WHERE NO MAN HAS GONE BEFORE

A History of Apollo Lunar Exploration Missions

"The Hammer and the Feather" (48" × 37½", *acrylic on masonite*)
Copyright © 1986 by Alan Bean. Used by permission.

Apollo 15 commander David R. Scott confirms Galileo's hypothesis that in the absence of air resistance all objects fall with the same velocity. A geologic hammer in Scott's right hand and a falcon feather in his left hand reached the surface of the moon at the same time (see chapter 13). The demonstration was performed before the television camera on the lunar roving vehicle, and no photographs were made.

NASA SP–4214

WHERE NO MAN HAS GONE BEFORE

A History of Apollo Lunar Exploration Missions

William David Compton

The NASA History Series

National Aeronautics and Space Administration
Office of Management
Scientific and Technical Information Division
Washington, DC   1989

Library of Congress Cataloging-in-Publication Data

Compton, William David.
  Where no man has gone before.

   (The NASA historical series) (NASA SP ; 4214)
   Supt. of Docs. no.: NAS 1.21:4214
   1. Project Apollo (U.S.)—History. I. United
States. National Aeronautics and Space
Administration. Scientific and Technical
Information Division. II. Title. III. Series.
IV. Series: NASA SP ; 4214.
TL789.8.U6A528 1989 919.9'104 88-600242

For sale by the Superintendent
of Documents, U.S. Government
Printing Office, Washington, DC 20402

PREFACE

The purpose of this book is only partly to record the engineering and scientific accomplishments of the men and women who made it possible for a human to step away from his home planet for the first time. It is primarily an attempt to show how scientists interested in the moon and engineers interested in landing people on the moon worked out their differences and conducted a program that was a major contribution to science as well as a stunning engineering accomplishment.

When scientific requirements began to be imposed on manned space flight operations, hardly any aspect was unaffected. The choice of landing sites, the amount of scientific equipment that could be carried, and the weight of lunar material that could be brought back all depended on the capabilities of the spacecraft and mission operations. These considerations limited the earliest missions and constituted the challenge of the later ones.

President John F. Kennedy's decision to build the United States' space program around a manned lunar landing owed nothing to any scientific interest in the moon. The primary dividend was to be national prestige, which had suffered from the Soviet Union's early accomplishments in space. A second, equally important result of a manned lunar landing would be the creation of a national capability to operate in space for purposes that might not be foreseeable. Finally, Kennedy felt the need for the country to set aside "business as usual" and commit itself with dedication and discipline to a goal that was both difficult and worthwhile.

Kennedy had the assurance of those in the best position to know that it was technologically possible to put a human on the moon within the decade. His political advisers, while stressing the many benefits (including science) that would accrue from a strong space program, recognized at once that humans were the key. If the Soviets sent men and women to the moon, no American robot, however sophisticated or important, would produce an equal impact on the world's consciousness. Hence America's leadership in space would be asserted by landing humans on the moon.

This line of reasoning was convincing enough for most congressional leaders, who would have to provide the money, and was accepted by a majority of Americans. But Apollo was conceived and developed in an era when the scientific community was emerging as a political force in the country. Scientific research was becoming a big business in the late 1950s and early 1960s, sustained by unprecedented financial support from the federal government. Science had constituted the major portion of NASA's early space program and was the rationale for the space program in the first place; hence scientists considered space to be their province. The investigations of the pioneer space scientists did not require a human's presence; hence man had, in their view, no important role in space. Since Apollo was not a scientific project, it was unnecessary; because it would be expensive, it probably would be detrimental to the legitimate space program already under way.

In one respect the critics were right: Apollo was not primarily a scientific project. The engineers charged with accomplishing the lunar landing *within the decade* had far too many problems to solve to give much thought to secondary matters—a category to which they relegated scientific experiments. Some may have felt that the landing itself was enough; the President had called for nothing more. Most probably reasoned that someone else would specify what the astronauts would do on the moon. For the engineers responsible for carrying it out, Apollo required a strict ordering of priorities: be sure we can get the crew there and back; then provide for other objectives. Risks abounded, and the glare of publicity surrounding Apollo made it certain that the loss of a single astronaut's life could imperil the project's very existence.

Objections by the scientists had no effect on the nation's determination to carry out the manned lunar landing. Science would, however, considerably affect the conduct of the lunar missions that followed the first landing. By the time the project ended in December 1972, engineers and scientists had developed a mutual respect and a commonality of aims. No one lamented more strongly than the scientists—not, for the most part, the same ones who had objected so vigorously to the program in the early days—the cancellation in 1970 of three planned lunar exploration missions.

Apollo might have been considered complete when the crew of the spacecraft *Columbia* came aboard the U.S.S. *Hornet* in the Pacific Ocean on July 24, 1969. Indeed, for some years after President Kennedy proposed it, the first lunar landing was regarded as the objective of Apollo. Planning for manned space flights to follow the lunar landing began in 1963 with studies on how the Apollo spacecraft could be modified to extend the duration of the missions and increase the payload that could be carried to the moon. In 1965 a program called "Apollo Applications" emerged, which included long-duration earth-orbital flights as well as lunar exploration. It would build and launch a few Apollo spacecraft and Saturn rockets each year, sustaining the nation's manned space capability and producing useful information while the nation decided what the next major step in manned space exploration would be.

By late 1967, however, it was clear that that decision would not come early, and that post-Apollo programs other than lunar exploration required more thought. An Office of Lunar Exploration Programs was opened in NASA Headquarters to direct the continued exploration of the moon under the Apollo banner, and the earth-orbital portion of Apollo Applications, which would evolve into Skylab, was split off.*

The development of the spacecraft, rockets, and launch facilities necessary to accomplish the primary goal of Apollo has been described in three prior volumes in the NASA history series.** The story of the lunar spacecraft and the flight program up to the return of Apollo 11 is detailed in *Chariots for Apollo*. The present

---

*See W. David Compton and Charles D. Benson, *Living and Working in Space: A History of Skylab*, NASA SP-4208 (Washington, 1983).

**Courtney G. Brooks, James M. Grimwood, and Loyd S. Swenson, Jr., *Chariots for Apollo: A History of Manned Lunar Spacecraft*, NASA SP-4205 (Washington, 1979); Roger E. Bilstein, *Stages to Saturn: A Technological History of the Apollo/Saturn Launch Vehicles*, NASA SP-4206 (Washington, 1980); Charles D. Benson and William Barnaby Faherty, *Moonport: A History of Apollo Launch Facilities and Operations*, NASA SP-4204 (Washington, 1978).

volume is both a parallel and a sequel to *Chariots*; it traces the development of the Apollo science program from the earliest days and continues the history of the Apollo program, laying major emphasis on the scientific exploration of the moon conducted on the later flights, Apollo 12 through Apollo 17.

One issue of great concern to scientists and engineers alike was whether a trained scientist should be included in the crew of the lunar module. If the point of man in space was to make best use of the unique capabilities of humans, would not manned space science require a professional scientist? Or were the intricacies and potential hazards of flying the spacecraft so great that only test pilots could be trusted to lead the missions? Could one professional be sufficiently trained in the other's skills to be an adequate surrogate? The qualifications of candidates for selection and training as astronauts, and the choice of crews for each mission were points that were never really settled during Apollo and remained a point of contention throughout the project.

Another issue concerned the scientific study of the samples returned from the missions, which was complicated by the U.S. Public Health Service's insistence on quarantining everything returned from the moon until it could be shown that no exotic microorganisms had been accidentally imported. The Lunar Receiving Laboratory and its role in the storage, dissemination, and preservation of the lunar samples are important to the scientific story of Apollo.

Finally, this history must make a first cut at answering the following questions: what have scientists made of the data produced by Apollo? Do we understand more about the origin and history of the moon and the solar system as a result of Apollo's six voyages? Any answers to these questions can only be provisional since, like all scientific questions, they are subject to revision as new investigators apply new techniques to the samples. By the time this history was written, scientists had reached consensus on very few answers to the questions that lunar exploration hoped to clarify, but I have tried to summarize their tentative conclusions.

A program as complex as Apollo is not easily handled by a simple chronological account. In the early stages, from 1961 to roughly the end of 1966, the several phases of the program had to be hammered out more or less independently and many complex relationships had to be built. For those reasons I have organized the early chapters of the book topically, the better to deal in some detail with these early developments.

By early 1967 most of the separate elements were in place; then on January 27, 1967, the program was shaken to its foundations by the command module fire that killed three astronauts in training for the first manned Apollo mission. The fire was a watershed for Apollo, setting back operations by a year or more while NASA and its contractors examined every detail of spacecraft design, manufacture, and management. It had almost no negative effect on the science program; in fact, lunar exploration probably benefited by the delay, which gave some much-needed time to the development of the lunar surface instruments and the lunar receiving laboratory. From the fire to the first lunar landing in July 1969, a basically chronological account of development is somewhat more manageable.

The other accident in the Apollo program, the aborted Apollo 13 mission, was deliberately mentioned only briefly in the main text. To have covered that flight in detail, dramatic as it was, would have lengthened the story unacceptably; and since other authors have dealt with it in detail (see n. 78, chapter 11), I decided to treat the essentials of the accident, the management of the flight, and the results of the investigation in appendix 8. The safe return of the crippled spacecraft and its three crewmen is a monument to the skill and determination of hundreds of dedicated individuals; the subsequent investigation was a masterly piece of engineering detective work, lacking a "body"—the failed spacecraft itself—to provide clues; but space did not permit full treatment of the flight in all its aspects. As for its impact on lunar science, Apollo 13 created a delay of a few months, giving scientists a little breathing space, which they welcomed, to refine plans for later missions. The loss of one load of lunar samples and the data from one more set of surface experiments was of small importance in the end. More than that was lost a few months later when two of the remaining six missions were canceled.

In writing this history I was granted unrestricted access to the extensive Historian's Source Files at the Lyndon B. Johnson Space Center in Houston, to the files in the History Office at NASA Headquarters, and to documents stored at the various Federal Archives and Records Centers. To the extent possible within the time constraints of my contract, participants reviewed my drafts and offered comments. They also corrected factual errors where they found them. Their comments were given thoughtful consideration and incorporated into the history whenever the documentation seemed to support them or when an insider's viewpoint yielded insights the historian cannot glean from the documents alone. The interpretations of events here recorded, as well as any errors that remain, are my responsibility.

I have tried to define fairly and accurately the arguments on both sides of the scientific and technological issues that bore on the conduct of the Apollo exploration missions. I found it somewhat difficult to treat the opposition to the Apollo program voiced by many prominent scientists. Of course the scientists who spoke against Apollo and later criticized NASA's management of it were merely exercising their right to political expression. One reviewer who took exception to my treatment noted that this was the only avenue open to the scientists, who were put off by the engineers and had no choice but to "go public" with their objections, in the hope that political pressure would gain what their efforts within the system had not. Nevertheless, their objections, usually based on the unspoken assumption that purely scientific projects were entitled to privileged treatment, often smacked of intellectual arrogance. More irritating still—even to an outsider— few of those who criticized the project wanted to assume any responsibility for managing it. On the whole the objectors preferred to remain outside, where they could pursue their rewarding scientific careers while freely criticizing NASA— often in ignorance of the political, operational, and cost restrictions within which the space agency had to operate. Those scientists who made a commitment to the program and stayed with it to the end, establishing close relationships with mission planners and scaling their objectives to the capability of the system, deserve more credit than they usually get for the ultimate scientific productivity of Apollo.

The reader who suspects that I have a lingering bias in favor of Apollo's engineers is probably correct. I would not dispute the scientists' assertion that the engineers in charge of Apollo often seemed to be throwing roadblocks in the way of science. I suggest, however, that the engineers' reluctance to stretch the missions as far and as soon as the scientists wanted grew out of a healthy respect for the limitations of their equipment and procedures. However easy it may have come to seem, landing on the moon and returning to earth was not a bit less hazardous on the last mission than on the first. When lives are at risk, the line between boldness and recklessness can seem narrow to those who carry the responsibility. NASA and the nation paid the price of haste on January 28, 1986, when the space shuttle *Challenger* and its crew of seven were lost 73 seconds after launch.

One word about terminology. Throughout the text I have used generic terms like "scientific community" as convenient shorthand, which may be misunderstood. Obviously no single, homogeneous "scientific community" exists now, or ever did, and I use the term only to indicate the source of comments or criticisms. Terms like "manned space flight enthusiasts," "Headquarters officials," or "MSC engineers" only categorize the source of a comment and do not imply that all persons in that category agreed with the statement or point of view thus attributed.

W.D.C.
Houston, 1987

ACKNOWLEDGMENTS

The onerous task of acquiring, sorting, and culling the Apollo documentation at Johnson Space Center had been all but completed when I undertook this history. Chief among those responsible for acquiring these records and putting them in order are James M. Grimwood, historian at JSC from 1962 to 1979, and the late Sally D. Gates, editor and archivist in the JSC History Office until her death in 1980.

By helping me learn to find my way through this morass while I worked on an earlier project, Jim Grimwood and Sally Gates earned my lasting gratitude. Similarly the staff of the Headquarters History Office—Monte D. Wright, director, Frank W. Anderson, Jr., assistant director, Carrie Karegeannes, editor, and Lee D. Saegesser, archivist—provided moral support, critical evaluation, and substantial assistance in my early days. Sylvia D. Fries, who took over as director in 1983, was no less helpful and encouraging on the present project than her predecessor. In the later stages of research and writing I was fortunate to have a sympathetic and helpful technical monitor in William E. Waldrip of the Management Analysis Office at JSC. Not only did he help in working with the NASA bureaucracy; he also took a deep and genuine interest in the organization and content of the book and offered perceptive comments while I was writing it.

Mrs. Sarah C. Arbuckle deserves mention for the hundreds of hours of tedious labor she applied to preparing the computerized index to the Apollo files, which was one of the tasks required by my contract. Although the results of her efforts do not appear in this book, the index was of great help in research in the later stages of its preparation. Researchers who use these files in the future will surely be grateful for her work.

Captain Alan L. Bean, USN (Ret.), artist and former astronaut, generously provided the frontispiece: a reproduction of his picture, "The Hammer and the Feather," which depicts the demonstration performed by David R. Scott on Apollo 15. No photograph of this demonstration was taken on the moon; the only ones available were taken from the television screen on which earth viewers saw it performed.

Finally, my thanks go to the participants in Apollo who provided interviews and criticized the manuscript.

CONTENTS

1

AMERICA STARTS FOR THE MOON:
1957–1963

When the crew of Apollo 11 splashed down in the Pacific Ocean on July 24, 1969, Americans hailed the successful completion of the most audacious and complex technological undertaking of the 20th century: landing humans on the moon and returning them safely to earth. Just over eight years before, when President John F. Kennedy proposed the manned lunar landing as the focus of the United States' space program, only one American—Lt. Comdr. Alan B. Shepard, Jr.—had been into space, on a suborbital lob shot lasting 15 minutes. At the end of the first lunar landing mission, American astronauts had logged more than 5,000 man-hours in space. To the extent that any single event could, the first successful lunar landing mission marked the National Aeronautics and Space Administration's development of the capability to explore space by whatever means were appropriate for whatever purposes seemed to serve the national interest.

To many, Apollo 11 demonstrated that the United States had clearly won the "space race" with the Soviet Union, which had been one of the space program's major purposes. By the time that was done, other issues dominated the scene. National interests were not the same in mid-1969 as they had been in 1961. Of the public reaction after Apollo 11, a congressional historian has written,

> The high drama of the first landing on the Moon was over. The players and stagehands stood around waiting for more curtain calls, but the audience drifted away. . . . The bloody carnage in Vietnam, the plight of the cities, the revolt on the campuses, the monetary woes of budget deficits and inflation, plus a widespread determination to reorder priorities pushed the manned space effort lower in national support.[1]

Project Apollo encompassed more than simply sending men to the moon and back. It reflected a determination to show that humans had an important role to play in exploring space, as they had in exploring the unknown corners of the earth in earlier centuries. That proposition was not universally accepted. From the time the space agency determined to put humans into space, many Americans argued vigorously against manned space flight on the grounds that it was unnecessary and inordinately expensive. Space scientists had already shown how much could be done with instruments, and planners were designing spacecraft that would revolutionize communications, weather forecasting, and observation of the earth, all without requiring the presence of people in space.

These arguments were difficult to refute. Only when it came to exploring other planets did humans seem superior. For all of their limitations, humans were far more flexible than the most sophisticated robot, capable—as preprogrammed instruments were not—of responding creatively to the unexpected. If people had a place in space exploration, surely it would be on the surface of the moon.

Man's place in space exploration was decided, however, on other grounds. President Kennedy chose to send humans to the moon as a way of demonstrating the nation's technological prowess; and Congress and the nation endorsed his choice. That demonstration made and the tools for lunar exploration developed, Americans would go back to the moon five times, to explore it for the benefit of science.

Organizing for Space Exploration

The Soviet Union's launch of the world's first man-made satellite (*Sputnik*) on October 4, 1957, concentrated America's attention on its own fledgling space efforts. Congress, alarmed by the perceived threat to American security and technological leadership, urged immediate and strong action; the President and his advisers counseled more deliberate measures. Several months of debate produced agreement that a new federal agency was needed to conduct all nonmilitary activity in space. On July 29, 1958, President Dwight D. Eisenhower signed the National Aeronautics and Space Act of 1958 establishing the National Aeronautics and Space Administration (NASA).[2]

When it opened for business on October 1, 1958, NASA consisted mainly of the four laboratories and some 8,000 employees of the government's 43-year-old research agency in aeronautics, the National Advisory Committee for Aeronautics (NACA).* Within a few months NASA acquired the Vanguard satellite project, along with its 150 researchers from the Naval Research Laboratory; plans and funding for several space and planetary probes from the Army and the Air Force; and the services of the Jet Propulsion Laboratory (JPL) outside Pasadena, California, where scientists were planning an unmanned spacecraft (Ranger) that would take close-up television pictures of the lunar surface before crashing into the moon.[3]

Vanguard and JPL brought a strong scientific component into NASA's activities. Many of the Vanguard scientists became administrative and technical leaders at NASA Headquarters and at its new space science center (Goddard Space Flight Center**) at Greenbelt, Maryland. JPL's contributions to the space program would be strongest in instrumented spacecraft for the planetary programs. It also shared with Goddard major responsibility for development and operation of the tracking and telemetry network used in deep space operations, including Apollo.[4]

---

*NACA' s installations were Langley Memorial Aeronautical Laboratory, Langley Field, Va., with its subsidiary Pilotless Aircraft Research Station at Wallops Island, Va.; Ames Aeronautical Laboratory, Moffett Field, Calif.; Lewis Flight Propulsion Laboratory, Cleveland, Ohio; and the High-Speed Flight Station at Edwards Air Force Base, Calif. Langley, Ames, and Lewis became "Research Centers" under NASA, and the High-Speed Flight Station was renamed the Flight Research Center, later the Hugh L. Dryden Flight Research Facility (honoring NASA's first Deputy Administrator and long-time Director of NACA, who died in 1965).

**Named in honor of the American pioneer of liquid-fueled rockets, Dr. Robert H. Goddard (1882-1945).

These new acquisitions were grafted onto NACA, an organization that had played a leading role in the development of aircraft technology since 1914. After World War II, new aerodynamic and control problems had to be solved as the demand for military aircraft to perform at greater speeds and higher altitudes increased. By 1957 the X-15, one of a series of rocket-propelled piloted aircraft, was on the drawing boards. It was intended to be capable of exceeding Mach 6 (six times the speed of sound) and of climbing beyond 107,000 meters (67 miles)—above nearly all the sensible atmosphere. NACA was, in fact, approaching the conditions of space flight by extension of the operational limits of manned aircraft.

Other NACA engineers were working on other space-related problems. At Langley's Pilotless Aircraft Research Division, aerodynamicists were acquiring important data on aerodynamic heating at speeds of Mach 10, unattainable in the wind tunnels of the time, by flying models of aircraft and missiles mounted on rockets.[5] When *Sputnik* went up, many of these engineers were already talking about the problems of putting humans in an earth-orbiting spacecraft.[6]

The necessity for thinking about humans in space was made apparent when, less than a month after *Sputnik*, the Soviets orbited *Sputnik II*, a 500-kilogram (1,100-pound) satellite carrying a living passenger—a dog named Laika. With this clear evidence that the Russians intended to send men into space, both the Army and the Air Force resurrected dormant schemes to follow suit. Neither could produce a credible mission for humans in space, and both lost out to the new space agency in 1958, when President Eisenhower assigned all manned space flight projects to NASA.[7] Before NASA was a month old, Administrator T. Keith Glennan chartered a Space Task Group (STG) at Langley and charged it with managing the United States' first project to put man in space: Project Mercury. In 1961 STG was redesignated the Manned Spacecraft Center, a connotation of its newly expanded responsibility for all manned projects, and located on 1,660 acres (6.5 square kilometers) of flat Texas pasture land 22 miles (35 kilometers) southeast of downtown Houston.[8]

Crucial to any ambitious program in space was the ability to launch large payloads into earth orbit and to send instrument payloads to the planets. Rockets far exceeding the capacity of existing launch vehicles were required, but only one was being seriously pursued. At the Army's Redstone Arsenal just outside Huntsville, Alabama, the Free World's most experienced rocket engineers—Wernher von Braun and the team built around the hundred-odd Germans who developed the V-2 rocket during World War II—were about to undertake construction of a vehicle called Saturn I, five times as powerful as the biggest then available. By 1959, however, the Army had lost its last tenuous foothold on space flight and had no use for Saturn—nor could it provide any other pioneering work for the ambitious von Braun. On July 1, 1960, rocket development at Redstone Arsenal followed some earlier Army space programs into NASA when von Braun and 4,600 employees, along with many of the facilities at Redstone, became the George C. Marshall Space Flight Center.[9]

Thus by the end of 1960 NASA had the elements of a comprehensive space program in place. Marshall Space Flight Center would design, test, and launch*

---

*Marshall maintained a subsidiary Launch Operations Directorate at the Air Force's Eastern Test the rockets and oversee their production by industry. The Manned Spacecraft Center would manage spacecraft design and testing, conduct flight operations, and train the astronauts. Goddard and JPL would be responsible for tracking, communication, and data management. At Headquarters, a triumvirate comprising the Administrator, Deputy Administrator, and Associate Administrator managed the overall program, determining policy, preparing budget requests, and defending the program and the budgets before congressional committees. Agency programs—science, manned space flight, advanced research—were managed by directors of Headquarters program offices. The field centers reported to the Associate Administrator, who coordinated the program offices and allocated resources to the centers.

Project Apollo: The Decision

During NASA's first two years, manned space flight managers struggled with the problems of organizing extremely complex and technologically demanding projects. The established space science programs continued to produce new data on the earth and its space environment. President Eisenhower, among others, favored continuing the productive (and comparatively inexpensive) unmanned science programs and withholding judgment on manned programs. In his departing budget message to Congress, the retiring president noted that more work would be needed "to establish whether there are any valid *scientific* [emphasis added] reasons for extending manned spaceflight beyond the Mercury program."[10] In early 1961, a committee of scientists appointed by newly elected President John F. Kennedy recommended that *"we should stop advertising Mercury as our major objective in space activities* [emphasis in the original]," and instead try to "find effective means to make people appreciate the cultural, public service, and military importance of space activities other than space travel."[11] So problem-ridden did Mercury seem that Kennedy's advisers felt the new president should not endorse it and thereby risk being blamed for possible future failures; better, the scientists believed, to emphasize the successful science and applications programs and the tangible benefits they could be expected to produce.

In spite of Mercury's early problems, manned space flight enthusiasts were thinking far beyond manned earth-orbital flights. NASA's engineers were confident that they could send people to the moon and back. A moon flight was an obvious goal for the manned programs. It would be an end in itself, needing no justification in terms of its contribution to some larger goal, and it would demonstrate the nation's superiority in space technology to all the world. Preliminary work and discussion during 1959 turned up no insurmountable obstacles, and in mid-1960 NASA announced its intention to award contracts to study the feasibility of a manned lunar mission. The project even had a name: Apollo. On October 25, study contracts were let to three aerospace firms.[12]

NASA might conduct studies to show that man could go to the moon, and scientists might argue that manned space flight was of doubtful value, but Con-

---

Range at Cape Canaveral, Fla., which was made autonomous in 1962 as the Launch Operations Center (renamed the John F. Kennedy Space Center in December 1963), responsible for final assembly, checkout, and launching of manned space vehicles.

gress and the president would have to make the commitment, and the decisive stimulus was still lacking. Then on April 12, 1961, the Soviets once more spurred a major advance in the American space program by sending Major Yuri A. Gagarin into space for one orbit of the earth. Congressional advocates of an all-out effort to "beat the Russians" renewed their cries; influential media organs saw a challenge to America's world leadership, as did many high government officials. President Kennedy called on Vice-President Lyndon Johnson, chairman of the National Space Council, to survey the national space program and determine what project promised dramatic results that would show the United States' supremacy in space. Johnson immediately began consultations with NASA and Defense Department officials and with key members of Congress.[13]

Kennedy's desire for "dramatic results" did not coincide with what others had in mind for the space program—especially the scientists. Neither Eisenhower's nor Kennedy's science advisers believed that any results from manned space flight could compare with those expected from space science and applications programs. During the debate on the creation of a space agency, the President's Science Advisory Committee (PSAC) issued an "Introduction to Outer Space," which asserted that "scientific questions come first" and that "it is in these [i.e., scientific] terms that we must measure the value of launching satellites and sending rockets into space."[14] Eisenhower's chief scientific adviser, James R. Killian, former president of Massachusetts Institute of Technology and chairman of PSAC, said after leaving his White House position in 1960 that the Soviets' space exploits were attempts "to present spectacular accomplishments in space as an index of national strength." He deplored the tendency to design American programs to match the Soviet Union's and urged that the United States define its own objectives and pursue them on its own schedule, not indulge in costly competition for prestige in space exploration—by which he apparently meant manned space flight. "Many thoughtful citizens," Killian said, "are convinced that the really exciting discoveries in space can be realized better by instruments than by man."[15] His views were shared by many scientists, including Jerome Wiesner, a member of PSAC since its formation who became principal scientific adviser to John Kennedy. What the scientists could not, or would not, recognize was that their excitement was neither understood nor shared by any substantial majority of the people.

Some scientists, however, believed the space program should include elements with strong public appeal. The Space Science Board of the National Academy of Sciences,* NASA's officially designated source of scientific advice, discussed the question of man in space early in 1961 and later that year adopted a position paper on "Man's Role in the National Space Program." The board asserted that the goal of the nation's space program should be the scientific exploration of the moon and the planets but recognized that nontechnical factors were vital to public acceptance of a space program. Human exploration of the moon and planets

---

*The National Academy of Sciences is a private, nongovernmental body chartered in 1863 to promote the advancement of science and to provide advice, when asked, to the government on scientific matters. Membership in the Academy is regarded as recognition of eminence in research and is the highest honor an American scientist can be awarded short of the Nobel prize. See Daniel S. Greenberg, "The National Academy of Sciences: Portrait of an Institution," *Science* 156 (1967): 222–29, 360–64, and 488–93. In June 1958 the Academy created a Space Science Board to advise the government on the space science program.

"I believe that this nation should commit itself to achieving the goal, before this decade is out, of landing a man on the moon and returning him safely to the earth. . . ." President John F. Kennedy issues the challenge of Apollo, May 25, 1961. Behind him, Vice-president Lyndon B. Johnson (left) and House Speaker Sam Rayburn.

would be "potentially the greatest inspirational venture of this century and one in which the world can share; inherent here are great and fundamental philosophical and spiritual values which find a response in man's questing spirit. . . ." Thus the space exploration program must be developed "on the premise that man will be included. Failure to adopt . . . this premise will inevitably prevent man's inclusion," presumably because of the costs involved. "From a scientific standpoint," the paper went on, "there seems little room for dissent that man's participation in the exploration of the Moon and planets will be essential, if and when it becomes technologically feasible to include him."[16] This endorsement of man's participation in space exploration was at variance with a substantial body of opinion in the American scientific community, as events of the next two years would show; and the board's adduction of nonscientific values to justify manned space flight would later draw pontifical rebuke from an influential scientific organization.[17]

On May 8, 1961, Lyndon Johnson's survey of the space program culminated in a lengthy report drafted by NASA and Defense Department officials. The report recommended strengthening the civilian space program in all areas. Particularly pressing was the need for new and much more powerful launch vehicles. As for the best way to put the nation ahead of the Soviets, the report chose a manned lunar landing: "It is man, not mere machines, in space that captures the imagination of the world." However small its value in military or scientific terms, such a project would not only recover the country's lost prestige, it would stimulate advances in every phase of space technology and give the nation the means to explore space in whatever way best suited the circumstances.[18]

With this strong endorsement of a lunar landing project, and after Alan Shepard's successful suborbital Mercury flight on May 5, Kennedy put together a message to Congress on "Urgent National Needs," which he delivered in person on May 25, 1961. While the speech covered many issues, its major impact was on the space program. In it Kennedy expressed his belief that a manned lunar landing, "before this decade is out," should be the principal goal of the American space effort. Stressing that this meant a long and costly development program to reestablish the nation's world leadership in technology, he cautioned that "if we are to go only halfway, or reduce our sights in the face of difficulty . . . it would be better not to go at all."[19] It was a call for the country to commit itself wholeheartedly to a long-term project that required sustained effort, substantial cost, and determination to see it through to a successful conclusion.

If congressional reaction was less than enthusiastic, as Kennedy is reported to have felt afterwards,[20] events of the following summer proved that Congress was solidly behind the venture. The supplemental budget request to get Apollo under way—$675 million over Eisenhower's proposed $1.1 billion—carried both houses with large majorities after little debate and suffered only minor reduction by the House Appropriations Committee.[21] Congress and the nation were eager to see Apollo succeed; but NASA engineers, while confident that it could be done, better understood the magnitude of the task. Robert R. Gilruth, head of the Space Task Group, recalled later that he was simply aghast at what NASA was being asked to do.[22]

Project Apollo: The Debate

Support for the Apollo commitment was not unanimous, either in Congress or among the public. The public opposition most often questioned the wisdom of spending so much money on space when so many domestic problems confronted the country. Those who spoke for science often shared this concern, but their special objection was Apollo's distortion of priorities within the space program. One unidentified astronomer was reported to have complained to Senator Paul Douglas of Illinois that the space program was becoming "an engineering binge instead of a scientific project."[23] Petulant as that comment may sound, it epitomized what many space scientists most feared about the lunar landing project. Space science was a rapidly expanding field, offering almost limitless possibilities for exploitation by ambitious investigators. It had been generously supported by NASA for three years and had produced a rich harvest of scientific knowledge, much of it unfamiliar to the public. Manned space flight, merely because of man's participation, drew attention that gave it prominence far out of proportion to its scientific value. The pioneers of space science were what one historian has called "sky scientists"—mainly astronomers and physicists interested in studying the sun and stars and particles, fields, and radiation in near-earth space.[24] Sky scientists could well have believed that their projects would suffer as lunar and planetary science gained support. Lunar science, which stood to gain the most from Apollo, counted only a few practitioners who did not yet have the influence of the established space science programs.

American science generally was still riding a wave of public esteem and government subsidy that had begun in the early 1950s and had swelled again after *Sputnik*. Basing their arguments on the indisputable contributions made by scientists to the war effort during World War II, American scientists had worked long and hard after the war to convince the public and the Congress that America's standard of living and position in world affairs depended on a strong scientific base, which in turn depended on generous funding of basic research. By the mid-1950s government support of basic research had risen to a level that prewar researchers could not have dreamed of. This new status had not been easily achieved and often had to be defended; many congressmen would have preferred to support practical projects rather than pure research, which often seemed pointless. (Indeed, congressmen and journalists frequently enjoyed making fun of research projects that had absurd-sounding titles, such as the reproductive physiology of the screwworm fly.[25]) But scientists had grown increasingly influential in governmental affairs. Prominent scientists found their counsel being sought more and more frequently by government at all levels, and science had enough influential friends in and out of government to ensure the continuity of a substantial level of support throughout the postwar years.[26]

Nonetheless, the most prominent and influential spokesmen for science seemed to feel uneasy about the viability of their favored status. Their behavior was characterized by one critical observer of the science-government interaction as "not unlike [that of] a *nouveau riche* in a fluctuating market." Every threat, real or imagined, to reduce the support of science—or even to reduce its rate of growth—was regarded as a potential catastrophe.[27] So the space scientists may have perceived Apollo as a threat. No one could accurately predict its ultimate cost—estimates ranged upward from $20 billion—but it would be expensive enough that Congress might trim other programs to provide its funds.

Scientists' misgivings about Apollo were expressed intramurally in the summer of 1962 at the Space Science Board's first summer study of NASA's science programs. Convened at the request of NASA, the six-week summer study brought together more than a hundred participants from universities and industry to evaluate NASA's past activities and recommend future policies and programs. The final report of the study, noting that "there is considerable confusion about the Apollo mission and its proper justifications," stated that Apollo was just what Kennedy had said it was: a program to put America first in space, with no necessary commitment to science. Until the success of the lunar landing could be clearly foreseen, Apollo was, and must be, an engineering effort, "and the engineers must be protected in their ability to do their jobs." Scientific investigations would be phased into the program later; still later, assuming intermediate success, "scientific investigations will become the primary goals." It was evident that these considerations were not well understood—and perhaps not accepted—by the scientific community, for the report urged NASA to work harder to make them clear.[28]

This section of the report was addressed primarily to the scientific community rather than to NASA, but whether it allayed any fears is debatable. If it did, events of the following fall could well have raised stronger ones. In November, manned space flight projects were severely cramped by lack of funds, and Brainerd Holmes, director of Headquarters's Office of Manned Space Flight, wanted to ask Congress for a $400-million supplemental appropriation to cover unanticipated costs. NASA Administrator James Webb, unwilling to risk undermining congressional support, did not agree. Holmes then proposed to transfer money to Apollo from other NASA programs, including science, but again Webb refused. When the question was taken to the White House, Webb told the President he would not take responsibility for a program that subordinated all else to the lunar landing. The extra funds could wait, he said, until NASA went to Congress with its fiscal 1964 budget. President Kennedy accepted this compromise.[29] Webb's stand for a balanced program should have surprised no one, for both he and his Deputy Administrator, Dr. Hugh L. Dryden, had repeatedly stated their view that the lunar landing was not in itself the ultimate goal of the space program; it was a project which, to be successful, required the advancement of space technology and science on a broad front.[30]

Webb went to Capitol Hill in March 1963 asking for $5.712 billion—$2.012 billion more than the previous year's budget request. Nearly 80 percent of the increase was for manned programs, but funding for space science was also substantially raised, by 50 percent over the previous year's budget.[31] For the first time NASA met significant resistance to its presentation. The sudden drastic increase in the total budget (54 percent in one year), the growing awareness of the probable total cost of Apollo (estimated at $20–$40 billion), and the increasing dissatisfaction in the country with the administration's priorities all combined to raise opposition to the manned space program to a peak during the spring and summer of 1963.

As hearings on the administration's budget proceeded, the space program drew fire from many sources. Retired President Eisenhower reiterated his conviction that Apollo was not worth the tax burden it would create.[32] The Senate Republican Policy Committee published a report questioning the Democratic administration's expenditures on space rather than on other urgent national needs.[33] Two years before, Kennedy had warned that the cost would be high and that careful consideration by Congress and the public was essential. It was useless to agree that the country should bid for leadership in space, he had said, unless "we are prepared to do the work and bear the burdens to make it successful."[34] Under the pressure of Soviet achievements, the commitment had been endorsed. When the bills began to come due, the country was not so sure.

In the debate that spring and summer, many scientists spoke from their peculiar point of view concerning the space program. On April 19 Philip H. Abelson, editor of *Science*,* summarized the case against Apollo in an editorial. It did not deserve the priority it had been given in the space program, Abelson believed. Its scientific value, small at best, would be even less if (as seemed likely) a trained

---

Science, the weekly journal of the American Association for the Advancement of Science (AAAS), probably reaches more scientists in different disciplines than any other single scientific publication. (In 1963, AAAS numbered about 76,000 members.) Besides publishing technical papers, *Science* provides news and analysis of many subjects of interest to the scientific community. Acerbic and outspoken, Abelson characterized himself as "a damned maverick" in testimony before the Senate space committee later in the year.

scientist was not sent on the first landing mission. More and better data could be obtained by unmanned probes, at about one percent of the cost. In Abelson's view, neither the military advantages nor the "technological fallout" cited by advocates could justify the cost of sending men to the moon.[35]

After Abelson's editorial, many other scientists expressed their reservations concerning the space program, and a general debate ensued in the press.[36] Criticism focused on several points: the lunar landing program had almost no scientific value and science would be advanced much more by spending the same money on unmanned projects; the space program lured promising young talent away from other worthwhile research, creating an imbalance in the nation's overall scientific effort; and the money spent on Apollo could be better invested in educational, social, and environmental programs. Some seemed to feel that Apollo had been promoted as a scientific program and to resent the confusion in the public mind. Hugh Dryden reminded the critics that "no one in NASA had ever said [Apollo] was decided upon solely on the basis of its scientific content."[37] Other scientists agreed with Dryden and expressed their acceptance of the lunar landing on its own terms.[38]

Senator Clinton P. Anderson, chairman of the Senate Committee on Aeronautical and Space Sciences, which was then considering NASA's authorization bill for fiscal 1964, reacted to this debate by inviting several prominent scientists to present their views to the committee. During two days of hearings, ten scientists (including Philip Abelson, who was the first to be heard) ranged over most of the ground covered in the public debate. If there was any general agreement, it was that the time limit set for Apollo was probably conducive to waste, and that many national problems deserved equal attention; but there was no agreement that American science was being skewed by so much attention to space. The strongest protest against the program was a written statement provided by Warren Weaver, vice-president of the Alfred P. Sloan Foundation, who listed many good things that could be bought with $30 billion*—a price he said was "undoubtedly an underestimate"** of Apollo's ultimate cost.[39]

Senator Anderson got what he wanted from the scientists—a variety of views that improved his perspective of the space program.[40] The disapproving witnesses' doubts were echoed in Congress. NASA's budget did not go through unscathed, but the cuts actually made were less than some in Congress would have liked. When the House approved NASA's authorization bill on August 1, support for the space program was still strong: the majority was six to one.[41] After three more months of debate and cuts totaling $612 million, NASA's appropriation ($5.1 billion) passed both houses by large majorities.[42] Many opponents of expensive

---

*Thirty billion dollars, Weaver said, would give every teacher in the U.S. a 10 percent annual raise for 10 years; give $10 million each to 200 small colleges; provide 7-year scholarships at $4,000 per year to produce 50,000 new Ph.D. scientists and engineers; give $200 million each to 10 new medical schools; build and endow complete universities for 53 underdeveloped nations; create 3 more Rockefeller Foundations; and leave $100 million over "for a program of informing the public about science."

**The official estimate provided to Congress in 1973 was $25.4 billion. House, Subcommittee on Manned Space Flight of the Committee on Science and Astronautics, *1974 NASA Authorization*, Hearings on H.R. 4567, 93/2, Part 2, p. 1271.

manned space flight programs would express their objections over the next decade, but Apollo would go forward, carrying—as some thought—the rest of the space program with it.[43]

What might have been, if there had been no lunar landing project, can be (and was) long debated. Over and over, advocates of the manned programs pointed out the reality of the situation: the nation could afford whatever it valued enough to pay for. Social welfare and other desirable programs had to win support on their own merits and would not necessarily be given Apollo's $3 billion a year if it were canceled. The politics of a technological project with a clear goal and self-evident success or failure were much simpler than any plan to conquer poverty, rebuild the cities, or clean up the environment.

No proof is possible that space science (or science generally) would have been better supported if Apollo had not been claiming such a large fraction of the space budget. In fiscal 1964, when NASA's budget request first encountered serious resistance in Congress, space science was authorized $617.5 million; its spending authority grew to $621.6 million and then to $664.9 million in the next two fiscal years.[44] (The entire Mercury project, from 1958 to 1966, cost about $400 million.[45]) But manned space flight budgets were three to four times those amounts, and Homer Newell, director of NASA's science programs for the first nine years, would later recall that space scientists never hesitated "to complain about not getting their fair share of the space budget." Newell, a space scientist himself and as active an advocate as space science had, understood and accepted the overall priorities of the space program, as the scientists apparently did not, and they sometimes tried his patience. He would later remark that "whatever complaint there might have been about either the absolute or relative level of the space science budget . . . , there can be little doubt that it represented a substantial program."[46]

Project Apollo: Prospects, 1963

Apollo survived the debate of 1963, as it would survive worse troubles later, but the cut in NASA'S budget request (more than 10 percent) left its mark. The following spring Administrator James Webb would not assure Congress, as he had in the past, that he was confident the lunar landing would be accomplished within the decade—only that it was possible, if everything went well.[47]

And much could yet go wrong. Spacecraft design and the basic mission operations plan had been settled and the major contracts had been let. Years of testing and design refinement lay ahead. An entire project, Gemini, was still to be conducted, to establish the feasibility of rendezvous—bringing two spacecraft together in orbit—on which the success of Apollo depended. In terms of technical milestones, the lunar landing was still a long way off. The science community had registered its objections to Apollo, as had other concerned citizens, and the nation had reaffirmed the commitment asked of it by its late president. Those same objections would continue to be voiced, but the lunar landing would remain the major driving force behind the national space program.

One thing that could be clearly seen at the end of 1963 was that manned space flight had an important interest in reaching some kind of accommodation with science. Over the next four years NASA officials and members of the science community worked to establish a program of scientific exploration that would become the primary purpose of the later Apollo missions.

2

LINKING SCIENCE TO MANNED SPACE FLIGHT

Scientific exploration of the moon required close cooperation between two quite different organizations within the space agency. Space science was an active field of research when NASA was created, with a well-organized constituency and established procedures for generating and developing experiments. The Office of Space Sciences, which managed these projects, relied heavily on scientists outside the agency for advice on policy and regarded itself as an operational arm of the nation's scientific community, providing opportunities for that community to conduct the research it deemed important. The Office of Manned Space Flight, on the other hand, had no interested constituency outside of the space agency. Having been handed their primary assignment by the President in 1961, engineers of the manned space flight organization reported to the NASA Administrator and to Congress on the progress of their projects.

To get these two offices working together on exploration of the moon was not simple. Starting in 1962, Homer Newell, director of the Office of Space Sciences, began to lay the organizational foundations on which eventual collaboration would be built. The Office of Manned Space Flight, feeling the pressure of the Apollo deadline, was at first reluctant to spend much time preparing for science. By the end of 1963, however, much of the preliminary work had been done and the broad outlines of a lunar science program were taking shape.

The Moon and the Space Science Program

NASA's initial space science programs were largely defined by the projects transferred from other agencies and were mainly concerned with the study of phenomena in near-earth space. But shortly after taking over direction of space sciences, Homer Newell established a Theoretical Division to support programs in planetology and lunar science.[1] Unlike space physicists and astronomers, those interested in the moon and planets had little hard data to work with. Lunar and planetary science in 1960 was a field for theoreticians, and few scientists devoted their entire attention to it. So when Robert Jastrow, whom Newell appointed to head the new division, set out to learn all he could about current theories and research in that area, he had a very short list of sources to consult. High on the list was the name of Harold C. Urey, professor at large at the University of California at San Diego.

Urey, a chemist whose scientific career spanned four decades, had won the Nobel Prize in chemistry in 1934. During the second World War he had directed one of the major projects for concentrating uranium-235, the fissionable material of the first atomic bomb. A scientist of catholic interests, Urey became fascinated by the distribution of the chemical elements within the earth and in the solar system. Noting that there had been an extensive separation of iron from the rocky materials of the earth and meteorites, he began to consider possible mechanisms for the accretion of the planets out of the primordial matter of the solar system. In 1952 he published a book on the origin of the planets, in which he asserted his belief that the moon might provide the key to understanding the formation of the solar system. On retiring from the University of Chicago in 1958 at age 65, he continued to teach and conduct research in California, devoting considerable time to cosmology.[2]

Urey brought a chemist's approach to a subject that had previously been the province of astronomers and astrophysicists. Like almost any chemist of his era, he would have preferred to have samples that he could study in the laboratory. Lacking lunar samples, he used information from meteorites, plus such physical data as were available concerning the moon, to construct working hypotheses. When Apollo was created, Urey supported it for the contributions it could make to his own research interests, but he was conscious of its nonscientific value as well. In 1961 he thought that the lunar landing was too expensive for its potential scientific return, but on reflection he decided that if the money were not spent on Apollo it might well go to less productive projects and changed his mind.[3] Urey never failed to criticize NASA's practices when he felt criticism was justified, but on the whole he was a dependable supporter of the lunar landing program.[4]

Impressed by Urey's exposition of his theories and the potential they held for space investigation, Jastrow brought him to Headquarters to confer with Newell about possible NASA programs for lunar exploration. Their enthusiasm convinced Newell that space science should make room for a program in lunar and planetary sciences, and in January 1959 he appointed an ad hoc Working Group on Lunar Exploration to coordinate the efforts of NASA and academic scientists and to evaluate proposals for lunar experiments.[5]

Such interest as there was in lunar missions in early 1959 was at the Jet Propulsion Laboratory (JPL).* Even there, however, many scientists favored missions to Venus and Mars rather than to the moon, partly because the best opportunities for launch to the near planets occurred less frequently.[6] The moon—in earth's back yard, so to speak—offered an optimum launch opportunity once every lunar month, and if one were missed because of problems with a launch vehicle the delay was only four weeks, whereas a mission to Mars would have to wait two years if an optimum launch date were missed. While JPL was developing a plan for 12 deep-space missions, including 5 moon probes, Jastrow was urging Newell to accelerate NASA's lunar exploration programs.

Once again, however, the Soviets' eagerness to achieve space "firsts" exerted its pernicious influence on American space programs. Even before the Working

---

*JPL's director William Pickering had proposed an unmanned lunar probe as a response to *Sputnik* but had found no support for it.

Group for Lunar Exploration could finish drawing up a list of recommendations for lunar missions, the Russian *Luna I* swung by the moon and into solar orbit, measuring magnetic fields and particles in space. A month after JPL submitted its plan to Headquarters on April 30, 1959, orders went out to Pasadena to reorient the program to concentrate on lunar orbiting and soft-landing missions. (Apparently Headquarters felt that the more frequent opportunities for lunar missions offered the best chance to beat the Russians to their apparent target.) As the year progressed, the Soviets sent two more *Luna* spacecraft to the moon; one crash-landed, the other photographed the hidden side of the moon for the first time. In December Headquarters killed JPL's planetary exploration plan, in part because of problems with the proposed Atlas-Vega launch vehicle, and substituted a program of seven lunar missions using the Atlas-Agena B. Emphasis was on obtaining high-resolution photographs of the moon's surface, but some space science instruments would be carried as well. JPL would also investigate the feasibility of sending a hard-landing instrument package to transmit data about the moon. This project, called "Ranger," was explicitly recognized as a high-risk project geared to very short schedules and intended to capture the initiative in lunar exploration from the Soviet Union.[7]

Since lunar and planetary exploration seemed to have a promising future, Homer Newell established a Lunar and Planetary Program Office at Headquarters in January 1960 to manage it.[8] Initially, Ranger was the the new office's only lunar project. In July 1960 a second, Surveyor, was approved. More ambitious than Ranger, Surveyor had the objective of soft-landing a large (2,500 pounds, 1,100 kilograms) instrumented spacecraft on the moon's surface to gather physical and chemical information about the lunar soil and return it to earth by telemetry.[9]

Both Ranger and Surveyor were technically ambitious projects, requiring improvements in spacecraft stabilization, navigation and guidance, and telemetry. Both encountered technical and management problems that pushed back their completion dates to the point where rapidly changing events made their original objectives obsolete. In 1960, neither Ranger nor Surveyor was primarily intended to support the manned lunar landing, which at that time was still only an idea in the minds of NASA's planners, although both, if successful, would yield information useful to that project. But the pressures generated by the needs of Apollo between 1961 and 1963 forced Ranger and Surveyor into supporting roles for the manned space flight program, to the intense chagrin of the space scientists.

Manned Space Flight and Science

When the engineers of Robert R. Gilruth's Space Task Group began work on Project Mercury in 1958, they could not—as the space scientists could—draw on 10 years of experience in designing their spacecraft and conducting their missions. Aviation experience was helpful in some aspects of manned space flight, but in many others they faced new problems. Apollo posed many more. The engineers did not lack confidence that the President's goal could be met, but they knew only too well how much they had to learn to achieve it. A sense of urgency pervaded the manned space flight program from the beginning right up to the return of Apollo 11—an urgency that determined priorities for engineers at the centers. Every ounce of effort went into rocket and spacecraft development and operations planning. Science was considerably farther down the list, and for the first five years they gave it little thought.

Manned space flight projects were ruled by constraints that were less important to science projects. One was safety. Space flight was a risky business, obviously, but the risks had to be minimized. No matter that the astronauts themselves (all experienced test pilots in the beginning, accustomed to taking risks) understood and accepted the risks. From the administrator down to the rank-and-file engineer, everyone knew that the loss of an astronaut's life could mean indefinite postponement of man's venture into space. Moreover, NASA was in a race, competing against a competent adversary and working in the public eye, where its failures as well as its successes were immediately and widely publicized.

Reliability was one key to safety, and spacecraft engineers strove for reliability by design and by testing. With few exceptions, critical systems—those that could endanger mission success or crew safety if they failed—were duplicated. If redundancy was not feasible, systems were built with the best available parts under strict quality control, and tested under simulated mission conditions to assure reliability.[10] The measures taken to ensure reliability and safety contributed to the fact that manned spacecraft invariably tended to grow heavier as they matured, making weight control a continuing worry.

Those constraints were not so vital in the unmanned programs. Instruments needed no life-support systems and required no protection from reentry heat; scientific satellites were usually expendable. Being smaller than manned spacecraft, they required smaller and less expensive launch vehicles. Furthermore, those vehicles could be less reliable. More science could be produced for the money if experimenters would accept less than 100-percent success in launches, and space scientists were content with this.[11] The loss of a scientific payload, though serious to the investigators whose instruments were aboard, did not cost a life.

On the whole the engineers were content to go their way while the scientists went theirs. But the scientists were not (see Chapter 1), and their protests seemed to require a response. Manned space flight enthusiasts spoke of the superiority of humans as scientific investigators and of the benefits to science that would result from putting trained crews in space or on the moon to make scientific observations. No existing instrument, they said, could approach a human's innate ability to react to unexpected observations and change a preplanned experimental program; if such an instrument could be built, it would be far more expensive than putting people into space.[12]

This argument did not move the space scientists, most of whom worked in disciplines where human senses were useless in gathering the primary scientific data. The role of a person in space science was not to make the observations but to conceive the experiment, design the instruments to carry it out, and interpret the results.[13] Cleverness in these aspects of investigation was the mark of eminence in scientific research. The early manned programs offered space scientists no opportunities that could not be provided more cheaply by the unmanned programs.

The relationship between the manned and unmanned programs—essentially one of independence—took quite a different turn with the Apollo decision. Within two weeks of President Kennedy's proposal to Congress, NASA Deputy Administrator Hugh L. Dryden told the Senate space committee that Apollo planners would have to draw heavily on the unmanned lunar programs for information about the lunar surface. Knowledge of lunar topography and the physical characteristics of the surface layer was vital to the design of a lunar landing craft. Ranger was the only active project that could obtain this information, and to provide it, NASA asked Congress for funds to support four additional Ranger missions. The day after Dryden testified, NASA Headquarters directed the Jet Propulsion Laboratory to examine how to reorient Ranger to satisfy Apollo's needs.[14]

This directive was received with mixed feelings by the participants in Ranger. JPL's project managers favored a narrower focus, because the scientific experiments were giving them technical headaches that threatened project schedules. They proposed to equip the four new Rangers with high-resolution television cameras and to leave off the science experiments, using the payload space to add systems that would improve the reliability of the spacecraft. Scientists who had experiments on the Ranger spacecraft, however, were upset by the proposed change. When they complained to Newell, he and his Lunar and Planetary Programs director reasserted the primacy of science in Ranger and did everything they could to keep the experiments on all the flights. But the difficulties with the Ranger hardware and the pressure of schedules proved too much. In the end, the problem-plagued Ranger carried no space science experiments on its successful flights, but did return photographs showing lunar craters and surface debris less than a meter across.*[15]

Apollo could command enough influence to affect the unmanned lunar programs, but science had no such leverage on manned flights. For that matter, scientists had little interest in Mercury; its cramped spacecraft and severe weight limits, plus the short duration of its flights, made it unattractive to most experimenters. Still, the Mercury astronauts conducted a few scientific exercises, mostly visual and photographic observations of astronomical phenomena.[16] Comparatively unimportant in themselves, these experiments pointed up the need for close coordination between the scientists (and the Office of Space Sciences) and the manned space flight engineers. After John Glenn's first three-orbit flight on February 28, 1962, the Office of Space Sciences and the Office of Manned Space Flight began to look toward the moon and what humans should and could do there.[17]

Apollo managers had spent the second half of 1961 making the critical decisions about launch vehicle and spacecraft design; in the spring of 1962 they were wrestling with the question of mission mode. Should they plan to go directly from earth to the moon, landing the whole crew along with the return vehicle and all its fuel on the lunar surface? Or would it be better to assemble the lunar vehicle in earth orbit—which would require smaller launch vehicles but would entail closely spaced multiple launches, rendezvous of spacecraft and lunar rocket, and the unexplored problems of transferring fuel in zero gravity from earth-orbiting tankers to the lunar booster? Or was the third possible method, lunar-

---

*Ranger's results came too late (1964–1966) to affect the design of the Apollo lunar module; they did confirm that the designers' assumptions about the lunar surface were satisfactory and that the lunar module needed no modification.

orbit rendezvous, preferable: building a separate landing craft to descend from lunar orbit to the moon, leaving the earth-return vehicle circling the moon to await their return?[18] Apart from its essential impact on the booster rocket and space-craft, the mission mode would determine how much scientific equipment could be landed on the moon, how many men would land to deploy and operate it, and how long they would be able to stay. Until the decision was made it was pointless to try to design equipment, but by early 1962 the mission planners needed to know in general terms what the scientists hoped to do on the moon and some important questions of responsibility and authority had to be settled.

Coordinating Science and Apollo

The question of science on Apollo was far down the list of priorities in the Office of Manned Space Flight during 1961, behind such overriding questions as the choice of mission mode and the configuration of the launch vehicle. Elsewhere, however, stirrings of scientists' interest in the lunar mission began to appear. The director of MIT's Instrumentation Laboratory, which was designing the navigation system for Apollo, proposed that at least one Apollo crewman should be a scientist, since the major interest in the moon would be scientific. Furthermore, he said, it would be easier to train a scientist to pilot the spacecraft than to make a scientist out of a test-pilot astronaut.[19] A similar suggestion was made by a group of scientists working with the Lunar Exploration Committee of OSS, who asserted that the scientist should be a geologist, or at least a scientist well versed in geology and geophysics. They further proposed that NASA begin to recruit astronaut trainees from the ranks of professional scientists.[20]

That the Office of Manned Space Flight and the Office of Space Sciences would have to coordinate their efforts became evident early in 1962. When Homer Newell appeared before the Space Sciences Subcommittee of the House Committee on Science and Astronautics to defend NASA's authorization request, he was pointedly questioned about the support his office was providing for Apollo.[21] This line of questioning evidently perturbed Newell, for he subsequently wrote a personal letter to the subcommittee chairman explaining that the specific information needed by Apollo would become available through the normal course of lunar scientific investigations. Newell acknowledged the importance of the lunar landing, but could not agree that concentrating on the immediate needs of Apollo's engineers would best serve the overall space program. Newell's attitude gave rise to a feeling, even within OSS, that "space sciences was rather unbending in not getting scientific data which would assist the manned program," in the words of a Langley official. Langley had proposed that future Ranger missions should carry an experiment to measure the load-bearing capacity of lunar soil intended to assist in the design of the Apollo lunar landing craft, but the proposal had been rejected in favor of a purely scientific exercise in lunar seismometry.[22]

Newell, far from being indifferent to the needs of Apollo or unconscious of its importance, was simply trying to conduct his programs in the best interests of space science. Mindful of space scientists' increasing discontent over Apollo and its effect on NASA's budgets, he was trying to avoid alienating his major constituency. As nearly as Newell could make it, the Office of Space Sciences was run along lines that suited the scientific community. Advice on policy—the general lines that OSS programs should follow—was provided by the Space Science Board of the National Academy of Sciences and reflected the best consensus the space science community could reach. The content of space science projects was determined (within rather broad limits) by the interests of individual investigators, evaluated and endorsed by the Space Sciences Steering Committee and conducted under the direction of the investigator who proposed it. No one in OSS would have dreamed of telling scientists what experiments they should conduct with the expectation of having their instructions followed or their advice appreciated. Indeed, had anyone in OSS attempted to direct the course of a scientist's experiment, he would have brought down the wrath of the entire scientific community on the space science program. The prerogative of individual scientists to explore problems of their choice, with the endorsement of their scientific peers, is one of the hallmarks of basic research, and probably the most jealously guarded.[23] While Newell might have been able to find a way to supply the data Apollo needed, he risked losing the confidence of the scientists in doing it. At a meeting of OSS field directors in June 1962, Newell's director of Lunar and Planetary Projects reiterated that "pure science experiments will provide the engineering answers for Apollo."[24]

The issue was focused more sharply in mid-June, when Brainerd Holmes issued a document specifying the information Apollo required: the radiation environment in cislunar space, the physical properties of lunar soil, and the topography of the moon, including photos and maps to permit selection of a landing site. All groups conducting lunar investigations were asked to give top priority to obtaining the specified data.[25] The curt wording of the document and its assumption of overriding importance for the lunar landing project were not calculated to win friends in the space science programs, where managers had been struggling with the Jet Propulsion Laboratory over the same issue on Ranger and Surveyor. As events developed, it was JPL's director, William C. Pickering, who forced the issue later in the summer by urging Associate Administrator Robert C. Seamans, Jr., to seek an agreement between the Office of Space Sciences and the Office of Manned Space Flight that would allow JPL to define Ranger's objectives more clearly. Pickering's inclination was to support Holmes, because the science experiments were among Ranger's prime sources of difficulty.[26]

While unintentionally stirring up resentment on one front, the Office of Manned Space Flight was more cooperatively seeking assistance on another. In March 1962 the Space Sciences Steering Committee, at OMSF's request, established an ad hoc working group to recommend scientific tasks to be performed on the moon by the Apollo crews. Not less important, the group would recommend a course of scientific instruction for astronauts in training. Chaired by Charles P. Sonett of the Lunar and Planetary Programs Office, the working group initially included five members from the Office of Space Sciences and one from the Office of Manned Space Flight; membership was later expanded to thirteen and a roster of some dozen consultants was added.[27]

The Sonett committee first met on March 27, 1962, to hear William A. Lee, assistant director for systems in OMSF, explain what his office wanted it to do.

19

As the minutes of the meeting recorded it, Lee's approach was far from peremptory and demanding; rather, he explained,

> the Office of Manned Space Flight now wants very much to have on Apollo moon flights experiments which are of fundamental significance scientifically. They would like the Working Group to consider these experiments without constraints set in advance by OMSF. OMSF will attempt to tailor flights and flight equipment to meet these needs. As examples, it is expected that the duration of stay on the moon may be largely determined by space science requirements and that if necessary, one or more members of the crew could be professional scientists trained as test pilots. . . . They welcome suggestions from the Group as to astronaut selection procedure with respect to scientific background; OMSF will handle physiological and psychological factors in astronaut selection. . . . Unmanned and earth-based scientific experiments which tie in with or prepare for manned experiments will be undertaken by OMSF as part of Apollo; suggestions by the Group are desired.[28]

At the committee's third meeting on April 17, Lee presented the engineering guidelines to be considered by the group in proposing experiments. OMSF expected to fly the first mission "before 1970" and subsequent missions at intervals of about six months. The mission mode, not yet selected, was not to be considered by the committee, nor were engineering and operational details of the flights. Landing sites would be chosen by manned space flight officials, but the Sonett committee was encouraged to indicate the desirability of particular areas. OMSF considered it "possible that one of the flight crew might be a professional scientist trained to perform flight operations. . . . [But this] would significantly complicate our selection and training program, and should not be done unnecessarily." Lee urged the committee to assume no additional constraints. Even if some experiments required difficult engineering development, OMSF wanted the committee to list them so that the requirements could be considered in designing the Apollo spacecraft.[29]

The guidelines Lee provided stated that the Office of Manned Space Flight would consider the committee's report "a major factor" in determining some characteristics of the proposed missions. For example, the planned number of missions ("more than one but less than ten") and time to be spent on the lunar surface ("between 4 and 24 hours") might be strongly influenced by scientific considerations. If enough worthwhile scientific work could be done, "stays up to 7 days are not impossible." Similarly the planned payload (100 to 200 pounds, 45 to 90 kilograms) might be increased and the mobility of the astronauts on the surface might be extended, for example by providing a motorized vehicle and tailoring the space suit for increased ease of manipulation, if the increased scientific return justified the added expense. A soft-landing unmanned supply vehicle carrying up to 30,000 pounds (13,600 kilograms) of support equipment and supplies was under active consideration; this vehicle might carry additional heavy or bulky scientific equipment.[30]

Lee's presentation indicated that the Office of Manned Space Flight was willing to be as accommodating as possible in providing for scientific exploration of the lunar surface. But the qualifiers in Lee's guidelines suggest that OMSF left itself many escape clauses that could have important effects on the scientific program.

With these guidelines in mind, the Sonett committee began to collect suggestions for lunar surface experiments. Their criteria, established early, were simple: experiments must be scientifically feasible and important, capable of being performed only on the moon, significantly improved by having a human aboard, and likely to lead to additional scientific and technological progress. Three basic types of experiments were suggested: measurements and qualitative observations to be made by the astronauts on the lunar surface; experiments to be performed on samples selected and brought back by the crews; and instruments to be emplaced by the astronauts and left on the moon to transmit data to earth.[31]

A point that seriously concerned the Sonett committee was the background and training of the lunar explorers. As a basic requirement, the group suggested sufficient scientific judgment and maturity to recognize and act appropriately upon unexpected phenomena. They noted that scientist-astronauts might be caught in a serious conflict between acquiring proficiency in spacecraft operation and maintaining their skills in research[32]—a question that was to complicate relations between the science community and NASA's operations experts for the next 10 years.

Sonett's committee submitted a draft report in early July 1962; its recommendations, considered in the following weeks by the first of NASA's Summer Studies and endorsed by the external scientific community, would form the basis for the initial planning of Apollo's lunar exploration program. Meanwhile, the committee's work, together with the controversy over the content of the Ranger missions, emphasized the desirability of continuous contact between the offices of Space Sciences and Manned Space Flight on scientific matters. The experience in developing the scientific exercises carried out by the Mercury astronauts showed that overlapping responsibilities requiring close supervision would develop when science went aboard manned spacecraft, and some kind of formal liaison needed to be established.

Homer Newell moved to provide coordination in September 1962 when he proposed to establish a Joint Working Group to replace the ad hoc arrangements that were proving cumbersome. The basic tasks of this group would be to recommend to manned space flight planners a detailed program of scientific exploration, and to suggest to the Office of Space Sciences a program of data acquisition to ensure that Apollo's needs were met. It would also keep the field centers and outside scientists informed of the science programs planned for manned space flight. In carrying out its functions, the group would have the implicit responsibility of assuring that NASA's interface with the scientific community remained where it belonged—within the Office of Space Sciences—and to assure that each program office's projects were maximally responsive to the interests of the other. The group's chairman was to be assigned to OSS for administrative purposes but functionally would be a member of both offices, reporting to the Lunar and Planetary Programs Office in OSS and the Office of Systems Studies in OMSF.[33] Newell would later remark that this official "had two bosses to try to satisfy, which is universally recognized as unsatisfactory"[34]; but no better alternative seemed available.

To chair this Working Group, Newell appointed Eugene M. Shoemaker, a geologist from the U.S. Geological Survey who had recently been assigned for a one-year stint to NASA from the Survey's Astrogeology Branch. Shoemaker had received his doctorate from Princeton under Harry H. Hess, chairman of the Space Science Board from 1962 to 1968 and an enthusiastic promoter of space science.[35] Since October 1961 Shoemaker had been a co-investigator for the television experiment on Ranger.[36] Having been a space enthusiast since pre-Sputnik days, he had consciously fashioned his professional career with an eye to becoming a scientist-astronaut.[37] Although he never realized that ambition, for seven years starting in 1962 he would contribute to the design of Apollo's lunar surface activities and help to train the men who would be his surrogates on the moon.

For the first several months Shoemaker's new job took him into a tangle of uncharted responsibilities and unclear jurisdictions. At Headquarters, Newell's Office of Space Sciences had asserted its responsibility for all of NASA's science, manned or unmanned; in the opinion of many scientists in OSS, neither the Office of Manned Space Flight nor the Manned Spacecraft Center had a staff qualified to manage scientific experiments, but both were considering adding their own science people.[38] Newell and his staff expected to plan and develop the manned science program and to help select and train the astronauts. The Office of Manned Space Flight, foreseeing the many complex interactions that such divided responsibility would require, felt that it was simply not feasible to allow anything—even the science—to be managed by another office.[39] When scientists showed interest in the Mercury project, the Manned Spacecraft Center (MSC) had quickly moved to establish a Mercury Scientific Experiments Panel to screen proposed scientific observations and prevent any interference with the program's primary purposes.[40]

Shoemaker spent much of 1962 devising a structure for the working group and soliciting support from the centers.[41] Toward the end of the year he recommended to Newell that the group should comprise an executive board and two panels: one on data requirements generated by OMSF, the other on scientific missions recommended by OSS for manned space flight. The executive board would recommend projects to the respective directors (Holmes and Newell).[42] It would, however, neither initiate nor manage specific scientific projects.[43]

In the first months of 1963 Homer Newell moved steadily to bring manned space science under his office's direction. He also began a campaign to persuade the Manned Spacecraft Center that science was something more than a means of acquiring the data required to design their spacecraft.[44] Shoemaker, who had been urging MSC to start training astronauts in geology, had found a few receptive minds at the Houston center—among them Max Faget, director of engineering and development—who understood that "it wouldn't look very good if we went to the moon and didn't have something to do when we got there."[45] In late March, Faget drafted a letter for MSC Director Robert R. Gilruth's signature, formally requesting the U.S. Geological Survey to assign six of its scientists to the Houston center. Newell had paved the way for this cooperation in informal discussions with the Survey's director, who agreed to Gilruth's request.[46] Newell saw no need to establish a geological competence in NASA, and both he and Shoemaker preferred to have the Geological Survey staff the Apollo science program rather than geologists hired by MSC.

Meanwhile Newell and Holmes, anticipating a need for increased cooperation as Project Gemini progressed, were negotiating a formal division of responsibility for manned science projects between their offices. Newell insisted that the

Office of Space Sciences had the responsibility to solicit, select, and approve all of NASA's scientific experiments; Holmes similarly asserted the Office of Manned Space Flight's responsibility to approve any hardware that went into a manned spacecraft and any procedure that affected the flight plan. Between the selection of an experiment and its execution on a flight was a large area where responsibility for funding and management of design, fabrication, and integration into the spacecraft of the scientific instruments had to be worked out.

After considerable negotiation, deputies for Newell and Holmes signed a memorandum of agreement on July 25, 1963, spelling out the responsibilities of their two offices. Planning and development of all manned space science projects was assigned to the Office of Space Sciences, as well as any research and development necessary to support them. This meant that OSS would select the experiments and principal investigators and undertake preliminary development of the necessary instruments, in consultation with the Office of Manned Space Flight and the appropriate field center. OMSF agreed to develop and integrate flight hardware and work the scientific objectives into mission plans, usually acting through the Manned Spacecraft Center. OSS took on the job of formulating a science training program for the astronauts, which OMSF would conduct. Newell's office would also establish the scientific qualifications for scientist-astronauts (though none had yet been selected) and would participate in selection of scientists for astronaut training. Each office would budget the funds for its part of experiment development. Fabrication, testing, and installation of the flight instruments would be supervised by the same team that monitored its design and early development.[47] This agreement proved to be about as workable an arrangement as possible, given the division of responsibility inherent in the Headquarters organization. With minor changes, it governed OSS-OMSF relations in manned space science throughout the Apollo program.

While the new agreement was being worked out, Newell reorganized Shoemaker's working group as the Manned Space Science Division on July 30, 1963. The new division reported to both program offices as before, but at a higher level, and it would be the focus of OSS's management of manned science experiments. Shoemaker, however, did not stay to direct the lunar science program from this position. When his tour of duty with NASA ended in November 1963 he returned to the Geological Survey to continue his work with Ranger and Surveyor. He was succeeded by Willis B. Foster, who had served in the Pentagon as Deputy Assistant Director for Research in the Directorate of Defense Research and Engineering.[48]

Reorganizing Around Apollo

The establishment of the Manned Space Sciences Division coincided with a major reorganization of Headquarters. From 1961 to November 1963 the field centers reported to the Associate Administrator but might be conducting projects under the authority of any or all of the Headquarters program offices. Under the new organization each center reported to one of three program offices, the heads of which were now designated Associate Administrators for Manned Space Flight, for Space Science and Applications,* and for Advanced Research and Technology. This change greatly simplified the lines of authority between the program offices and the centers, as well as freeing NASA's three top managers to attend to matters of broader concern. The greater autonomy of the program offices, however, tended to make them more self-sufficient and parochial and to make interoffice cooperation more cumbersome.[49]

In September 1963 Brainerd Holmes, who had steered the lunar landing program through some of its most critical decisions, left NASA to return to private industry. His replacement was George E. Mueller. Headquarters scientists did not know what to expect from the new chief of manned space flight, but they could hardly anticipate less consideration than Holmes had given their projects. One historian has characterized Holmes as "masterful, abrasive, and determined to get what he needed to carry out his assignment, even at the expense of other programs."[50] Considering the state of the manned space flight organization when Holmes took charge and the development problems that had to be solved to put men on the moon by 1970, it is hardly surprising that he gave science such a low priority; but the science community was not inclined to accept that as an excuse for his indifference to their suggestions.

Mueller, like Holmes, was an electrical engineer. He had managed Air Force missile and space development projects for Space Technology Laboratories, Inc., for five years before joining NASA. But he also held a Ph.D. in physics, and his experience included several years of teaching and research at Ohio State University;[51] it was at least conceivable that his attitude toward science might be different from Holmes's. No less committed than Holmes to the success of Apollo and no less determined to have his own way,[52] Mueller came to Washington about the time the public debates concerning Apollo and the space science program had subsided (see Chapter 1), so he was in a position to see that it would be politically advantageous to accommodate the scientists if he could.

Mueller moved quickly to reorganize his office and strengthen its lines of communication with the centers. He also brought in several high-ranking Air Force officers familiar with his management style to fill key positions in OMSF. One of Mueller's first priorities was to evaluate the general health of the Apollo project. He detailed two experienced Headquarters officials to study the progress of its components and estimate its chances for success by 1970. After surveying current plans in light of recent progress and anticipated funding, they reported that in their judgment the project had about one chance in ten of meeting its stated goal. Mueller then imposed some drastic changes on the Saturn program: he canceled the scheduled earth-orbital test flights of the Apollo spacecraft on the Saturn I and ordered that the Saturn V be tested "all up"—with all stages and systems complete and functioning from the first test flight—rather than proving each stage's readiness separately.[53] Then, while the centers adjusted to their new boss's ways, he turned his attention to the science program.

Mueller believed that his office had to have ultimate control over every part of every manned space program, including the science experiments; and by the

---

*Newell's responsibilities were extended to encompass the space applications program (communications, weather, and navigation projects), which since 1961 had been managed by the Office of Applications (abolished in the reorganization).

end of the year he had decided to set up a Manned Space Flight Experiments Board (MSFEB) to review all experiments proposed for manned missions. The board's charter established four categories of experiments (scientific, technological, medical, and Department of Defense*) and the channels through which they were to be submitted. Assessment of the scientific merit of a proposed experiment was left to the sponsoring agency. The board would assess the operational feasibility of each proposal by referring the experiment plans to the appropriate field center (usually MSC) for review. In case the board could not agree on whether to fly any particular experiment, Mueller, as chairman, could make the final decision. Experiments accepted by the MSFEB and assigned a priority comprised a list from which those to be flown on a particular mission were to be taken. Mueller assigned responsibility for developing the flight hardware for each experiment to the field center involved (again, normally MSC), which appointed a technical monitor to work with the experiment's principal investigator in developing the flight-qualified instrument.[54]

Not everyone was happy with the new experiments board; some space science officials protested that Mueller was usurping their prerogatives to select and evaluate experiments.[55] Experimenters complained that Mueller's system took too much time and paperwork and unduly increased the cost of manned space science. Nonetheless, Mueller had to ensure that a proposed experiment was compatible with the spacecraft in all respects and that it could be carried out without interfering with mission operations. As long as OMSF was still learning how to get people to the moon and back, science took second place when it came aboard at all.

1963: Progress and Prospects

As 1963 ended, manned space flight officials could look back at two and a half years of intense activity and considerable progress. Project Mercury had flown four earth-orbital missions, the last one remaining aloft for 34 hours. A new project, Gemini, was under way; its objectives were to explore some of the problems posed by the lunar landing mission, especially rendezvous.[56] Apollo managers had made several critical decisions, including the choice of lunar-orbit rendezvous, the basic design of the two lunar spacecraft, and the configuration of the three-stage Saturn V lunar launch vehicle. Nine of the 12 men who would ultimately walk on the moon were in training for their part in the program. A new NASA Administrator, James E. Webb, had taken control, and the agency had been restructured for better support of Apollo. The new president, Lyndon Baines Johnson, had more experience with the space program than any other politician in Washington and would remain committed to it despite severe criticism and unforeseen tragedy.

---

*For the involvement of DoD in the early manned space flight programs, see Barton C. Hacker and James M. Grimwood, *On the Shoulders of Titans: A History of Project Gemini*, NASA SP-4203 (Washington, 1977), pp. 117–22, and W. Fred Boone, "NASA Office of Defense Affairs: The First Five Years," NASA Historical Report HHR-32 (Washington, 1970).

The dominant position Apollo occupied in the space program was indicated by the changes imposed on Ranger and Surveyor in 1962 and 1963. Space science officials tried hard to preserve the priority of science in both projects, but with limited success. After thorough examination of the lunar exploration projects, the Space Science and Space Vehicle Panels of the President's Science Advisory Committee came down on the side of the Office of Manned Space Flight. In a mid-October report they concluded that Ranger and Surveyor must be planned so that the first few successful missions would provide the information engineers needed to design the lunar landing craft, and that those projects' technical management and funding should be integrated with Apollo's. Notwithstanding that concession, the panels insisted that after the first couple of landings science should determine the content of lunar missions and that the lunar surface activities must be planned with the full participation of the scientific community.[57]

Lunar science planning had begun, in the broadest sense, by the end of 1963, but how much science could actually be done during Apollo was still a question. Planners were contemplating as many as 10 lunar missions; what could be done on each one would have to await events. If the first lunar landing attempt did not succeed, and several attempts had to be made, science might have to wait. As 1964 began, Headquarters was preparing to deal with the details of the lunar missions: choosing landing sites, deciding what specific experiments would be done on the moon, and selecting the experimenters.

3

APOLLO'S LUNAR EXPLORATION PLANS

The years from 1963 to 1966 were the busiest of the lunar landing project, no less for science planners than for the spacecraft and rocket builders. After the major decisions made in 1962 provided a basis for planning, the Office of Space Sciences called on its academic advisory bodies to define first the broad outlines of a program of lunar and planetary exploration, then more specific plans for the moon, based on the clearer definition of the Apollo project that was emerging during the period.

Scientific Interest in the Moon

In the two and a half centuries since Galileo first turned his crude telescope on the moon, astronomers have mapped its surface features, measured the height of its mountains and the depth of its craters, and calculated its size, mass, and orbital parameters with increasing accuracy over the years. Even so, the best optical telescopes, under the best observing conditions, could not show any detail smaller than about 300 meters across (1,000 feet, about the size of the capitol building in Washington); so information concerning the nature and texture of the surface was limited.

To the telescopic observer, the moon's surface shows two distinct types of regions: 1. the maria (Latin = "seas"; singular *mare*)—dark, apparently smooth, and roughly circular areas extending over hundreds of kilometers; and 2. the highlands, mostly mountainous regions much lighter in color. Among the most striking features are the craters—tens of thousands of circular depressions ranging in diameter from 180 miles (290 kilometers) down to the limit of telescopic resolution. Most large craters have high walls and a depressed floor, and many have a central peak or ridge. Some are centers from which streaks of light-colored material (rays) extend for considerable distances. Besides the craters, the lunar surface shows domes, ridges, and rilles—long, narrow channels resembling dry watercourses—that run for several kilometers.[1] Measurements of reflected light indicate that the lunar surface is covered with a layer of finely pulverized material. Airless and arid, the lunar surface is subjected to temperature changes of more than 200°C (360°F) during the course of a lunar day.[2]

For decades these features fascinated astronomers and cosmologists, who have generated volumes of speculation as to the moon's origin and history. Three hypotheses attracted adherents: one, that the moon was spun off from a molten proto-earth; a second, that the moon was formed in a separate event and later

THE MOON

The earth-facing side of the moon, depicted in an artist's rendition based on scientific findings. North is at the top; east is to the right. Especially noticeable is the contrast between the dark, relatively smooth areas (the "seas" or "maria"), mostly in the upper half of the picture, and the lighter, heavily centered areas in the lower right. The large circular feature in the upper left quadrant is Mare Imbrium, bounded on the southeastern edge by the Apennine Range and on the southwest by the Carpathian Mountains. East of Mare Imbrium lies Mare Serenetatis, with the Taurus Mountains on its eastern edge. Mare Tranquilitatis joins Mare Serenetatis on the southeast. The near-circular area in the upper right is Mare Crisium; below it lies Mare Fecunditatis. Spreading irregularly over the area west and south of center is Oceanus Procellarum. Two smaller maria, Mare Humorum and Mare Nectaris, lie to the south and southeast.

The lighter-colored areas are marked by an abundance of craters, in contrast to the maria; this has led scientists to believe that the maria are younger features, perhaps filled with lava. Lunar craters range in size from hundreds of kilometers to less than a meter. Many of the larger ones have prominent central peaks while others have relatively flat floors. A prominent crater visible here is Copernicus, west-southwest of the center, from which streaks of light-colored material ("rays") extend into Mare Imbrium. Farther west is Kepler, a rayed crater surrounded by a light-colored blanket. In the southwest quadrant, two prominent parallel rays point toward Tycho. Relatively inconspicuous in this view, Tycho is a medium-sized crater with a central peak about one-fourth of the way from the bottom of the photograph to the center.

captured by the earth; and the third, that the earth and moon formed at about the same time in the same region of space (this hypothesis was considered somewhat less likely than the first two). Those who believed the moon was formed by accretion of smaller bodies generally supposed that during its evolution the moon, like the earth, went through a molten stage in which its components were chemically fractionated, with iron and nickel being separated from the rocky minerals as the moon cooled. A smaller body of opinion held that the moon never grew large enough to be completely melted by the heat generated as its component particles coalesced, and would not be chemically differentiated.[3]

All of these assumptions led to difficulties when examined in light of accepted celestial mechanics. If the moon separated from the earth at some early stage in its formation, its orbital plane should lie in the plane of the earth's equator, but it does not. If, on the other hand, the moon was formed elsewhere in the solar system, its capture by the earth would be highly improbable and its present orbit difficult to account for. Finally, if the earth and moon were formed out of the same primordial matter by any mechanism, it is hard to explain the fact that the earth is nearly half again as dense as the moon.[4]

None of the hypotheses could be proved or disproved on the basis of the evidence available from visual observation alone. Both led to logical conclusions concerning the chemical composition and internal structure of the moon that could not be tested. Scientists who held that the moon was once molten regarded the maria as massive lava flows from volcanoes or fissures in the lunar crust, similar to known examples on earth, and pointed to the domes and certain craters that are much like well known terrestrial volcanic structures. Advocates of the "cold moon" theory considered the maria to be the solidified remains of large bodies of molten rock created by collision with meteorites or asteroids, or perhaps by localized heating due to some other cause.[5] The origin of lunar craters was a matter for debate. Some had undoubtedly been produced by impacts of cosmic debris, but strong arguments could be made that others were volcanic in origin. It was generally accepted that the moon's surface, unaffected by wind and water, preserved a record of cosmic events which the earth must also have undergone; on earth, however, the effects have been obliterated by erosion.[6]

Crucial to the confirmation of either hypothesis, or to the creation of an alternative, was information that could be obtained only by direct examination of the moon. Analysis of samples of the lunar surface would show whether the moon was chemically similar to the earth and whether the lunar material had ever been extensively melted. Measurement of the flow of heat from the moon's interior to the surface would show whether it was still cooling from a molten state. Other important investigations included the seismic properties of the moon, which could reveal its interior structure. Some of this information could be provided by instruments, perhaps including remotely controlled samplers capable of returning lunar material to earth. But some tasks, such as examining the moon's surface and selecting samples on the basis of that examination, could better be done by humans. With the advent of the space age, lunar scientists could look forward to sending instruments to the lunar surface; only after the decision to land people on the moon could they hope to send a trained explorer.

Planning for Lunar Exploration

Homer Newell set lunar exploration planning in motion in 1962 with the appointment of the Sonett committee on Apollo scientific experiments and training (see Chapter 2). After three months of consultation with leading experts in the scientific fields related to lunar exploration, the committee outlined its conception of the scope of Apollo lunar science. The primary objectives were examination of the lunar surface in the immediate area of the landed spacecraft, geologic mapping of the landing area, investigation of the moon's interior by means of emplaced instruments, studies of the lunar atmosphere, and radio astronomy from the lunar surface.[7] No specific experiments were recommended, but the criteria used in developing the objectives limited the possibilities. The experiments should be scientifically important and feasible, possible only on the lunar surface and only with a human on the mission, and likely to lead to further scientific and technological development.[8]

A week after the Sonett committee issued its draft report, NASA committed Apollo to lunar-orbit rendezvous. This decision, probably the most thoroughly debated of the entire program, directly affected the scope of lunar science. The mission mode determined how many men and how much equipment could be landed on the moon. Whereas the other possible modes (direct ascent and earth-orbit rendezvous) contemplated a single large spacecraft that would land with its entire three-person crew on the moon, lunar-orbit rendezvous would put down a separate specialized landing craft and a crew of two.[9]

With that decision made, Associate Administrator Robert C. Seamans, Jr., asked Newell to present several questions to NASA's outside scientific advisers for discussion. What should be the preferred scientific objectives of the earliest lunar landings? Would the acquisition of scientific data require more than two persons on the moon at the same time? Were there sufficient scientific reasons to establish a semipermanent station on the moon for extended exploration? Advice on these questions would be important in establishing policy with respect to lunar science and could have direct influence on the design of the lunar landing craft.[10]

An appropriate forum for such a discussion was already in session that summer—the first summer study conducted by the Space Science Board, meeting on the campus of the State University of Iowa (see Chapter 2). Manned space science programs were not the study's primary concern, but two working groups, one on lunar and planetary exploration and one on the scientific role of humans in space, provided opportunities to examine the plans for Apollo's experiments and put some of the scientists' concerns on record.[11]

At the start of the summer study the chairman of the Space Science Board, Lloyd V. Berkner, instructed the participants that their advice on the existence of a national space program and the division of effort among its projects was not sought. Nonetheless, because of mounting fears for what it would do to NASA's space science budget, Apollo was never far below the surface, and objections to the manned programs were repeatedly voiced. Berkner, Newell, and others tried to steer the discussions into more productive channels, such as how the capability being developed for Apollo might be exploited for scientific purposes; but many of the participants could not be dissuaded from protesting the priority assigned to the lunar landing project.[12]

Seamans's questions were aired, in substance at least, and the summer study's final report contained recommendations concerning the points he had raised. The working group on the scientific role of humans in space endorsed the findings of the Sonett committee but recommended that experiments that only used the moon as a base for observation (e.g., radio astronomy) be relegated to later missions. It found that there was indeed a valid scientific requirement for a lunar surface laboratory for long-term investigations. The working group on lunar and planetary exploration reached similar conclusions and called for increased emphasis in the Ranger and Surveyor programs on providing necessary engineering information for Apollo.[13]

Of much more concern to the summer study participants was the scientific competence of the men who would land on the moon. Acknowledging a continuing need for astronauts whose primary skills were in spacecraft operation, the study report nonetheless urged the inclusion of a professional scientist among the crew of the first lunar landing mission. To ensure that such a person would be ready for the first landing, NASA should recruit qualified scientists at once and begin training them as astronauts. The science community insisted that these scientist-astronauts be given the means to maintain their scientific competence while acquiring piloting skills. To that end, a space institute should be established convenient to the astronaut training center, a facility "of the very highest scientific calibre [maintaining] liaison with major centers of research in the space sciences [and functioning] as a graduate school offering advanced degrees in various fields of science." It should either be operated "under contract with a major university" or administered by "that office of NASA responsible for scientific research and planning,"[14] (i.e., the Office of Space Sciences). This "space university" proposal—which Newell later remarked never had the slightest chance of being accepted by NASA[15]—was soon abandoned, but the demand for scientist-astronauts was reiterated by the science community right down to the end of the Apollo project.

With the modifications made by the Iowa summer study and with the imprimatur of the National Academy of Sciences, the Sonett committee's recommendations formed the framework of Apollo's initial lunar science planning. Shortly after its establishment in July 1963 the Manned Space Science Division promulgated the first scientific guidelines for Apollo, which reflected this preparatory work. The primary scientific activity was to be the study of the moon itself. First priority among the various lunar studies was given to geologic mapping, followed by collection of samples for return to earth and emplacement of instruments to return data by telemetry.[16]

Headquarters-Center Relations in Science

While working out arrangements for cooperation with the Office of Manned Space Flight, Newell and his staff also had to establish working relationships with the Manned Spacecraft Center (MSC). In early 1963 MSC's only experience with science had been the visual and photographic experiments made on the first three Mercury orbital missions.[17] These somewhat hurriedly improvised "experiments" produced some useful data, but mainly they served to show that people could conduct scientific exercises in orbit.[18] They also showed that both scientists and engineers had much to learn about each other's objectives and methods. One result was that MSC acquired a reputation among space scientists of being at best indifferent and at worst hostile toward scientific investigations.[19] Eugene Shoemaker, always an enthusiastic supporter of manned space flight, was "utterly dismayed" by the attitude of the MSC representatives at the Iowa summer study: "we don't need your help; don't bother us." This experience led Shoemaker to agree to spend a year at NASA Headquarters to try to establish a lunar science program; unless someone concentrated on that task, lunar science might never get done at all, because "there was no planning for it [and] no program for it."[20]

Some of MSC's indifference to science was the predictable consequence of the formidable development tasks the center faced and its intense concentration in 1962 on learning to operate in space. But Newell felt that the Houston center's engineers and managers (also engineers, for the most part) simply did not appreciate what space science and manned space flight could do for each other. Houston had no scientific research under way; the few scientists who worked there were mostly inexperienced in research and served almost entirely in support roles, providing data to the engineers. Newell spent considerable effort in 1963 trying to create a more receptive attitude toward science at MSC.[21] One of Shoemaker's first accomplishments after he came to Headquarters was to persuade MSC to expand its small Space Environment Division, a branch of the Engineering and Development Directorate that existed mainly to collect environmental data affecting the design of spacecraft and mission plans. One geologist joined the division in 1963, and a team of specialists from the U.S. Geological Survey was assigned later that same year. Their functions were to set up a research and training program in geology, develop a model of the lunar surface for use by the spacecraft designers and mission planners, assist in the evaluation of lunar scientific instruments, and develop plans for geologic field work on the moon.[22]

Lunar surface science, though important, was only part of the larger question of manned space science, and Newell, looking ahead to the earth-orbital flights of Gemini and Apollo, wanted to establish a place for science on those missions as well. The agreement worked out between Newell and Brainerd Holmes for developing experiments called for the appropriate manned space flight center (usually MSC) to oversee the development of experiment hardware once the basic design had been worked out by the Office of Space Sciences. As OSS saw it, this would require more scientific competence than the Houston center had. Up to mid-1963 MSC's concern for experiments had been limited to assuring that they fit into the spacecraft and the flight plan and did not compromise a mission, a task carried out by an Experiments Coordination Office in the Flight Operations Division.[23] Now, OSS saw a need to establish what would amount to a space sciences division at Houston. Discussions with MSC produced agreement that the Space Environment Division would be the nucleus of the prospective science branch.[24]

The Office of Space Sciences and the Manned Spacecraft Center spent the rest of 1963 defining their relationship. Newell firmly maintained his office's responsibility for all of NASA's science programs, while MSC occasionally displayed reluctance, to say the least, to accept direction from Headquarters.[25] In the old days of NACA the field laboratories had enjoyed considerable independence in the conduct of their programs, and all of MSC's top managers were old NACA hands. At times they seemed inclined to insist on running their programs their way, including the science. But by the end of the year MSC had agreed in principle to set up a scientific program manager on Director Robert Gilruth's immediate staff, and Headquarters and center elements were beginning to work out a description of that person's responsibilities.[26]

Lunar Surface Experiments

Early in 1964 the Office of Space Science and Applications (OSSA) began to define the Apollo lunar science project more narrowly. Rather than issue a request for experiment proposals to the scientific community at large, which was the usual procedure for soliciting experiments, Newell and his manned space flight counterpart George Mueller agreed that initially a few selected scientists should be called upon to identify the most important experiments within the areas agreed on by the Sonett committee and the Iowa summer study. A representative of Headquarters's Manned Space Science Division and the chairman of the Space Science Board then compiled a list of experts who met on January 30 to begin the process.[27] The first investigations agreed on by this planning group comprised geology (field geology, petrography and mineralogy, and sample collection), geochemistry, and geophysics (seismology, magnetic measurements, heat-flow measurements, and gravity measurements). For each area the group suggested several prominent scientists to work out detailed experiment plans.[28] In April the first Apollo science planning teams, groups of experts in subdisciplines of the earth sciences, began meeting to define more specifically the lunar surface experiments and instruments. Later, teams would be established for lunar atmospheric studies and biosciences.[29]

At Houston, meanwhile, mission planners had started defining the lunar landing mission in detail. In late 1963 the first set of operational ground rules was established, listing mission objectives and constraints and identifying critical points where a flight might have to be aborted or diverted to an alternate mission in case of failure of some essential system. Using these ground rules and the latest available weight and performance data for the launch vehicle and spacecraft, planners then prepared the "reference trajectory," a detailed description of the mission from liftoff to splashdown.* For planning purposes the reference trajectory listed 10 landing sites within a zone 85 miles (137 kilometers) wide and 1,480 miles (2,380 kilometers) long stretching along the equator on the moon's

---

*The reference trajectory was one of the most important mission planning documents. It enabled flight planners to evaluate the effect of changes in spacecraft weight, propulsion capability, mission requirements, etc., on every phase of the mission. An initial reference trajectory was often based on many assumptions, but as test data and operational experience accumulated, the trajectory was updated.

visible side. It assumed that the lunar landing module would stay on the moon for 24 hours.[30]

Science and mission planning came together in mid-June of 1964 when Houston hosted a lunar landing symposium to give each group a look at the other's preliminary plans. MSC representatives presented their concepts of a lunar landing mission, described the two spacecraft and what they could be expected to carry to and from the moon, and outlined the current astronaut training course. Members of the Apollo science planning teams sketched out their tentative plans for surface activities and experiments. After four days of discussion, both groups went home to refine their concepts.[31]

The June symposium defined the operational bounds within which the lunar science planning teams had to work, and after five more months of deliberation the planning teams' report went to the Office of Space Science and Applications. Distributing the report for comment, Headquarters noted that its recommendations would eventually become the basis for a lunar exploration science program definition document on which final instrument and experiment designs would be based. First, however, OSSA was planning to have the report discussed at another summer study, scheduled for 1965.[32]

As the planning teams saw it, the ultimate objective of the lunar science program was simply the same as that of all science: to add to human knowledge. More specifically, lunar studies could lead to a better understanding of the solar system and its origin, of primary forces that shaped the earth, and of geological processes whose effects have been obscured on earth by erosion and other secondary processes not operating on the moon. (Making the obligatory gesture toward practical results, the report mentioned that lunar studies "may have direct bearing towards more intelligent search for mineral resources on Earth," though it did not specify how.) The larger objectives, however, could scarcely be met in the short times the early Apollo missions would stay on the lunar surface. Really productive lunar studies required more time on the moon, greater mobility for the astronauts, and logistic support.[33]

For the approved Apollo missions the planning teams stressed field work, sampling, and emplacement of instruments, all planned to yield the maximum information in the time available. Instruments relaying data by telemetry were preferred for measurement of gravity, magnetism, magnetic phenomena, and seismic studies. The geochemistry and bioscience planning teams emphasized the importance of bringing back the greatest possible weight of lunar material in the form of carefully selected and documented samples for laboratory study. Eugene Shoemaker's field geology team stressed visual observation and panoramic photography from the lunar module, followed by surface traverses during which the astronauts would collect samples, emplace instruments, and describe the important geologic features of the landing area. Both the geology and geophysics teams listed the tools and instruments that should be taken along, as well as preliminary estimates of weight, volume, power, and telemetry requirements. The report included tentative operations plans for several lunar surface missions, taking into account operational constraints and certain contingencies that might require changing plans during the mission.[34]

With these recommendations in hand, Headquarters and MSC began making plans for managing the experiments. In mid-January 1965 Houston's Space

Environment Division appointed interim coordinators for lunar surface experiments.[35] At the end of the month manned space flight director George Mueller conducted a program review at which he called for a series of studies to evaluate several program management alternatives for Apollo science, from experiment development to handling of lunar samples.[36] Discussions between Headquarters and MSC continued during the spring; Newell's Office of Space Science and Applications continued to hope for the establishment of a separate science organization at Houston.[37] Progress was slowed somewhat by OSSA's apparent difficulty in working out its internal lines of authority.[38]

In late February Newell and Mueller met to formulate policy for managing and funding science experiments in manned space flight. An afternoon's discussion produced agreement on a division of responsibility only slightly different from that adopted by Newell and Brainerd Holmes two years earlier (see Chapter 2). Newell's Office of Space Science and Applications would publicize the science opportunities in the manned programs, solicit experiment proposals, and make the initial selection. After further evaluation, which would include constructing a "breadboard" model of the instrument to demonstrate its feasibility, OSSA would select experiments and experimenters and arrange for construction of prototype instruments. Mueller's Office of Manned Space Flight would then develop flight-qualified instruments, integrate them into the spacecraft, work the experiments into the flight plan, and collect the data produced in flight. OSSA would arrange for distribution, analysis, and dissemination of the data. Deputies for Newell and Mueller formalized details of this agreement later in the year.[39]

MSC and OSSA worked for several months on a procurement plan for Apollo's lunar surface experiments, submitting it to Mueller in May 1965. Mueller decided that a two-phase procurement was advisable, the first phase to better define the instrument package and the second to build the instruments. Several contractors would be selected to conduct the definition studies and one of them would be picked to build the instrument package.[40] In June Houston sent out requests for proposals to conduct six-month definition studies for a lunar surface experiments package. Nine companies responded, and three* were selected in early August.[41]

Woods Hole and Falmouth Conferences, Summer 1965

In the summer of 1965 the Space Science Board again convened a conference to consider NASA's plans for space research. The Iowa summer study three years before had reviewed current programs and recommended changes of emphasis; the 1965 study at Woods Hole, Massachusetts, was charged with recommending directions and priorities for the next 10 to 20 years. The major topics considered were lunar and planetary exploration, astronomy, and the role of humans in space exploration.[42]

The Woods Hole conference report made many detailed recommendations in individual areas of research. Space research, the study found, required the use

---

*Houston Division of Bendix Corp., Ann Arbor, Mich.; Space-General Corp., El Monte, Calif.; and TRW Systems Corp., Redondo Beach, Calif.

of ground-based observations as well as satellites, sounding rockets, and balloons. Throughout, the report stressed *balance*: between manned and unmanned programs, between lunar and planetary exploration, and between ground-based studies and research in space. The salient conclusions of the study were easily summarized. Planetary exploration was judged to be the most rewarding scientific objective for the post-Apollo period, with Mars being the most interesting target. Concerning humans in space, the participants agreed that "The distinction between manned and unmanned programs is an artificial one; scientific objectives should be the determining factors." But before people could be dispatched on missions to the planets, "an orbiting research facility for the study of long-term effects of space flight is essential."[43]

The 1965 conference was somewhat more sanguine than the 1962 study concerning the role of humans in space science and somewhat more tolerant of the Apollo program. The working group on the role of man noted that

> Few . . . scientists would attempt to justify the entire cost of developing manned space flight solely on the basis of its "scientific value;" however, most scientists would agree that this capability, when developed, should be utilized for scientific purposes whenever it seems possible to do so.

The group did not concede that people could be replaced by instruments in every imaginable case:

> Manned intervention [in an unmanned system] increases reliability through the possibility of extending the lifetime of spaceborne equipment almost indefinitely by means of repair and replacement. . . . Man greatly increases the flexibility of the system, for he can decide [how best] to use an instrument [and] can make alterations or improvements in the instrument itself. Data transmission can . . . be virtually eliminated for manned experiments, and the design of an instrument can be simplified. . . .

Not only that, but

> if the presence of man in the system were already considered essential, the scientist would assign many scientific tasks to man because of the greater reliability and flexibility he would bring to the system.[44]

On the question of selection and training of scientists for space missions, the scientists were of much the same mind as those in 1962:

> For some tasks—those requiring scientific insight—it would seem better to have a scientist possessing judgment, experience, and imagination and to train him as an astronaut to the extent required.

Again conceding that test pilots had the edge in some areas, the working group nonetheless concluded that

> the astronaut selection and training program [should] take into account progress in manned space flight. Successes so far[*] may be viewed as evidence

---

*Four successful manned flights had been made in the past three years, including two of the two-man Gemini spacecraft.

that it may be possible to relax the present high physical standards at a pace faster than has yet been contemplated.[45]

Concerning the lunar exploration program, the Woods Hole study report confined itself to defining the major scientific questions in lunar exploration. Three basic problems should be explored: structure and processes of the lunar interior, the composition and structure of the moon's surface and the processes that have modified it, and the sequence of events by which the moon has arrived at its present configuration. As guidelines for exploring these problems, the report listed 15 specific scientific questions that should direct lunar exploration, both manned and unmanned (see Appendix 3).[46]

One question that assumed increasing importance as planetary exploration became more realistic was the existence of life forms or their precursors on the moon or the planets. The working group on biology concluded that the evolution of organisms or prebiotic materials was most likely to have occurred on Mars. However, the group affirmed the necessity to avoid contaminating any celestial body, including the moon, with terrestrial organisms or organic materials that might invalidate later experiments attempting to detect life. As to the need for protecting earth from biological contamination by material from the moon, the group believed the hazards were small but that NASA should be safe rather than sorry: "the consequences of misjudgment are potentially catastrophic."[47] Opinion on this point was not unanimous throughout the conference, however, for the working group on lunar exploration noted only a "minor possibility of finding prebiotic material, either buried or in sheltered locations."[48]

The Woods Hole conference was, as far as Apollo was concerned, a policy-setting meeting. When it adjourned, many of the participants stayed on at nearby Falmouth for a two-week conference dealing with details of lunar exploration and science on Apollo. The Falmouth conference elaborated on the previous year's work of the Apollo lunar science planning teams in light of plans for manned space flight for the 10 years following the first Apollo landing. As plans then stood, the first few landings—the exact number was not certain—comprised the Apollo program; it was assumed that these landings would be minimal missions to establish confidence in the Apollo systems. They were to be followed by flights using improved spacecraft that could carry larger payloads and stay longer on the lunar surface.* Ultimately, perhaps in the last few years of the post-Apollo period, a lunar base might be available for scientific studies. Falmouth planners structured their discussions around this idealized schedule.[49]

At Falmouth, members of the lunar science planning teams collaborated with NASA and academic scientists to take the first steps in detailed scientific operations planning. As specifically as they could, disciplinary working groups laid

---

*In 1965 plans for manned space flight after Apollo were still in the formative stage. OMSF had conducted several studies on "Apollo Extension Systems" to determine how far the propulsion, life-support, and electrical power systems on the Apollo spacecraft could be upgraded without major redesign. These "extended Apollo" components would be used in missions—earth- and lunar-orbital flights as well as lunar surface missions, lasting from 10 to 45 days—that would sustain the manned program until the next major program could be defined. See W. David Compton and Charles D. Benson, *Living and Working in Space: A History of Skylab*, NASA SP-4208 (Washington, 1983), Chaps. 1, 3, and 5; also George Mueller's presentations to the House and Senate space committees in the NASA authorization hearings for fiscal 1966 and 1967.

out their requirements for procedures and equipment. For the first time, an astronaut in training described the mission's essential constraints from the astronaut's point of view, stressing the limitations of space suits, life-support equipment, and operational contingencies (such as an aborted landing).[50] For some scientists this was the first time they had had to consider their own experiments in the context of other scientific work and operational restrictions, and one participant reported a noticeable change in attitudes as the discussions progressed.[51]

In its final report the Falmouth conference summarized its recommendations for the early landings, the "post-Apollo" (advanced) missions, and the more distant future when a lunar base could be contemplated (see Appendix 3). Highest priority on the early landings was assigned to returning the greatest number and variety of samples as feasible; emplacement of long-lived surface instruments was next, followed by geologic exploration of the landing area by the astronauts. The early missions could only sample isolated areas of the lunar surface. A survey of the entire moon, plus detailed studies of the equatorial belt, should be the objective of later missions. These advanced missions, five or six landings flown at the rate of one or two per year, should be supported by an unmanned logistics system that would land additional consumable supplies and scientific equipment. Crews might stay as long as 14 days and explore as far as 15 kilometers (9 miles) from the landing site. The additional equipment should include some analytical instruments for on-the-spot discriminatory tests on lunar material, so that astronauts could select a wider variety of samples. Surface transportation should be provided—a wheeled vehicle with a range of 8 to 15 kilometers (5 to 9 miles) and a flying vehicle that could carry 135 kilograms (300 pounds) of instruments from point to point over a 15-kilometer range. With the flying unit, astronauts could secure samples from otherwise inaccessible locations, such as a crater wall.[52]

Falmouth provided the best scientific advice NASA could get at the time, and its recommendations formed the basis for the earliest mission planning. It was the first of several iterations of scientific planning that would take place during the rest of the Apollo program. NASA would make every effort to carry out as much of the program as could be done within a changing context of available resources. Progress often seemed intolerably slow to some scientists,[53] but in the end a gratifying proportion of the Falmouth recommendations would appear in mission plans.

Site Selection: Ranger, Surveyor, Lunar Orbiter

The visible face of the moon offered scores of interesting sites for scientific exploration. As far back as 1961, Harold Urey, responding to a question from Homer Newell, listed five general regions of high scientific interest: high latitudes, to determine whether water might exist where temperature extremes were less marked; two maria, to determine whether they were of different composition; inside a large crater; near one of the great wrinkles in the maria; and in a mountainous area. Assuming that equatorial sites would draw the most attention anyway, Urey offered no suggestions concerning them.[54] Eugene Shoemaker noted not long afterward that scientific objectives would be subordinate to operational requirements, at least on the early missions, and that the sites most operationally suitable for an Apollo landing—level, featureless maria, most likely—would be the least suited to determination of geologic relationships.[55]

Operations planning for the earliest Apollo missions was designed to assure the safe return of the astronauts. The spacecraft, still attached to the Saturn upper stage, would be inserted into an earth-circling "parking orbit" so that mission control could verify that all systems were working properly. Lunar landings were to be made in direct sunlight and at specific times of the lunar day chosen for optimum visibility; return to earth must take place in daylight; and allowance would be made for possible interference with communication caused by solar activity.[56] These rules, along with constraints on the lunar orbit of the command module, defined a landing zone along the moon's equator within which the first mission would have to land.

Equally important for selecting a landing site was the nature of the lunar surface—how much weight it could support and how many rocks and small craters were present. For this purpose, high-resolution photographs of the lunar surface were required, which were to be supplied initially by the Ranger spacecraft (see Chapter 2). To supplement Ranger's photographs and also to provide direct information about the surface, a second unmanned spacecraft, Surveyor, would be built. One version would soft-land on the moon and transmit scientific data and television pictures; a second version would be an orbiter that would circle the moon, photographing large portions of its surface. But when Apollo engineers specified the detail required for their purposes, the Surveyor orbiter as then defined could not meet them.[57]

To supply the high-resolution photographs needed to certify landing sites, both OMSF and OSSA endorsed a new lunar-orbiting spacecraft. Preliminary studies indicated that the project was feasible, and after briefly considering giving the orbiter to the Jet Propulsion Laboratory, Headquarters officials decided to assign it to Langley Research Center. By the end of August 1963 Langley had drawn up specifications for the spacecraft and camera system and called for bids. Before the year ended, project officials had selected the Boeing Company as the prime contractor. Boeing proposed a solar-powered satellite, attitude-stabilized in three axes, carrying a film camera system designed by the Eastman Kodak Company. Film was to be developed aboard the spacecraft and the images were to be transmitted to earth by an optical scanning and telemetry system. Five photographic missions were planned; the first Lunar Orbiter would be ready for flight less than three years after the contract was let.[58]

In Ranger and Surveyor, the two projects that contributed most to early Apollo site selection, the Manned Spacecraft Center had worked closely with the Jet Propulsion Laboratory to coordinate Apollo's requirements with mission plans (see Chapter 6).[59] As late as 1964, no single organization was responsible for collecting the data needed to evaluate Apollo landing sites and recommending the final choice for each mission.[60] In mid-1965, acting on a recommendation from

Bellcomm, Inc.,* George Mueller formally established an Apollo Site Selection Board to evaluate and recommend landing sites for the Apollo missions. Chaired by the Apollo program manager in the Office of Manned Space Flight, the board would weigh all available scientific and operational considerations and recommend landing sites to Mueller.[61]

As planning for lunar science progressed during the middle 1960s, three key issues emerged: management of the returned lunar samples, selecting and training astronauts for scientific exploration, and the choice of landing sites.

---

*Bellcomm, Inc., was a systems-engineering organization created in 1962 by the American Telephone & Telegraph Co. at NASA's request for the purpose of conducting independent analyses of many aspects of the Apollo program. It employed about 500 people at peak strength (1969). In 1972, its work for NASA completed, Bellcomm was merged with Bell Laboratories. J. O. Cappellari, Jr., "Where on the Moon? An Apollo Systems Engineering Problem," *The Bell System Technical Journal* 51 (5) (1972): 955.

HANDLING SAMPLES FROM THE MOON

The samples of rock and soil brought back from the moon would be a priceless scientific resource, and for scientists to be able to extract the maximum information from them, the samples would have to be carefully protected. Minute traces of earthborne contaminants could lead to completely erroneous interpretations of laboratory results. In 1964 scientists at the Manned Spacecraft Center proposed that NASA provide a laboratory in which lunar samples would be cataloged and subjected to preliminary examination, so that the requirements of principal investigators for specific types of material could be met. There were, as well, certain time-critical examinations that would have to be done as soon as possible after the samples were returned to earth.

To provide for these requirements would have required only a modest facility. But as plans for managing the samples developed, NASA came under pressure from space biologists and the U.S. Public Health Service to protect the earth against the introduction of alien microorganisms that might exist in lunar soil. What would have been a small laboratory designed to protect lunar samples against contamination grew into an elaborate, expensive quarantine facility that greatly complicated operations on the early lunar landing missions.

Early Plans for Lunar Sample Management

Preliminary definitions of the lunar science program noted the importance of laboratory studies on returned lunar material, but offered no suggestions as to how samples should be collected and handled.[1] Neither within nor outside NASA did anyone give serious thought to the details of preserving lunar samples in near-pristine condition until late 1963. Elbert A. King, Jr., and Donald A. Flory, two geoscientists who joined MSC's Space Environment Division that year, were among the first to propose action to protect valuable scientific information that could be lost unless the lunar samples were handled under carefully controlled conditions.

In February 1964 King and Flory put together a concept of a sample receiving laboratory and forwarded it to Max Faget, director of engineering and development at the Manned Spacecraft Center. Their plan called for a small (100 square feet, 9.5 square meters) laboratory in which sample containers could be opened and their contents repackaged under high vacuum (one ten-millionth of atmospheric pressure) for distribution to the scientists who would conduct most of the studies. Remotely controlled manipulators would be used to carry out operations within the chamber, which would be sterile, chemically clean, and used for no other purpose.[2]

Faget recognized the importance of the proposed facility to the lunar science program and encouraged King and Flory to expand their concept. The second version of the "sample transfer facility" was considerably larger and more sophisticated. A 2,500-square-foot (232-square-meter) clean room contained several analytical instruments for performing preliminary tests on the samples. Within this area was a high-vacuum system containing remote manipulators and a separate sterile laboratory for biological testing. The vacuum chamber was equipped to prepare mineralogical and petrological specimens as well as divide and repackage the samples. The atmosphere in the entire area would be closely monitored so that subsequent investigators would know what contaminants might be present in their samples.[3]

These preliminary studies received considerable support in June 1964 when the Apollo science planning teams (see Chapter 3) met at Houston. Both the geochemistry and mineralogy-petrology teams emphasized the importance of controlling the environment in which the sample containers were first opened and the need for extensive preliminary examination of the samples at the receiving site.[4] After discussions with members of these teams, King and Flory reworked their proposal and described an elaborate lunar sample laboratory. Projected at more than 8,000 square feet (740 square meters) of floor space, their third concept included offices for 30 visiting scientists as well as laboratories for chemical analysis, low-level short-lived radioactivity measurements, biological examination, and mineralogical and petrological preparations. This facility was not a mere sample-receiving and -packaging laboratory, but the center for much of the preliminary scientific work that would be done on the lunar samples.[5] They presented this concept to MSC's director, Bob Gilruth, on August 13. Gilruth approved, and Faget set about preparing to contract for design studies for it.[6]

MSC's plans required Headquarters approval and funding, and when Faget explained the project Willis Foster's Manned Space Science Division reacted cautiously, to say the least. While agreeing in general terms with the concept, Foster noted that it was a Headquarters responsibility and that the laboratory would be only a "receiving laboratory." The detailed preliminary studies Houston was proposing should be left to outside investigators. In response to Faget's request for $300,000 to conduct the design study, Foster replied that he could allot only $100,000.[7] In view of the alarm with which some of his people viewed the size of the project MSC was proposing, Foster appointed an ad hoc group of Headquarters and MSC scientists to review it.[8]

The group's first meeting in early November was, from Houston's point of view, disappointing. Few of the participants had given much thought to the requirements for a receiving laboratory, and the discussion was long and inconclusive. The group's chairman seemed determined to keep the size and cost of the proposed lab to the absolute minimum. Most members seemed to feel that a facility such as MSC was proposing would take much of the lunar science program out of the hands of academic investigators. In spite of MSC's insistence that time was short, the group adjourned without taking any useful action.[9] But the second meeting, a month later, produced enough progress that MSC's representative felt Houston could go ahead with initial engineering studies.[10]

While Foster's ad hoc group ruminated on the need for a receiving laboratory, Homer Newell—probably sensing that it would entail a considerable increase in the costs of lunar exploration—felt that an independent assessment by the scientific community was needed. Early in December he wrote to Harry H. Hess, chairman of the Space Science Board, requesting the board's judgment on the kinds of analysis that should be performed on the lunar samples as soon as they were returned, the facilities needed to do that work, and the staffing that would be required.[11] A five-man committee—three members of the Space Science Board and two academic scientists—met in Washington on January 14, 1965, to discuss Newell's questions and to confer with members of the ad hoc group.

Three weeks later Hess reported to Newell that a sample receiving laboratory having a relatively restricted mission was indeed needed. The only critical examination was measurement of radioactivity induced in the lunar surface material by cosmic-ray bombardment, which would have to be measured as soon as possible because it quickly dropped to a very low level. The committee raised a question that Newell had not put to it: its members foresaw a need to quarantine the lunar samples until they proved biologically innocuous. A simple, general biological examination could be done at some existing Public Health Service or Army installation. Without specific information or plans to comment on, the committee could give only rough estimates of staffing requirements and probable costs. A minimal quarantine facility with a radiation-counting laboratory might be built for $2.5 million; it would require between 12 and 30 professional scientists plus a supporting staff.[12]

Hess's committee emphatically asserted that the studies MSC was proposing should not be done in the receiving laboratory—or, for that matter, by any single group, inside or outside government—but should be entrusted to the scientific community at large. Neither did they see a compelling need to locate the laboratory at Houston, although "it may seem desirable that MSC have a part in the activity by virtue of its Apollo role." If the Houston center could properly staff such a laboratory, however, it might "add to the overall environment at MSC." On the other hand, since the radiation-counting laboratory would have to be built deep underground to shield it from natural radiation, the waterlogged soil of the Texas coast might make construction more expensive there, and thus some other site might be preferable. The committee also cautioned that such a laboratory would have to operate continuously to conduct worthwhile research; it could not be geared up to operate whenever a new set of samples was available and then shut down until another lunar mission was flown,[13] which, they evidently suspected, was the way MSC was likely to operate the laboratory in light of its minimal scientific capability.

Manned Spacecraft Center officials, meanwhile, were anxious to get their preliminary engineering studies under way. Preliminary studies, design studies, contractor selection, and contract negotiations had to be disposed of as quickly as possible. Cost estimates and justifications had to be prepared for inclusion in the center's budget proposals for fiscal 1967, when construction would have to start. According to early 1965 estimates, the laboratory would have to be operational by January 1969 to support the first lunar mission. Many critical and complex operations in the laboratory would have to be checked out beforehand, and managers estimated that a 9- to 12-month shakedown would be needed.[14]

The sense of urgency felt at MSC was not shared in Washington, however. Faget wrote to Foster in mid-January 1965 urging him to release funds for preliminary engineering studies for the laboratory and suggesting that Headquarters' ad hoc committee be replaced by a standing committee to oversee the incorporation of scientific requirements into the laboratory during construction. A month later Foster replied that study funds could not be released until the ad hoc committee made its report. He concurred with Faget's desire for a standing committee and urged him to appoint one of his staff to it, pointing to the value of "a greater understanding by MSC of the scientific objectives of the laboratory," which would enhance Houston's chances of getting the facility. "Other NASA centers," Foster said, "are submitting 'bids' for the laboratory."[15]

For the next several weeks, MSC and Headquarters discussed management of the laboratory, finally reaching an understanding as to future activity on the project. Foster would appoint a standing committee that would be given free access to design reviews and relevant program materials, so that Headquarters could be assured that the science requirements levied on the laboratory were being met. At Faget's insistence, however, the committee would have no right to approve plans; their advice would be made available through the committee chairman to MSC's point of contact for the receiving laboratory. Except for the specialized radiation-counting equipment, the cost of the laboratory would be included in MSC's construction of facilities budget for fiscal 1967.[16] Some discrepancy still existed between Headquarters's and Houston's cost estimates. Foster's office seemed be thinking of a $1- to $2-million facility; the figure included in MSC's preliminary 1967 budget was $6.5 million.[17]

The Specter of "Back-Contamination"

Houston's planning for the sample receiving laboratory was vastly complicated the following summer by a question Hess's committee had emphasized in its February report. They stated a clear requirement for quarantine of the lunar material until biologists could ascertain that it harbored no living organisms that might threaten the earth's biosphere.[18]

The possibility that life exists or has existed elsewhere in the universe—even within our solar system—evolved from a science-fiction fantasy to a serious scientific question within a few decades. Although no positive evidence has ever been found to indicate that even the simplest living organisms exist elsewhere, a considerable accumulation of evidence that life *might* appear, under the right conditions, has led to a widespread conviction that it *has* appeared, somewhere.*[19]

---

*The argument runs roughly as follows. Simple organic molecules related to the substances out of which living matter is made have been detected in space. Other organic material, possibly derived from living organisms, has been found in meteorites. Conclusive experiments have shown that precursors to living matter can be built from chemically simple substances under conditions presumed to have existed on the ancient earth. Given the vast number of galaxies observable in the universe (a billion billion, according to one estimate) it seems probable that solar systems like our own exist somewhere in those galaxies, and that conditions favoring the origin of life exist on an appreciable number of planets similar to earth.

As early as 1960 the Space Science Board had advised that NASA and other concerned government agencies (e.g., the Public Health Service) should establish an interagency committee on interplanetary quarantine to formulate a national policy for handling spacecraft and material returned from other planets.[20] Two years later, the working group on biology of the Iowa summer study (see Chapter 1) noted that

> the introduction into the Earth's biosphere of destructive alien organisms could be a disaster. . . . We can conceive of no more tragically ironic consequence of our search for extraterrestrial life.

Acknowledging that scientists by no means unanimously agreed on the existence of extraterrestrial life, the group nonetheless recommended that NASA employ

> appropriate quarantine and other procedures . . . when handling returned samples, spacecraft, and astronauts [in order to] make the risk as small as possible.[21]

These cautions had little effect on NASA's plans—in part because the danger seemed remote in the early 1960s, but also because no one in the space agency spoke for the life sciences.*[22] In the early days of space flight few biologists were interested in the space environment; the frontiers of biology were on earth. The life-science community created no demand for NASA support comparable to that created by space physics and astronomy.[23]

As the Apollo program progressed, however, and the prospect of people returning from the moon with boxes of lunar rocks and soil became increasingly likely, concerned biologists continued to call attention to the need for precautions against contamination of the earth by organisms from the moon. On July 29, 1964, the Space Science Board convened a conference of representatives from the Public Health Service, the Department of Agriculture, the Fish and Wildlife Service, the National Academy of Sciences, and NASA to assess the back-contamination problem and recommend courses of action. The conference concluded that

> the existence of life on the moon or planets cannot . . . rationally be precluded. At the very least, *present evidence is not inconsistent with its presence. . . . Negative data will not prove that extraterrestrial life does not exist; they will merely mean that it has not been found* [emphasis added]."**

To contain any alien life forms, astronauts, spacecraft, and lunar materials coming back from the moon should be placed immediately in an isolation unit; the

---

*See Newell, *Beyond the Atmosphere,* chap. 16 ("Life Sciences: No Place in the Sun"). Of all the life sciences, space medicine—the effects of the space environment on human physiology—was the only one of prime concern to manned space flight; but it was only a subsidiary effort in Mercury and Gemini and was concerned mainly with settling some crucial operational questions. See John A. Pitts, *The Human Factor: Biomedicine in the Manned Space Program to 1980,* NASA SP-4213 (Washington, 1985).

**This statement is quite correct; experimental proof of a negative postulate, such as "life does not exist on Mars," is, in any practical sense, impossible. But many must have felt like one anonymous reader at MSC, who penciled opposite the sentence in the margin of his copy of the report: "Like witches."

astronauts should be held in rigid quarantine for at least three weeks; and preliminary examination of the samples should be conducted behind "absolute biological barriers, under rigid bacterial and chemical isolation." NASA should immediately take steps to work out the operational details of these procedures.[24]

When Harry Hess's committee, speaking for the Space Science Board, reported its position on back-contamination the following February, both Headquarters and the Manned Spacecraft Center realized that quarantine was a more serious concern than they had anticipated.[25] Although the director of Biosciences Programs in the Office of Space Science and Applications had kept in contact with the National Academy of Sciences and the Public Health Service, the emphasis on the possible dangers of lunar material came as a surprise to him.[26] Most speculation about extraterrestrial life excluded the moon. At Houston, the report portended serious complications in the design of the receiving laboratory and probably in flight operations as well, and Faget's organization took steps to clarify the quarantine requirements.[27] Action at higher levels was slow in coming, however. Only in May did the NASA Administrator and the Surgeon General (chief of the Public Health Service [PHS]) discuss the matter, agreeing to set up an interagency advisory committee to deal with back-contamination.[28]

By the end of July 1965, MSC had incorporated a general requirement for quarantine into its justification for building the receiving lab, but time was growing short and detailed specifications were needed. Then-current plans required the laboratory to be in operation by January 1, 1969. Allowing a year or more for engineering and design studies, another six months for checking out the equipment and procedures, and six months to correct deficiencies uncovered in the shakedown, just over a year would be left to build the laboratory and install the special scientific equipment. Procedures and space requirements for quarantine had to be settled as soon as possible. MSC already had an outside engineering firm working on a preliminary engineering survey, which would define the laboratory's special requirements and determine what additional studies might be needed to specify its specialized scientific equipment.[29]

Others at MSC were working to formulate a center policy on quarantine in the hope that it could be simplified as much as possible. In late July, Headquarters arranged for an informal meeting of PHS and manned space flight representatives to discuss quarantine and the lunar sample receiving laboratory.[30] Knowing that MSC had little chance of convincing the PHS that no hazard existed, Elbert King consulted with center medical experts and prepared a statement asserting that only subsurface samples should be treated as potential hazards and that quarantine could be terminated as soon as returned samples were found free of exotic organisms. He also set down several important policy questions concerning quarantine to serve as the basis for discussions with the PHS.[31]

When the two groups met at Houston on September 27, 1965, it soon became apparent that MSC's evaluation of the hazard presented by lunar material was not shared by PHS officials. King argued that the lunar surface could be considered sterile: it was in a high vacuum, devoid of water, exposed to intense ultraviolet radiation and subatomic particles from the solar wind, and subjected to severe temperature changes. "If you really wanted to try to design a sterile surface," King later summarized his argument, "this was it." Thus surface material and the astronauts who came in contact only with it should not require such rigorous isolation as samples taken from greater depths. The PHS representative, Dr. James Goddard, chief of the Communicable Disease Center in Atlanta, was unmoved by these arguments. He asked whether anyone could be certain that no microorganisms could survive anywhere on the moon—in sheltered areas, for example. When no one could offer such assurance, Goddard insisted that quarantine must be strict. He and other PHS officials were upset, in fact, that MSC had taken such a casual view of the biological hazard. Even if it cost $50 million to implement an effective quarantine, Goddard said, the importance of the issues justified the added expense.[32] When MSC asked whether the PHS's immigration officers would allow the Apollo astronauts to enter the United States if they were handled in the same way the Gemini crews had been, the reply was emphatic: they would not.[33]

When the conference was over MSC officials knew that the lunar receiving laboratory would have to be even larger and more expensive than they had expected. Quarantine would require astronauts and a fairly large support staff to live in isolation for at least three weeks. The numerous postmission debriefings would have to be conducted through the biological barrier. Not only that, but recovery operations would be much more difficult. PHS officials wanted to prevent exposure of the command module's interior to the earth's atmosphere from the moment it splashed down in the ocean, and the astronauts would have to be isolated at once—even in the rubber rafts used by the recovery crews. On reaching the recovery ship they would be led straight to a mobile quarantine chamber that could be flown back to the mainland. Passing the word to all branches of the Systems Engineering Division at MSC, Division Chief Owen Maynard directed them to show what measures were being taken to comply with these requirements. "Rather than assume the standard answer that no changes can be made within present weight, cost and schedule limitations," he said, "you should assume that [we are] morally obligated to prevent any possible contamination of the earth." Initial examination should be based on the ground rule that "no [command module] components can be exposed to the earth's atmosphere following entry, except those components external to the pressure shell which cannot be contaminated by the cabin environment." While conceding that the first look might show the problems to be insurmountable, Maynard noted that the hazards should be completely documented so that action could be taken as needed.[34]

Maynard's instructions to his division probably reflected a widespread mood of resignation to working around the difficulties that would result from imposition of strict quarantine, which, in the view of some, was unnecessary.[35] But the question of back-contamination had been raised by the scientific community and recognized as important by the Space Science Board and thus had the potential to become a political issue that could create much adverse publicity for the Apollo program.

Toward the end of 1965 it was generally accepted that crew and samples would have to be strictly quarantined. On November 15 NASA's Deputy Administrator, Hugh L. Dryden, wrote to the Surgeon General proposing that a formal liaison office with the Public Health Service be established, that a NASA-PHS advisory committee be set up to establish guidelines for back-contamination control and oversee NASA's efforts to avoid infecting the earth, and that the PHS recommend the kind of facilities and staff required to carry out those efforts.[36] The Surgeon General's reply a month later paved the way for establishing formal cooperation in managing quarantine in the Apollo program.[37]

Congress Objects to the Receiving Laboratory

MSC had initially projected the cost of building the lunar receiving laboratory at $6.5 million, but providing for quarantine would increase that considerably. Before going ahead with plans to enlarge the laboratory, George Mueller, no doubt with a view to minimizing the cost increase, directed MSC to conduct a quick survey of quarantine facilities around the country that might serve to isolate crews and support staff following Apollo missions.[38] Houston evaluated 12 hospitals and research installations that were equipped for biological containment against the requirements imposed on Apollo, and concluded that none would be satisfactory. Only one, in fact, an Army hospital at Fort Detrick, Maryland (which provided care for personnel working with highly dangerous microorganisms), even came close. Converting it to accommodate Apollo, however, would require major new construction and would interfere with the Army's research programs.[39] If the Apollo crews were to be quarantined, NASA evidently would have to build its own isolation ward for the purpose, and in spite of its cost Mueller believed he had adequate justification for it.

But lean years were beginning for manned space flight. President Lyndon Johnson pressed hard for expanded domestic social programs while announcing his intention to keep the federal budget under $100 billion. At the same time the nation's involvement in southeast Asia deepened; troop commitments increased eightfold as the United States sent combat units to Vietnam. Other programs, including space, would be squeezed hard to keep costs down.[40]

Administrator James Webb went to Congress in the spring of 1966 with an authorization request for $5.012 billion, 60 percent of it for manned space flight. It was an austere budget, Webb said, that provided "no margins of time or of resources to counter the effects of setbacks or failures.[41] Manned programs required $54.4 million for construction of facilities—a sharp increase from the previous year's request, but only one-fourth of what NASA had received two years before.[42] The largest single amount, $36.5 million, went to Kennedy Space Center, most of it for completion of the Saturn V launch complex. The Manned Spacecraft Center needed $13.8 million, of which $9.1 million was allocated to the lunar receiving laboratory.[43]

The House subcommittee on manned space flight was generally sympathetic to Apollo's other requests, but it gave the lunar receiving laboratory unusually close scrutiny. In response to questions submitted by the subcommittee after the initial hearings on February 24, 1966, the Office of Manned Space Flight supplied for the record a detailed history, including statements from the Public Health Service, of the requirements imposed by the scientific community and NASA's efforts to satisfy them.[44] But when William Lilly, Headquarters Apollo Program Control Director, and George Low, MSC's Deputy Director, faced the subcommittee on March 1, some members apparently had not had time to study OMSF's responses. Congressman Donald Rumsfeld suspected the receiving lab was NASA's attempt to get a foot in the door for substantial expenditures later, and he saw no valid reason to build the lab at MSC.[45] Ranking minority member James Fulton of Pennsylvania, like Rumsfeld, did not see why the lab had to be at Houston; but he also faulted NASA for not planning to use existing facilities and questioned NASA's assertion that the laboratory was needed.[46] Whatever the reasons for the congressmen's antagonism to the receiving lab, Lilly and Low had an unexpectedly rough day. Olin Teague, chairman of the subcommittee and an unswerving supporter of manned space flight, did not preside at these hearings, and his absence undoubtedly deprived them of a friendly interrogator who would have helped them make their case. The following week the subcommittee, not convinced that NASA had justified the receiving laboratory, struck it out of the authorization bill.[47]

The loss of the laboratory was a blow to MSC and to Apollo; without it, NASA could not meet the scientific requirements imposed on the lunar landing missions. Mueller ordered a more detailed study of existing facilities that might serve either to quarantine the crews and samples or to conduct the required scientific studies.[48] MSC immediately appointed a site survey board to evaluate possible alternate locations. From a list of 27 facilities a group of 8 was selected for detailed investigation.* The board then prepared a list of detailed criteria for the laboratory, based primarily on its requirements for two-way biological containment, handling of samples under high vacuum, and low-level radiation counting. Secondary but important considerations included space, administrative and technical support services, and availability of utilities. Finally, the board was to consider logistics, principally the problem of travel to and from the site by engineers and others who needed to debrief the astronauts.

Since time was limited, the board split up into two teams, one to survey the eastern sites and one the western. Each team had a member who could estimate the cost of modifying and operating the candidate facilities. During the week of March 16-23 the teams inspected the eight establishments, conferring with officials at each site to assess the impact of Apollo's requirements on local programs and their willingness to accept the project. The board then summarized its findings:

1. There is no single facility that will meet the criteria standards for any *one* of the major functional areas of the Lunar Receiving Laboratory, without extensive modification.

2. There is no single facility that can be economically modified or adapted to meet all the requirements of the Lunar Receiving Laboratory.

3. The use of any one of the sites investigated will result in reduction or reprogramming of some phase of nationally significant research effort.

---

*USPHS Communicable Disease Center, Atlanta, Ga.; Army Biological Center, Ft. Detrick, Md.; National Institutes of Health, Bethesda, Md.; Oak Ridge National Laboratory, Oak Ridge, Tenn.; USAF School of Aviation Medicine, Brooks AFB, Tex.; NASA Ames Research Center, Moffett Field, Calif.; Navy Biological Laboratories, Oakland, Calif.; and Los Alamos Scientific Laboratory, Los Alamos, N. Mex.

4. An integrated facility is necessary to minimize public health hazards.

5. Maximum operational effectiveness, as a part of the Apollo mission, and minimum operating costs indicate a Houston location as the most preferred site.[49]

The final report stressed the last point: efficient management of Apollo required the astronauts, spacecraft, and lunar samples to be as close as possible to Houston's engineers and physicians, especially in the first few weeks after recovery.[50]

While this survey proceeded, Headquarters gathered more supporting information in preparation for a rehearing.[51] On March 31 Mueller, armed with the site survey report and stacks of extra facts, appeared before the subcommittee. In great detail he explained the history of the receiving lab, in particular the emergence of the requirement for quarantine, stressing NASA's cooperation with the Public Health Service and the Space Science Board. Since some subcommittee members apparently felt that NASA had sprung a surprise in proposing the new facility, Mueller pointed out that chairman Teague had been informed of the probable need for a receiving laboratory in August 1965. Col. John Pickering, who had headed the site-evaluation survey, then explained that group's operation and reiterated its conclusions.[52]

This time the subcommittee was convinced and restored funds for the laboratory to the authorization bill.[53] Fulton, however, held out. He put his objections on record in the full committee's report to the House, discounting the danger of back-contamination, asserting that the site survey teams had been packed with members predisposed to choose Houston as the site for the lab, and disputing the need to centralize all the operations in one place. Contrary to NASA's own findings, many other facilities in the country could be used with minor modification, Fulton said.[54] When the authorization bill came to the House floor, it passed with the lunar receiving laboratory's $9.1 million intact, though Fulton tried once more to kill it. "We simply have no facts," he told his colleagues, "on which to build a practical foundation and a laboratory."[55]

The lunar receiving laboratory survived the authorization hearings but fared somewhat less well in the Appropriations Committee,* which cut out $26.5 million in construction funds. The committee did not eliminate or reduce any specific projects; it was simply not convinced that they should all be started immediately. Some projects (presumably the receiving laboratory, although it was not mentioned by name), the committee report pointed out, would not be needed until after the lunar landing.[56] The Senate Committee on Aeronautical and Space Sciences was more specific, reducing funds for the laboratory by $1 million and warning NASA to keep a tight rein on its costs and provide only the necessary minimum of specialized facilities.[57]

In the final appropriations bill passed by Congress, NASA's budget request for fiscal 1967 was cut by $44 million, bringing funding for space below $5 billion

---

*Manned space flight (and the Manned Spacecraft Center) had lost a powerful friend in Congress a few weeks earlier. Representative Albert Thomas of Houston, chairman of the subcommittee that passed on NASA's appropriation bills, died on February 15, 1966.

for the first time since fiscal 1963.[58] Research and development, the category that covered most of the agency's expenses for space flight hardware and operations, suffered the smallest reduction, only $1.6 million out of $4,246.6 million requested. Construction of facilities was reduced by $18.2 million and administrative operations by $23.9 million;[59] both of these reductions would have a considerable impact on construction and operation of the lunar receiving laboratory.

Completing Design and Starting Construction

Throughout 1965, the Manned Spacecraft Center developed the architectural and engineering requirements for the lunar receiving laboratory (LRL), relying on Headquarters's Manned Space Science Division to define its scientific requirements.[60] By the end of the year Houston had completed its preliminary engineering study, requirements for quarantine of crew and support personnel had been incorporated, and in March 1966 NASA officials took to Congress a description for a building enclosing more than 86,000 square feet (8,000 square meters) of floor space. One-fourth of this was required for the crew isolation area, two-thirds for the sample-receiving laboratory and the biological facilities, and the rest for the radiation-counting laboratory—part of which was 50 feet (15 meters) below ground level and elaborately shielded against radiation from outside.[61]

Houston planned a two-phase construction program; the first, for the foundation and building shell, to begin in July 1966; and the second, to complete and equip the building, to start in September. Total construction costs were estimated at around $8 million. The vacuum system and its associated equipment were to be designed and built by the Oak Ridge National Laboratory of the Atomic Energy Commission. At the time of the congressional hearings MSC expected the entire facility to be ready for operation by the end of December 1967.[62] A program office and a policy board for the LRL were established in May 1966.[63]

Congress's unexpected opposition to the LRL delayed the schedule, however, and by midsummer Headquarters advised Houston that it could solicit bids but could not open them until NASA's authorization bill had passed. Both schedule and budget were tight, but it seemed that MSC would have to live with both, since Congress was far from favorable toward the project.[64] Clearance to open bids for the first phase of construction came on July 28 with the stipulation that no contract was to be awarded nor was the second contractor to be announced until Headquarters gave the word.[65] On August 1 MSC selected Warrior Constructors, Inc., of Houston for the initial phase of construction; on the 19th the contract for completing the laboratory went to a consortium of Warrior, National Electronics Corp. of Houston, and Notkin & Co. of Kansas City, Mo. The two contracts totaled approximately $7.8 million.[66]

Staffing the Lunar Receiving Laboratory

While most of the early planning for the lunar receiving laboratory necessarily concentrated on the building and its equipment, the scientists who were drawing up its functions continually pondered the problem of providing a competent staff to operate it. The Space Science Board's ad hoc committee (Harry Hess's group) saw merit in establishing a "small but competent [scientific] staff headed by a scientist of recognized stature" to maintain a modest research program, but felt that "the problem of attracting a highly competent small staff is serious,"[67] presumably because the environment at MSC did not appeal to scientists. The Houston center had only a few scientists, most of them younger professionals who had yet to establish their reputations and had little chance to do so in the roles they were assigned.[68] OSSA's ad hoc committee proposed a permanent staff to sustain the laboratory, supplemented by visiting scientists who would do much of the experimental work on the lunar samples.[69]

As plans matured, however, and particularly as the requirement for quarantine developed in 1965, the question of an organization and staff for the laboratory became increasingly pressing. In early September, James McLane, MSC's engineer in charge of the early planning, outlined the peculiar requirements of the new laboratory in a five-page memorandum. McLane noted that preliminary studies had pointed up the need for a minimum of three persons to serve as an administrative staff: a director, a technical director (chief scientist), and an assistant technical director (sample curator), all of whom should be MSC civil service employees. Major questions concerning the roles of these staff members needed resolution. Where would they fit in the MSC organization? What responsibility would they have in managing and distributing the lunar samples? Should they be eminent scientists to satisfy the scientific community? (At one time McLane had heard that a Nobel laureate was being suggested.) How would differences between visiting scientists and MSC's operations be settled? McLane urged that MSC and Headquarters work out these details, define the positions, and start recruiting." It would be highly desirable," he said, "to have at least one of these . . . positions filled reasonably early in the final design phase (late 1965)." Concerning quarantine, McLane felt that this might be better left to the Public Health Service; but whatever was decided, definite responsibility should be assigned as soon as possible.[70]

But while design studies went on under the pressure of the early-1969 schedule for beginning operations, laboratory organization and staffing languished. Discussions with the Public Health Service and definition of the PHS's role in quarantine took up most of the last half of 1965. MSC and the Headquarters standing committee for the sample receiving laboratory spent much of the first half of 1966 working over the lab design. And when Congress balked at Houston's plan for the receiving lab, considerable effort had to be given to justifying its existence and its location. All in all, the lunar receiving laboratory created a knotty management problem, which was probably complicated by the lack of a focal point for science at MSC.

Early in 1966 the Planetology Subcommittee of OSSA's Space Science Steering Committee,* worried by the lack of progress in defining the scientific role of the

---

*The Space Sciences Steering Committee, composed entirely of NASA employees, was responsible for recommending science programs and projects to the Associate Administrator for Space Science and Applications. Seven subcommittees, each having about half its members from the outside scientific community, advised the main committee on specific areas of space science.

receiving laboratory, recommended that the standing committee monitor that aspect of the planning and report periodically to the Steering Committee.[71] After some discussion within OSSA, the standing committee was replaced by a Lunar Receiving Laboratory Working Group chaired by Dr. Clark Goodman of the physics faculty at the University of Houston, a member of the Planetology Sub-committee. At its first meeting on May 5, 1966, the group reviewed progress on laboratory design and studied MSC's schedule for construction and activation of the lab. MSC's presentation made no mention of staffing, and when the question was raised it developed that no job descriptions were available and no recruiting was under way for key positions. While members of the working group were impressed by MSC's concept of the laboratory, they were concerned about the lack of definition of its organization. Their concern took the form of a resolution, in which they also recommended that at least 12 civil-service positions be made available for scientists in the LRL and stated the group's willingness to suggest candidates.[72] Later in the month MSC prepared an estimate of staffing requirements that showed 33 resident scientists (Ph.D. or equivalent) and provision to accommodate 15 visiting scientists working on grants from Headquarters.[73]

Houston's freedom to staff the receiving laboratory was restricted by the budget cuts imposed by Congress, which ultimately reduced the number of civil-service positions at the center by 84 in fiscal 1967.[74] The LRL Working Group continued to press for action while MSC struggled with the problem of fitting the laboratory into its organization.[75] A possible alternative came to light in midsummer when the University of Houston approached officials at MSC about managing the laboratory under contract.[76] The idea of using an outside management contractor had been considered by MSC, and Houston officials gave this initiative considerable thought. Early in September a plan was drafted for consideration by Headquarters, calling for the University of Houston, through its Houston Research Institute, to head a consortium of regional unversities to provide professional staff and technical support for operations in the receiving lab.[77] In further discussions with Headquarters MSC stressed the value of a connection with the academic world and the difficulty of staffing the laboratory under existing civil-service restrictions.[78]

Headquarters wanted to look more closely at the situation, however, and discussions continued through the rest of the year. NASA's top managers evidently had visions that the lunar receiving laboratory would become a national laboratory for lunar studies, in which case participation by the academic community on a nationwide scale would be desirable.[79] Meanwhile, MSC was directed to create a task force to take care of the receiving laboratory's basic needs, extend existing support contracts to meet immediate requirements, and await developments.[80] If additional scientific talent was needed, other NASA centers could be called on to provide it.[81]

During all these discussions, pressure from concerned groups mounted. The LRL Working Group especially emphasized the need to name a director for the laboratory. The Public Health Service felt that matters had reached a point where an organization and staff were critical to further progress, and suggested that the laboratory chief should be someone thoroughly familiar with biomedical science generally and quarantine in particular.[82]

In fact, MSC officials had finally come to recognize the need for more than just a staff and organization for the receiving laboratory. By mid-December the Houston center had decided to create a Science and Applications Directorate, organizationally on a par with Max Faget's Engineering and Development Directorate, in which all the center's scientific activity would be centralized (see Chapter 6). It would take over the activities of the Experiments Program Office and the Space Sciences Division as well as a number of other scientific functions scattered around the center. Robert O. Piland, chief of the Experiments Program Office, would act as director while a search was instituted for an established scientist to fill the position.[83] No director was named for the lunar receiving laboratory; pending a permanent appointment, Joseph V. Piland, project manager for LRL construction, was named acting manager.

To provide the laboratory technicians who would prepare the lunar receiving laboratory for operation, MSC turned to its support contractors. In mid-January 1967 Gilruth notified Headquarters that he intended to extend an existing contract with Brown & Root-Northrop, which would furnish technicians for the receiving laboratory under the direction of the project manager.[84] Between completion of the laboratory and the first lunar mission, Brown & Root-Northrop would train its employees in laboratory operations and maintenance and prepare for an operational readiness inspection and a rehearsal of a complete cycle of laboratory operation.

The Manned Spacecraft Center had recognized the need for a lunar sample receiving laboratory and was willing to support it, but many MSC engineers—and a few scientists as well—felt that the quarantine facility and its elaborate precautions were unnecessary impediments to Apollo operations. But however insignificant manned space flight officials believed the risk of back-contamination to be, it was a risk NASA could not afford to take. If they were wrong, the consequences would be, as the scientists said, disastrous.

5

SELECTING AND TRAINING THE CREWS

From the moment the lunar landing was proposed as the primary goal of manned space flight, NASA officials and outside scientists debated the qualifications of the people who would land on the moon. Scientists urged that at least one of the crew should be a scientist with enough experience to assess the significant features of the landing site quickly and accurately and to collect samples with discrimination. Those responsible for mission operations and crew training insisted that mission success and crew safety could be assured only if every crew member were a skilled pilot, preferably a test pilot, able to complete a mission alone if that unlikely situation should ever arise. Once the many unknowns in lunar landing operations were better understood, it might be considered safe to take along a person with different qualifications.[1]

Finding crew members who combined the experience of a scientist with the skills of a test pilot proved impossible. For the first five years, during the experimental and developmental phase of the Apollo program, piloting experience took precedence over scientific training as a requirement for admission to the astronaut corps. Later, as missions devoted largely to scientific operations were contemplated, scientists were admitted and trained as pilots.

Test-Pilot Astronauts

Before Project Mercury, design concepts for manned spacecraft were constrained by a lack of hard facts about the human ability to function in the space environment. Many believed that electronic systems, controlled from the ground or programmed for specific contingencies, offered the only safe and practical means of operating a spacecraft. From this point of view, the person in the spacecraft was simply a passenger, an experimental subject whose main function would be to provide the physiological data needed to define the limits of a human's role in space. Mercury's designers did not completely agree with this philosophy; they conceived a spacecraft in which the pilot could take control of critical systems if necessary and would have some measure of control at all times.[2]

Within months of the start of Project Mercury NASA selected its first group of pilots for the early earth-orbital flights. The "Original Seven," all military test pilots* with strong engineering backgrounds, were volunteers picked from a list of more than a hundred men provided by the Pentagon (see Appendix 6).[3] In view of Mercury's experimental character, the choice of test pilots was appropriate; they were accustomed to dealing with emergencies under stressful conditions and were familiar with the physical and psychological stresses of high-speed flight in unproven aircraft.

In choosing test pilots as its first astronauts, NASA came down on the side of people as active participant in spacecraft operations. It was yet to be shown how much humans could effectively participate, but the astronauts in the first group insisted on giving the pilot as much responsibility for control of the spacecraft as feasible; if the pilot did not operate the spacecraft, what was the point of a person in space (especially a test pilot)? Most of the engineers in the Space Task Group agreed with this viewpoint. In light of their long NACA experience with piloted aircraft, they too inclined toward giving pilots all the control they could safely handle.[4]

When the first group of astronauts entered the program in April 1959, Project Mercury and the Space Task Group were organizationally in flux. As a result, the Original Seven defined the role of the astronaut for the entire Apollo program. Administratively they reported not to any project office but directly to Robert R. Gilruth, director of the Space Task Group. For most of Gilruth's career in NACA he had worked with pilots, and he took a special interest in this group. Shortly after the astronauts reported aboard he assured them that whenever they had serious concern with any aspect of spacecraft design or mission operations he would see that they were listened to.[5]

The first two astronaut classes (1963). Seated, left to right, the "Original Seven": L. Gordon Cooper, Jr., Virgil I. (Gus) Grissom, M. Scott Carpenter, Walter M. Schirra, Jr., John H. Glenn, Jr., Alan B. Shepard, Jr., and Donald K. (Deke) Slayton. Standing, left to right: Edward H. White II, James A. McDivitt, John M. Young, Elliot M. See, Jr., Charles Conrad, Jr., Frank Borman, Neil A. Armstrong, Thomas P. Stafford, and James A. Lovell, Jr.

---

*NASA originally intended to issue a general solicitation of applications for the position of "research astronaut-candidate," and considered that several occupations besides test pilot might qualify. President Eisenhower, however, directed the agency to select its astronauts from the ranks of military test pilots; this would simplify selection, keep out undesirable applicants, and eliminate the need to run security checks on the candidates.

Given this kind of autonomy the astronauts were considerably more than pilots in training to operate a new vehicle. While undergoing training they also took an active part in reviews of spacecraft design and operations planning, offered suggestions from the pilot's point of view, and contributed to the design of the flight simulators that soon became an important part of astronaut training. Each person was assigned an area of spacecraft systems or operations planning (e.g., attitude-control systems, communications, recovery operations) as a prime responsibility; in his specialty he closely followed developments and served as the point of contact between his astronaut colleagues and project engineers.[6] Training was strongly engineering- and operations-oriented, a pattern that would be carried into subsequent projects.

Scientists Call for Representation on Apollo Crews

NASA's scientific advisory community first addressed the role of the astronaut in space science at the Iowa Summer Study in 1962 (see Chapters 1 and 3). Concerning humans on the moon, the study report stated the belief that

> it is extremely important for at least one crew member of each Apollo lunar mission to possess the maximum scientific ability and training consistent with his required contribution to spacecraft operations.

This person should participate "in the earliest possible lunar missions," and, since the chosen mode of operations called for only two men on the lunar surface, "the maximum scientific return will be achieved only if the scientist himself lands on the Moon."[7]

A working group of the 1962 summer study considered in detail the role of people in space exploration, formulating the scientists' position with reference to science missions of many types, not merely lunar exploration. The group defined several combinations of scientific and astronautic skills that would be appropriate for different degrees of scientific participation in manned space missions. At the top of the scale was the "scientist-astronaut"; fully trained both as a scientist and as an astronaut, he could operate the spacecraft as well as make valid scientific observations. For the long term, the working group recommended creating an Institute for Advanced Space Study—a graduate-level institute with a unique curriculum in which candidates holding bachelor's degrees would be trained as scientist-astronauts (see Chapter 3). Meanwhile, aspirants to this position should be recruited from among qualified scientists and trained to achieve comparable qualifications as astronauts.[8]

The working group recognized that no such scientist-astronauts could be trained in time for a lunar landing within the decade. For the short term, they acknowledged that the best course was to give qualified astronauts as much training in science as possible, so that they could be useful observers for the scientist on the ground. These "astronaut-observers" were expected to play a role in lunar exploration even after fully trained scientist-astronauts became available. Others who would be important in conducting space science missions were the "ground scientist," a scientist thoroughly familiar with all the details of space flight operations, who would direct the activities of the astronaut-observer from the ground; and the "scientist passenger,"* a scientist physically qualified for space flight but not trained to operate the spacecraft.[9]

Not surprisingly, the summer study report repeatedly emphasized the need for the scientist-astronaut to keep up with his science; the scientist who does not maintain a continuous research program falls behind his colleagues who do and loses his standing in the scientific community. At the same time, the tone of the working group's findings implied that the techniques of operating the spacecraft could be learned by any intelligent person in a couple of years and were therefore of subsidiary importance. The level of comprehension of the astronaut's task was indicated by the responses—summarized in the report—to a questionnaire sent by the Space Science Board to a number of scientists. On the question of whether the first scientist on the moon should also be an astronaut, the consensus of those responding was:

> Of course. He should be familiar with all aspects of the spacecraft and be able to take over in an emergency. However, his qualification as a crew member would not depend so much on his ability as a space-pilot as on his scientific aptitude.

To the question of how astronaut-scientists should be developed, the scientists replied that graduate students or early postdoctoral fellows should be picked and trained

> for at least four or five years. *They should go through astronaut training for part of each year to become familiar with the problems of space flight. It is hoped that this would not involve too large a fraction of their time* [emphasis added], since emphasis should be on their development as scientists.[10]

In 1962, of course, few people fully understood the demands that would be made of Apollo crews; but if these statements reflected opinions widely held in the science community concerning the training required to become a proficient astronaut, it is not surprising that misunderstandings developed when the time came to choose crews for the lunar landing missions.

Organizing the Astronaut Corps

The first group of astronauts immediately became public figures, and as Mercury shifted into flight operations in the closing days of 1961, demands on their time for interviews and personal appearances multiplied. NASA welcomed the publicity for the space program, but this aspect of the astronauts' status often made impossible demands on their heavy training schedule. And when a second program, Gemini, was established late in the year, it was clear that more astronauts would be entering the program, further complicating training and flight preparations. When Space Task Group managers decided that someone should be

---

*Fifteen years would pass before NASA had scientist-passengers, now known as "mission specialists" and "payload specialists," who began flying on Shuttle missions in the early 1980s.

appointed to organize the astronauts' activities more efficiently, some of the astronauts suggested that they would prefer to have one of their own rather than an outsider in that job. As it happened, one was available.

Air Force Captain Donald K. ("Deke") Slayton, assigned to the second Mercury orbital flight, had been grounded a few weeks before the mission when physicians discovered a minor (and, as it turned out, apparently harmless) irregularity in his heartbeat. Although no one could definitely say that Slayton's condition would endanger him or the mission, neither would any medical expert assure NASA that *no* risk was involved. Prudence, a quality which NASA's high-level managers possessed in full measure, dictated that someone without any detectable abnormality should fly the mission instead, and Slayton was grounded until the physicians could be confident he was physically qualified to fly.[11]

Slayton was one of the most experienced of the original seven astronauts. He had flown combat missions in Europe and the Pacific in World War II and had been a test pilot assigned to fighter operations at the Air Force Flight Test Center at Edwards Air Force Base when he was selected as an astronaut. His personal commitment to the manned space program was complete. He had vigorously defended the cause of humans in space before the Society of Experimental Test Pilots when most test pilots, unfamiliar with the actual course that the project was taking, considered that Mercury offered them no future and little valuable experience. His disqualification, especially on physical grounds, was a shock to everyone in the project as well as to the public. It was personally devastating to him; besides losing out in the competition for a space flight assignment, he was forbidden by the Air Force to fly alone in high-performance aircraft.[12]

In September 1962 Slayton was appointed Coordinator of Astronaut Activities, reporting to Gilruth. Without complaint, he took over the largely administrative duties of scheduling training activities, visits to contractor plants, and public appearances and interviews with the news media. But the most important responsibility he assumed was that of assigning astronauts to specific missions. This responsibility he shared with no one else; although he had plenty of help in assessing each candidate's personal and professional qualifications and mastery of the spacecraft systems and mission plans, Slayton made the final decision. His decisions stuck: in the entire manned program, from the later Mercury flights through the Skylab missions, he could later recall only one instance in which higher authority challenged his judgment.[13]

When the Manned Spacecraft Center was reorganized the following year, Slayton's position was redesignated Assistant Director for Flight Crew Operations, organizationally on a level with the assistant directors* for engineering, flight operations, and administration.[14] For some time he was also chief of the Astronaut Office, the administrative unit that coordinated training and other astronaut activities, and he continued training with the first two groups as much as he could, hoping for eventual reassignment to flying status.

Under Deke Slayton the Astronaut Office was run much like a military unit—which for several years it effectively was, since almost all the astronauts were

---

*The title "Assistant Director for . . . ," meant "assistant to the director *of MSC* in charge of . . ." but was applied to the heads of directorates. This confusing designation was later dropped and chiefs of directorates were called simply "directors of. . . ."

or had been Air Force, Navy, or Marine officers. He encouraged open communication between himself and the astronauts, but expected that when a decision had been made the discussion was finished. As one of the astronauts characterized Slayton's management style, the astronaut's job was "to do what the commanding officer says, and if you (didn't) want to . . . , the door was always open"[15]—the door marked "this way out." Astronauts came into the program voluntarily and they could always leave the same way.

Slayton well understood the position manned space flight occupied in the national space program as a consequence of its prominence in the public eye: any failure, especially one that endangered or killed an astronaut, could set back the lunar landing for years, and might even kill the manned space flight program. His contribution to avoiding failure was to pick the best people for the crews, and as long as manned space flight entailed any hazardous uncertainties, the best people would be experienced test pilots.

More Missions, More Astronauts

While much was learned from Mercury, much more had to be learned before a lunar mission could be planned. Even before President Kennedy's decision to go to the moon was announced, Space Task Group engineers were planning the second phase of manned space flight. Project Gemini, approved in early December 1961, would test various techniques of rendezvous, determine whether men and systems could survive and function during long missions, investigate the radiation environment in near-earth space, and develop techniques for controlled landings. Twelve missions were planned, ten of them manned, to start in the spring of 1964 and fly at two-month intervals.[16]

Additional missions required additional astronauts, and on April 18, 1962, NASA announced it would accept applications for trainees. Once more test pilots were given preference, but the required number of flying hours was reduced, and civilians as well as military pilots were eligible. The upper age limit was reduced from 40 to 35 and the education qualification broadened to include degrees in physical or biological sciences as well as engineering.[17] A list of more than 250 applicants was cut to 32 by preliminary physical and psychological screening. After intensive evaluation in Houston, nine new astronaut trainees were chosen in September 1962: two civilians, four Air Force pilots, and three Navy officers, including some who had applied for the first group but had not been selected (see Appendix 6).[18] Selection of this group virtually depleted the pool of qualified candidates from the small corps of test pilots in the country, and it was the last group for which test-pilot certification would be a requirement.[19]

The new trainees reported at Houston in October 1962 to begin a two-year training course. A four-day work week was normally scheduled, the fifth day being reserved for public relations duties or for travel.[20] After two weeks of orientation to NASA's organization and familiarization with the near-complete Mercury project, the second class, joined by the first group, started on a three-month "basic science" course interspersed with briefings on Gemini and Apollo projects and systems. The classroom work covered astronomy, aerodynamics, rocket propulsion, and the physics of orbital flight and re-entry; it included lectures on computers, space physics, and the medical aspects of space flight. Almost one-third of the classroom time was spent on navigation and guidance. In mid-January 1963 the class flew to Flagstaff, Arizona, for a series of geology lectures and field trips conducted by Eugene Shoemaker.[21]

As the Apollo program came into clearer focus in 1962, MSC officials saw that they needed still more astronauts. At the end of the year projected manned flights included four development flights of the Saturn I, four of the Saturn IB (an "uprated" version of the Saturn I), and one of the Saturn V, starting in late 1964 and flying at three-month intervals until mid 1967.[22] The 16 astronauts in training would not be enough to staff the 10 Gemini missions plus the 9 scheduled for Apollo, and in April 1963 MSC announced its intention to recruit a third class of trainees.[23] On June 18 the Houston center issued its formal call for applications. For this group the requirement for flight experience was relaxed still further: 1,000 hours of jet time could substitute for test-pilot certification. The selection board might consider advanced degrees in engineering or science as offsetting some lack of flight experience. Industry, professional organizations, and the armed services were asked to recommend candidates.[24] Manned space flight chief Brainerd Holmes, acknowledging the Space Science Board's Iowa summer study recommendations, indicated to Congress that scientific qualifications would be taken into account in selecting this group.[25]

Of 271 applicants responding, 30 were selected for final screening. On October 18, 1963, MSC announced the names of the newest class of astronaut trainees (see Appendix 6). Again military officers outnumbered civilians, by 12 to 2.[26] (At the end of 1963 the astronaut corps comprised 26 military pilots and 4 civilians, all trained in military service.[27]) The new group was distinguished by a large num-

The third class of astronauts, selected October 18, 1963. Seated, left to right: Edwin E. (Buzz) Aldrin, Jr., William A. Anders, Charles A. Bassett II, Alan L. Bean, Eugene A. Cernan, Roger B. Chaffee. Standing, left to right: Michael Collins, R. Walter Cunningham, Donn F. Eisele, Theodore C. Freeman, Richard F. Gordon, Jr., Russell L. Schweickart, David R. Scott, and Clifton C. Williams, Jr. Four members of this class died during training: Bassett, Freeman, and Williams, killed in aircraft crashes, and Chaffee, killed in the AS-204 fire on January 27, 1967.

ber of advanced degrees: 8 of the 14 had master's degrees and one held a doctorate in astronautics. The two civilians were scientists actively engaged in
research. Most of the military officers held engineering degrees. In spite of MSC's
obvious preference for pilots, the scientific community raised no outcry about
the lack of scientists in the astronaut program. Harold Urey, however, publicly
reproved the agency, late in the year, for not recruiting geologists to explore the
moon.[28]

When the new group reported to Houston in January 1964, Slayton had 29 pilots
to look after, including 5 Mercury veterans.* The first four men who would walk
on the moon were in training, but at the time all attention focused on Gemini,
whose first manned launch was scheduled for November 1964.[29]

Classroom work began in February with a new basic science program, a 20-week
series of lectures, briefings, and field trips, strongly oriented toward Gemini but
also including substantial chunks of time devoted to geology, which was entirely
an Apollo concern. The veterans of the previous year's training skipped parts
of this course to spend time in the Gemini simulators, but the geology sessions
were required of everyone. Geologists from MSC and from the Geological Survey guided them through the equivalent of a one-semester college course in land
forms and land-forming geologic processes, minerals and their origin, and topographic and geologic mapping. Lectures and laboratory work were supplemented
by field trips to study the Grand Canyon, the Big Bend area of west Texas, and
the volcano fields near Flagstaff, Arizona, and Cimarron, New Mexico.[30] No one

*Astronauts Edgar Mitchell (right) and Alan Shepard (second from right) and three geology
instructors examine a small crater near Flagstaff, Arizona.*

---

*John Glenn resigned from the program in January 1964 to enter business (and later, politics). Another
Mercury astronaut, Scott Carpenter, would soon be devoting most of his time to the Navy's Project
Sealab, an experimental underwater habitat, although he would retain formal affiliation with the
astronaut program for another three and a half years.

expected the astronauts to become fully qualified field geologists as a result of this training, but they could at least learn to interpret what they would see on the moon in terms of its probable geologic history and to recognize important geological specimens if they found any. On the later field trips the geologist-instructors began simulating lunar exploration by sending their pupils into an area with a radio transmitter and instructions to note the geologic features they could see, describe what they considered important, and collect representative samples of rocks and surface material. Their commentary was recorded and the exercise was completed with a detailed critique of their performance.[31]

Geology was a new field for most of the engineer-astronauts, rather unlike anything in their experience. What impressed most of them was the large amount of specialized terminology they had to learn—new words having little relation to their accustomed vocabulary. Instructors found them willing enough students, for the most part, but highly variable in their response to the course. Some seemed to be born observers and quickly developed the knack of picking out the distinguishing geologic features of an area and describing them in geologist's terms; others had more difficulty acquiring the field geologist's eye. Apart from the problem of adjusting to a new discipline with a novel point of view, the astronauts faced the question of how heavily their performance in geology would count when the time came to select flight crews. Seniority and flying experience seemed to be of prime importance in determining who got the assignments for Apollo flights, and it was important to get picked as early as possible for a Gemini crew. Well aware that no one could completely master every aspect of training, the astronauts sought to shine in those aspects that were most likely to attract Slayton's attention. For the short-term future at least, geology seemed fairly far down the priority list.[32]

Moving Into Flight Operations

After a problem-filled 1963, Project Gemini looked toward better things in 1964. The first flight test of the spacecraft and its Titan II launch vehicle went off on April 11, raising hopes of a manned flight before year's end.[33] Two days later, the Manned Spacecraft Center announced the names of the first crews for the two-man earth-orbital missions. As might have been expected, the Commander of the first Gemini mission was one of the Original Seven, Virgil I. ("Gus") Grissom, who had ridden the second suborbital Mercury flight in July 1961. Paired with Grissom was one of the second astronaut group, John W. Young. Their backup crew likewise had a representative from each of the first two astronaut classes, Walter M. Schirra, Jr., pilot on Mercury-Atlas 8, and Thomas P. Stafford.[34]

A week after the announcement, the four Gemini crewmen headed into a full mission-specific training schedule. At the spacecraft builder's plant in St. Louis, at MSC in Houston, and at the Cape in Florida they put in long hours learning the design and function of the spacecraft systems, following the assembly and testing of their spacecraft, attending briefings on program and mission objectives, and practicing such tasks as getting out of a floating spacecraft. Simulators duplicated as closely as possible most of the conditions of launch, orbital flight, and recovery (weightlessness being a notable exception), and in these simulators the crews practiced normal operations as well as all the likely malfunctions their training officers could think of.* Occasional trips to the Navy's man-rated centrifuge in Johnsville, Pennsylvania, gave them practice in enduring the acceleration forces (''g-loads'') of launch and reentry. Training stretched from a planned 7 months to 11 when their flight was delayed by problems with the second unmanned test, and by the time their flight was ready for launch on March 23, 1965, the crews would have been hard to surprise with anything that might come up.[35]

When crews were named on July 27, 1964, for the second Gemini mission, Slayton broke the pattern of designating an orbital veteran as commander, choosing four inexperienced men for Gemini IV even though one Mercury astronaut had not yet been assigned.** James A. McDivitt and Edward H. White II were named commander and pilot of the prime crew, backed up by Frank Borman and James A. Lovell, Jr., all members of the second astronaut class.[36] Flight experience

At the end of a 50-foot arm, the man-rated centrifuge at the Manned Spacecraft Center spins a three-man gondola to simulate the g-forces during launch and re-entry.

---

*See Appendix 7 for a summary discussion of simulation and training.

**Of the Original Seven, Glenn had resigned, Carpenter was working in Project Sealab, and Slayton was medically disqualified. At the press conference naming the Gemini 3 crews it was also announced that Alan Shepard was suffering from a middle-ear inflammation that grounded him as well. (Later that year Shepard took over from Slayton as chief of the Astronaut Office.) Only Cooper remained unassigned at this time.

seemed clearly to be a factor in Slayton's choice of crews, but it was just as clearly not the only factor. L. Gordon Cooper, Jr., pilot on the 34-hour, 22-orbit Mercury 9 mission, was not assigned until the Gemini V crew was named in February 1965; his pilot was Charles ("Pete") Conrad, Jr., of the second group. Their backups, both from the second group, were Neil A. Armstrong and Elliott M. See, Jr.[37]

After Grissom and Young completed Gemini 3, Slayton announced that their backup crew, Schirra and Stafford, would be the prime crew for Gemini VI, backed up by Grissom and Young. To trainees eagerly seeking some clue to their prospects for flight assignment, this signaled that appointment to a backup crew was the key to flying a mission. The system Slayton followed, as long as circumstances permitted, was to promote each backup crew to prime crew of the next available mission after their own prime crew had flown. Each flight had different objectives, requiring different training, and the prime and backup crews had to train as a team to perform most efficiently. Almost to a man, the astronauts professed being in the dark as to exactly how Slayton chose crew members for their first assignment, but that did not matter once they perceived that when they were named to a backup crew they were, at last, in line for a flight assignment.[38]

Selection of the First Scientist-Astronauts

As the Office of Manned Space Flight began to consider post-Apollo possibilities in 1963–1964, science-oriented missions—lunar exploration and earth-orbital missions of many days' duration—appeared to be the most acceptable of a very few alternatives. The principal theme of George Mueller's expositions to Congress was the continued use of Apollo's rockets, spacecraft, and launch facilities to conduct scientific and technological investigations on the moon and in space—to produce a return on the nation's investment in manned space flight. Mueller's proposals were criticized as unimaginative and not conducive to the advancement of space technology, but none of NASA's top managers was willing to advocate bolder programs under the budgetary restraints that were becoming apparent in 1964.[39]

For any serious scientific work the crews in the spacecraft would have to include some scientists trained as astronauts rather than astronauts trained as scientific observers; and early in 1964 selection of scientists for the astronaut program began. MSC officials and representatives of the National Academy of Sciences met in February to draft a plan for recruitment and selection. Agreement was reached that the Academy would define the scientific qualifications desirable in the candidates while MSC would specify the physical and psychological requirements. On April 16 Homer Newell formally asked Harry Hess, chairman of the Space Science Board, to draw up a statement of the scientific qualifications for a scientist-astronaut.[40] In Mid-October, Headquarters announced that it would accept applications from scientists who wanted to become astronauts. The primary requirement was a doctorate in medicine, engineering, or one of the natural sciences. No applicant had to be a qualified pilot; those accepted by the Space Science Board and by NASA would be assigned to the Air Force for a year of flying training.[41]

Any doubt that scientists were interested in space flight was dispelled by the response: more than a thousand hopefuls sent in their applications. An ad hoc committee of the Space Science Board (chaired in Hess's absence by Eugene Shoemaker) rigorously scrutinized about 400 of those who passed NASA's preliminary screening, finally sending only 16 names to NASA for final evaluation.[42] MSC had hoped for a larger group to choose from; Slayton's selection board had much less information on the applicants' physical condition and psychological makeup than they had for military applicants, and the choices were consequently harder to make.[43] The Space Science Board, however, was evidently determined to pick only the most promising scientists. Shoemaker later recalled that the committee had been disappointed in the overall quality of the applications that came in. Not many of the scientists who applied came up to the rather high standards they set.[44] Whatever the reasons for this, NASA was able to pick only 6 scientist-astronauts instead of 10 or more, as it had initially planned.

On June 27, 1965, NASA announced the names of its first scientist-astronaut candidates: two physicians, Duane M. Graveline and Lt. Cmdr. Joseph P. Kerwin, MC, USN, and four Ph.D. scientists, F. Curtis Michel, Edward G. Gibson, Owen K. Garriott, and Harrison H. Schmitt (who was generally known by his nickname, ''Jack'').[45] Kerwin was a flight surgeon stationed at Cecil Naval Air Station in Florida; Graveline, a former Air Force flight surgeon, was working in the medical program at MSC. Gibson, a senior research scientist at the Applied Research Laboratories of Philco's Aeronutronics Division in San Diego, California, and Garriott, associate professor of physics at Stanford University, were both engineers engaged in research in solar and atmospheric physics. Michel was an

The first class of scientist-astronauts (1965). Left to right: Owen K. Garriott, Frank Curtis Michel, Harrison H. (Jack) Schmitt, Duane E. Graveline, Edward G. Gibson, and Joseph P. Kerwin.

assistant professor of physics at Rice University in Houston conducting research in the interaction of the solar wind with the earth's atmosphere. Schmitt, the lone geologist in the group, was working with Eugene Shoemaker at the Geological Survey's Astrogeology Branch. The only qualified pilots in the group were Michel, a former Air Force pilot, and Kerwin, a naval aviator. The other four were sent to Williams Air Force Base in Arizona to begin 55 weeks of flying training.[46] Within a few weeks Graveline resigned from the program, citing "personal reasons" for his action.[47]

Future Plans Require Yet More Astronauts

From 1964 through 1966 George Mueller, chief of manned space flight programs, worked hard to sell an ambitious post-Apollo program to his NASA superiors and to Congress. His Apollo Applications Program (AAP), established in August 1965, contemplated 29 lunar and earth-orbital missions between 1968 and 1971. Two-thirds of them would be manned flights, launched at the rate of eight per year. A manned program of that magnitude seemed to have little chance of becoming reality, but until circumstances forced him to back down from it, Mueller kept the pressure on the field centers to plan for big things.[48]

At Houston, Bob Gilruth and Deke Slayton—whatever their own views about the probability that such an ambitious schedule could be realized—had to be prepared to furnish crews for whatever emerged as the Apollo Applications Program. If the projected AAP missions should actually materialize, many crews would be needed in short order and training had to begin soon. While the recruitment of scientist-astronauts was under way, Gilruth and Slayton urged that additional pilot candidates be sought at the same time, but Mueller decided to wait until the scientist-astronauts had been chosen.[49] When only six trainees emerged from that selection process, he agreed to go ahead. On September 10, 1965, Headquarters announced it would accept applications for a new class of pilot-astronauts. Qualifications would be the same as they had been for the third group: a bachelor's degree in science or engineering plus 1,000 hours of jet flying time or qualification as a test pilot.[50]

The announcement yielded 351 applicants—the largest number of pilots ever to apply—of whom 159 met the basic requirements. Final screening during the next four months produced the fifth class of astronaut candidates in April 1966: 19 pilots, 4 civilians and 15 military officers (see Appendix 6). Eleven of the fifth group held advanced degrees, two of them doctorates.[51]

Headquarters Expands the Ranks of Scientist-Astronauts

In mid-1966, when the scientist-astronauts had completed their flying training and the third group of pilots had reported aboard, the astronaut corps numbered

The fifth group of astronauts, selected in 1967. Seated, left to right: Edward G. Givens, Jr., Edgar D. Mitchell, Charles M. Duke, Jr., Don L. Lind, Fred W. Haise, Jr., Joe H. Engle, Vance D. Brand, John S. Bull, and Bruce McCandless II. Standing, left to right: John L. Swigert, Jr., William R. Pogue, Ronald E. Evans, Paul J. Weitz, James B. Irwin, Gerald P. Carr, Stuart A. Roosa, Alfred M. Worden, Thomas K. Mattingly II, and Jack R. Lousma.

44 pilots* and 5 scientists (or 41 and 8, depending on how Cunningham, Schweickart, and Lind were classified)—a ratio that hardly supported the contention that NASA was interested in sending scientists into space. At the Manned Spacecraft Center, Deke Slayton and Bob Gilruth considered that they had quite enough pilots to carry out the programs they could realistically envision and that pilots could be trained to conduct the scientific work that was planned for the lunar landing missions. At Headquarters, however, both Homer Newell in the Office of Space Science and Applications and George Mueller in the Office of Manned Space Flight thought otherwise. Newell, representing the science community, wanted manned space flight to give more attention to science and less to the engineering and piloting aspects of space flight. Mueller, trying hard to sell an ambitious program of post-Apollo manned missions based largely on scientific research in space, could use more scientists in the astronaut corps to give credibility to his appeals to Congress and to gain political support from scientists outside NASA.

In spite of Houston's reluctance to take on astronaut trainees who would have little expectation of flying in space, Headquarters and the National Academy of Sciences announced on September 26, 1966, that applications would be accepted for a second group of scientists to be trained as astronauts. Selection would be

---

*Three astronaut trainees had been killed in flying accidents in the previous two years. Ted Freeman's T-38 hit a goose near MSC on Oct. 31, 1964, causing both engines to flame out; he ejected but was too low for his parachute to open. Elliott See and Charles Bassett, prime crew for Gemini IX, crashed on Feb. 28, 1966, after missing a landing approach at St. Louis Municipal Airport under marginal weather conditions.

made in about six months.[52] By the time they came aboard, however, post-Apollo manned space flight programs were in a precarious position and the future looked much less bright (see Chapter 7). The chances seemed good that any scientist who went to the moon would be one of the first five already in the program.

Scientists in the Astronaut Corps

During 1965 and 1966 the Manned Spacecraft Center was busier than it had ever been. Gemini flights were being launched from Cape Canaveral every other month, on average. The Apollo command and service module was progressing, not without difficulty, toward its first earth-orbital flight test. Mission planners were hard at work on lunar-mission trajectories and contingency planning. Others were studying photographs of the lunar surface from Ranger and Surveyor, looking for suitable landing sites and scrutinizing the barren surface for possible unwelcome surprises. Still to come were the extensive and detailed photographs from Lunar Orbiter.

The Astronaut Office was as busy as the rest of the Center. All of the remaining "Original Seven," plus the "Next Nine" and 10 of "The Fourteen" (third group) were training for and flying the Gemini missions. By the end of 1966 crews for the first four Apollo earth-orbital missions had been assigned and were spending much of their time in design reviews and flight simulations. Russell Schweickart, one of the two scientists picked as a pilot, was serving as a kind of ombudsman, mediating between the astronaut office and the experimenters who had projects on Gemini.[53] His compatriot Walter Cunningham was sent to the Falmouth conference in mid-1965 to explain to scientists some of the operational factors that so strongly constrained a lunar landing mission.[54] Two of the first five scientist-astronauts, Joe Kerwin and Curt Michel, did not start basic astronaut training during their first year, and so were assigned to represent the Astronaut Office in matters concerning space suits and Apollo Applications experiments, respectively. The other three, Owen Garriott, Ed Gibson, and Jack Schmitt, drew assignments to Apollo in-flight experiments when they returned from flight training in mid-year, as did Don Lind, the scientist who came in as a pilot with the fourth group.[55] Before long, however, Schmitt was working with academic and Geological Survey scientists to improve MSC's training course in field geology.[56]

Schmitt was fortunate in having a scientific specialty that was widely accepted as being important to Apollo. The other scientist-astronauts—except for Kerwin, whose medical training could be applied to a number of space-related questions—found themselves in an environment oriented almost exclusively to operational and engineering concerns. Independent research was all but impossible; only Curt Michel—whose academic home base was Rice University, less than an hour's drive from MSC—made an attempt to sustain his previous research program. Owen Garriott and Ed Gibson had to redirect their scientific interests into fields more closely related to NASA's needs and plans.[57]

Apart from the time they had to devote to mastering astronautic skills, the scientists had to spend long hours on chores that sometimes seemed distinctly subsidiary to the main objectives. Among the duties of the Astronaut Office were

The second class of scientists-astronauts, picked in 1967. Seated, left to right: Philip K. Chapman, Robert A. R. Parker, William E. Thornton, and John A. Llewellyn. Standing, left to right: Joseph P. Allen, Karl G. Henize, Anthony W. England, Donald L. Holmquest, Story Musgrave, William B. Lenoir, and Brian T. O'Leary.

making public relations appearances, participating in design reviews, and contributing the astronaut viewpoint to engineering decisions; the scientist-astronauts were expected to shoulder their share of these burdens just as the test pilots did. Precious little time was left for keeping abreast of scientific developments, but in Slayton's view this was a problem each astronaut had to solve for himself.[58] Nobody was told what he could *not* do, but it was understood that the astronauts' primary role was to become competent spacecraft operators, and whatever else they wanted to do had to be compatible with that; as long as it was, the Astronaut Office raised no objection to anyone's supplemental activities.[59] Those who made the adjustment gained the respect of their pilot colleagues; those who expressed annoyance at these ancillary duties and felt cheated out of scientific opportunities provoked some resentment.[60] After all, the door was always open.

When the first scientist-astronauts joined the program in 1965, it was not to be expected that science could simply force its way into Apollo, which had yet to fly its first test mission. Nonetheless, the scientific community wanted to make it clear that scientist-astronauts were entitled to consideration of their professional scientific requirements. In the fall of 1965 Headquarters's Manned Space Science Division commissioned a study group to look into the matter of astronaut training. After some weeks of discussion with MSC officials, the group concluded that the astronaut training program was much too short on science. More scientist-astronauts should be brought in as early as possible, to provide more scientific resources for the manned space flight program. The scientist-astronauts should be used as in-house tutors for other astronauts who wanted to improve their scientific background. The Astronaut Office should actively encourage the astronauts to develop their scientific skills by issuing a policy statement that ''after engineering evaluation flights are completed and a spacecraft is considered operational, scientific proficiency shall be a prime requisite for at least one member of each flight crew.'' Anything that seemed to increase their chances of flight assignment was of vital interest to every person in the corps, the group had learned. (If Slayton wanted someone on a crew who could speak Mandarin Chinese, one of the astronauts told the study group, they would all be studying Mandarin Chinese.) Therefore, if MSC made it clear that scientific proficiency was desirable for crew selection, even the pilot-astronauts could develop a passion for science.[61]

Perhaps the most difficult recommendation to implement was that the scientist-astronauts be encouraged to keep up their research activity by affiliating with an established research group. ''The minimum amount of time required to maintain scientific proficiency,'' the group concluded, ''is believed to be one day per week for discussions, seminars, etc.,'' plus ''one full week each month in which the scientist-astronaut can become completely immersed in his research.'' The group could not suggest how this could be squeezed into an already tight training schedule, but they noted that astronauts spent considerable time at seemingly trivial tasks in engineering design that might be relegated to others. Paradoxically, however, these time-consuming chores seemed an indispensable part of the program, since the astronauts were the only competent group having an overview of the whole operation, and were ''the only single group that another astronaut will trust.''[62]

Stressing as it did the importance of research to a scientist, the study group's report could have been read as calling for a division of the corps into a test-pilot group and a scientist group. The scientist-astronauts' need to spend more than one-third of their time in research was received with some skepticism by the pilots, whose reaction was later summarized by one of them:

> . . . [Some] of those guys came in figuring, ''I'll write my textbooks and my thesis and teach [university courses] and I'll come by twice a week and be an astronaut.'' Well, that didn't work. . . . We were devoting our lives to this whole thing, and you couldn't devote anything less, I don't care what your discipline was.[63]

The issue did not become divisive because the scientist-astronauts themselves accurately perceived the situation they were in and most of them did not try to make the system fit their unique needs. They saw the utility of the maxim, ''if you want to get along, go along.''

The study group's report was received politely but coolly at MSC.[64] If Mercury and Gemini had shown anything, it was that the unexpected may turn out to be the norm, and no one knew how well a scientist, however skilled and intelligent, would react to sudden operational emergencies. On the other hand, appropriate reaction to such situations was believed to be almost instinctive to a good test pilot. Slayton and Gilruth, pondering the problem of landing an exotic spacecraft on a strange and possibly dangerous surface, naturally adopted the view that piloting skills were essential to mission success. Slayton repeatedly expressed this view in plain language: nobody would benefit from a mission that left a dead geologist (*and* his colleague in the lunar module) on the moon[65] —implying that just such a thing might happen if the pilot of the lunar landing module could not cope quickly enough with a sudden emergency. So it was up to the scientists to prove that they could become competent astronauts, which most of them did. None would ever command an Apollo mission; none would ever pilot a lunar module to a moon landing or a command module through reentry; but they showed themselves able to tackle the training program and willing to share the less pleasant but essential duties of an astronaut. Of the first six scientists picked as astronauts, four eventually flew in space. Many of the others filled essential roles in science planning and mission operations during the later Apollo missions.

6

MISSION AND SCIENCE PLANNING: 1963–1966

Apollo's scientific objectives were always acknowledged, but scientists as well as engineers understood that the primary goal was, in John Kennedy's words, "landing a man on the moon and returning him safely to earth." As long as that remained to be done, what the man (or men) would do on the lunar surface was of secondary importance. The choice of a landing site, the time allowed for the astronauts to explore the landing area, the instruments to be taken to the moon, and the amount of lunar material to be brought back to earth, all were governed by operational factors: what the engineers considered prudent in light of the over-riding necessity to return the lunar explorers safely to earth.

Some scientists might object to the goal itself, but no one disputed the need to achieve it safely. So while the engineers worked out their operations plans, scientists concentrated on what they wanted done on the missions, on the under-standing that their plans would have to yield to operational constraints until NASA had accumulated some flight experience. By the time flight planners felt able to relax some of those constraints, scientists would have prepared a long shopping list of landing sites and scientific activities to answer some of their questions.

Input from the scientists in the early days was minimal. In 1961 Harold Urey, responding to a request from Homer Newell, listed the general areas he would like to see explored on the moon: high latitudes, where low temperatures might allow water to exist; inside a large crater; in two maria of different types; near one of the great wrinkles in the maria; and in a mountainous region.[1] Probably many lunar scientists would have agreed with Urey's choices; but as Eugene Shoe-maker recognized in a paper not long afterward, operational necessities would certainly militate against many of those choices, especially for the early missions.[2]

Operational Constraints on Landing Sites

Picking a spot where a lunar module could land was a complex exercise requir-ing tradeoffs among dozens of factors. Predominant among these were the topog-raphy and texture of the lunar surface and the requirements of the lunar module's guidance and navigation system. Other restrictions included the elevation of the sun at the landing site; the temperature of the lunar surface; the radiation environ-ment in space and on the moon; and the earth lighting conditions desired for launch and recovery.[3]

The most restrictive mission rule, so far as landing sites for the earliest lunar landing missions were concerned, was the requirement to place the spacecraft on a "free-return" trajectory—a flight path that allowed for failure of the service module's main engine. If the service module engine should fail to put the spacecraft into lunar orbit, the joined CSM and LM would loop around the moon under the influence of lunar gravity alone and head back to earth. This free-return trajectory required the spacecraft to leave earth orbit on a path that would bring it to the moon within five degrees of latitude of the lunar equator. As early as mid-1963 MSC's Space Environment Division selected four sites from a list compiled by various lunar scientists, balancing scientific interest against this mission rule. A few months later five more sites were picked for preliminary trajectory studies.[4]

The choice of lunar longitude for a landing site depended mainly on two considerations. To land at a predetermined site it was essential to determine the position and the flight path of the lunar landing craft as accurately as possible before beginning the final descent to the surface. Navigational sightings taken from the spacecraft on stars or on lunar surface landmarks provided the data from which ground-based computers determined the spacecraft's orbit and calculated the necessary course corrections. Accurate calculation of the orbit required the astronauts to take sightings on five prominent lunar landmarks some distance east of the landing site, and since the spacecraft went behind the moon on the west and was out of radio and radar contact with earth until it emerged around the eastern edge, those sightings could only be taken after earth contact was reestablished. The position of the navigational landmarks had to be known with an error of no more than 1,500 feet (450 meters). In mid-1963 this was not possible; at the eastern and western edges of the visible face, surface features might actually be as much as 6,000 feet (1,800 meters) from where the best lunar maps showed them. In light of these navigation requirements, early planning assumed that a landing could be plotted no farther east than 40 degrees east longitude. The "Apollo landing zone" thus defined extended to 40 degrees west longitude; other operational considerations made a more westerly landing undesirable.[5]

The second limitation on the longitude of the landing site was the elevation of the sun at the time of landing, which was the major factor considered in choosing the time of launch. To the pilot looking for a safe spot to land within the time the lunar module could hover, it was vital that the sun be high enough in the lunar sky to highlight the surface topography without casting long, confusing shadows, but not so high that all surface detail was washed out. After landing, the lunar explorers would also be hindered by a low or high sun. The moon has no atmosphere to scatter light and therefore shadows are completely black; at either low or high sun angles, visual observations can be difficult. Furthermore, solar heating of the lunar surface varies with sun angle, complicating the problem of protecting the astronauts and the spacecraft against extreme temperatures. Conditions would be best when the sun was 15 to 45 degrees above the horizon. Mission planners could choose a launch time so that the lunar module would land at a time when solar illumination was near optimum. Launch time was subject to the further constraint, however, that lunar missions had to leave the launch pad well before last light in case an aborted launch required emergency recovery operations.[6]

Two potential hazards to the lunar mission were more difficult to take into account: meteoroids and radiation. In 1963 no one knew how dangerous meteoroids were. It seemed prudent to avoid the predictable (and dense) swarms that recur annually, but the earth-moon system constantly encounters a smaller number of randomly distributed meteoroids. The last three test flights of the Saturn I launch vehicle carried meteoroid-detecting satellites into earth orbit to determine how serious this hazard might be. Radiation (subatomic particles, x-rays, and gamma rays) was more worrisome. Of special concern were the high-energy protons shot out from the sun during major solar flares, which could subject astronauts on the lunar surface to lethal doses of radiation.[7] Solar flares were more troublesome because they are completely unpredictable. Protection was extremely difficult and warning all but impossible: by the time a flare could be detected on earth and its magnitude assessed, the most energetic (and dangerous) particles would already have reached the moon.

Within the zone defined by all these constraints—roughly 185 miles (300 kilometers) wide, stretching 1,500 miles (2,400 kilometers) along the moon's equator—geologic factors would determine the choice of a landing site. Surface topography could not be known in any detail until Ranger and Surveyor provided information; but from lunar maps available in 1963, several landing areas* about 400 square miles (900 square kilometers) in size could be picked out where slopes apparently were not too great and craters not too numerous. Balancing the need to pick landing areas near the center of the moon (where lunar maps were most accurate) against the requirement to spread the areas as far as possible along the equator (which allowed maximum flexibility in launch dates), MSC's Space Environment Division found 10 landing areas that seemed promising enough to warrant reconnoitering by unmanned spacecraft and close scrutiny by mission planners. The areas were spotted in a zone extending from the southeastern edge of Mare Tranquillitatis (not far from where Apollo 11 would eventually land) to a point northeast of Flamsteed crater in Oceanus Procellarum (some 375 miles [600 kilometers] west of Apollo 12's touchdown point). Even among these, however, none was completely satisfactory with respect to all the known constraints.[8]

The effect of all these restrictions on the landing site was to reduce drastically the number of consecutive days per month on which a lunar mission could be launched. Considering only the two most important factors—sun angle and surface temperature—a given site could be reached only if the spacecraft were launched during a 2.3-day period each month. Experience to 1963 indicated that launch operations stood a good chance of being interrupted and launches postponed because of systems problems, and no one was willing to count on launching an Apollo mission within 2.3 days. But if flight planners could be prepared to land at more than one site for each launch—changing to a more westerly target if launch delays prevented reaching the first site—the number of consecutive days on which a launch was possible could be substantially increased. A Bellcomm study in early 1964 pointed out that choosing multiple sites for each launch would make the program considerably more flexible, though it would require certifying more sites through the Surveyor and Lunar Orbiter programs. That, however, might cost no more than postponing a few launches by a month each.[9]

---

*A landing *area* was a fairly large segment of the lunar surface which appeared sufficiently level and smooth to permit landing; a landing *site* was an ellipse a few hundred meters in size within which the lunar module would actually touch down. A site would be picked for exact targeting after the hazards of the landing area had been assessed.

The greatest uncertainty in the program at that time, pointed up by all these early studies, was the physical nature of the moon's surface. Astronomers held widely different views of what a lunar module would encounter when it touched down. Gerald Kuiper of the University of Arizona, one of the principal investigators in the Ranger project, was convinced that the surface was firm, though it might be unconsolidated and might be covered by a thin layer of dust. Cornell University astronomer Thomas Gold asserted, however, that the apparently smooth areas on the moon were likely to be covered with a layer of fine dust several meters thick, raising the prospect that the lunar module might sink out of sight with only a short-lived dust cloud to mark its disappearance.[10] There was the further possibility that the surface might be so cluttered with boulders and pitted with small craters that the lander would find no level spot large enough to land—or if it tried to land, would turn over or come to rest tilted at an angle that made return to orbit difficult.

NASA had been hoping that Ranger's television photographs would shed light on these questions, but by the end of 1963 Ranger had experienced its fifth failure in as many attempts and was undergoing a critical reappraisal (see Chapter 2).[11] Spacecraft engineers at Houston's Manned Spacecraft Center, meanwhile, in spite of their real need for this information in designing the lunar landing module, had to go ahead without it.[12] Lunar Orbiter, still in the early stages, would have to provide the information that mission planners needed for site selection. The spacecraft builders could only hope that data from Surveyor, when they got it, would not force them to revise their design too drastically.

Scientific Input to Site Selection

Early consideration of lunar exploration missions by the scientific community focused more sharply on what should be done on the moon than on where it would be done. The 1962 Iowa summer study paid particular attention to the scientific qualifications and training of the astronauts; the 1965 Woods Hole study formulated a list of 15 important scientific questions to be addressed by lunar exploration, which would define the experiments to be conducted (see Chapter 3 and Appendix 3). Neither conference expressed any preference for landing sites, although both pointed out the need to study highlands as well as maria. The Woods Hole study report concluded that really effective geophysical and geochemical studies would require investigations at several locations up to 1,000 kilometers (620 miles) apart.[13]

NASA had apparently made it quite clear to its outside scientific advisors that science would have to take its results from wherever it could get them—at least on the first few missions. The point was implicitly acknowledged at the Falmouth conference on lunar exploration immediately following the 1965 Woods Hole study. All of the earth-science study groups at Falmouth stressed the need to supplement manned exploration with detailed study by unmanned lunar-orbiting satellites to secure comprehensive scientific coverage of the lunar surface.[14] To a considerable degree this emphasis on unmanned studies was simply a recognition of the relative cost-effectiveness of the two modes of exploration, but it was also a recognition of the operational limitations of Apollo and the resultant restrictions on landing sites. Even looking ahead to longer stays on the moon, the groups urged development of surface vehicles and flying vehicles to increase the effective range of exploration but did not mention any need to extend the limited Apollo landing zone. It was assumed that operations could eventually be extended to higher latitudes and more difficult sites, such as highlands and craters—possibilities included in NASA's plans for extending Apollo,[15] which the Falmouth participants used as the basis for their recommendations.

The fact was that in mid-1965 it was simply too early for scientific priorities to be included in the selection of Apollo landing sites. Better information on the lunar surface was needed, and better understanding of the operational constraints on landing sites, before the scientific merit of any particular site could have any effect on the choice.

Site Selection: Early Results from Unmanned Programs

Throughout 1965, site selection was one of Apollo's biggest concerns. Early in the year the ill-starred Ranger project flew its second consecutive successful mission (in seven tries). Its photographs had enabled the Manned Spacecraft Center to draw some generally encouraging conclusions about lunar topography, but MSC's Space Environment Division concluded that Ranger photographs would never enable them to certify lunar landing sites. They would, perhaps, allow Apollo planners to rule out unsuitable areas, but picking a suitable landing site by eliminating the unsuitable ones could take years.[16] Besides, Ranger provided no information about the physical characteristics of the surface, which designers of the lunar landing craft needed.[17] Surveyor was expected to soft-land on the moon within a few months, and MSC was offering advice to Surveyor mission planners at the Jet Propulsion Laboratory as to where the missions should land to be of maximum use to Apollo. The Houston center was also working closely with Langley's Lunar Orbiter program office in planning for optimum use of its high-resolution cameras to locate sites for the lunar landing mission.

In efforts to satisfy the sometimes conflicting aims of lunar science and Apollo, the Ranger project had been subjected to a great deal of pulling and hauling (see Chapter 2). The result was that neither side was completely satisfied, and the lunar scientists were particularly upset when Apollo preempted their experiments. Intending to forestall similar problems with Surveyor and Lunar Orbiter, Homer Newell established an ad hoc Surveyor/Orbiter Utilization Committee in the spring of 1965, after the last Ranger mission (Ranger 9) successfully concluded that project. Newell appointed his deputy, Edgar M. Cortright, chairman of the committee, whose other members were chosen from Headquarters and center project offices.*[18]

---

*Other members were: Everett E. Christensen, mission operations dir., OMSF; Victor C. Clarke, Surveyor project office, JPL; Willis B. Foster, dir., Manned Space Science Div., OSSA; William A. Lee, asst. mgr., Apollo spacecraft program office, MSC; Urner Liddell, chmn., Planetology Subcommittee of the Space Sciences Steering Committee, OSSA; Benjamin Milwitzky, Surveyor project dir.,

Mueller, meanwhile, had discovered that no single organization was responsible for coordinating lunar data and evaluating candidate landing sites. Shortly after Newell created the Surveyor/Orbiter Utilization Committee, Mueller established an Apollo Site Selection Board in the Office of Manned Space Flight. Its primary responsibility was to select and recommend candidate sites after considering all data required to ensure the success of the lunar landing. If the board could not unanimously agree on a site, majority rule did not apply; Mueller wanted all dissenting opinions recorded, and as he had done when he created the Manned Space Flight Experiments Board (see Chapter 2), he reserved for himself the authority to make the final choice. Chaired by OMSF's Apollo program director, Air Force Major General Samuel C. Phillips, the board included members from the Office of Space Science and Applications and the three manned space flight field centers.*[19]

The overlapping membership of these two boards was yet another way of making sure that all interested parties were sufficiently involved in decisions when scientific concerns and manned space flight objectives converged (and sometimes conflicted) in a single program—which, some felt, had not been the case during Ranger. MSC Director Bob Gilruth expressed satisfaction with this arrangement and selected two of his engineers to serve on both boards.[20]

Since Surveyor and Lunar Orbiter were expected to fly their first missions within a year, the Surveyor/Orbiter Utilization Committee began work early. From late August 1965 until the first launches in mid-1966 the committee met as frequently as necessary to evaluate sites and set priorities. Lunar Orbiter project officials initially outlined four types of missions, from a general survey of the moon to photography of specific sites within the Apollo zone. Surveyor representatives presented 40 possible landing sites for their first spacecraft, chosen on the basis of independent evaluations by the Jet Propulsion Laboratory and the Geological Survey based on the best available geologic maps and telescopic observations. With little discussion the committee recommended that Lunar Orbiter's first mission give priority to Apollo site photographs. Surveyor, like Lunar Orbiter, had objectives independent of its usefulness to Apollo, but its project managers seemed less willing to allow their science goals to be subordinated to those of the lunar landing program. After some discussion, the committee agreed on a set of ground rules for selecting Surveyor landing sites. It was more important to have well-lit photographs of some part of the lunar surface as soon as possible than to have photographs within the Apollo zone; however, if a daylight landing was possible within that zone as well as outside it, the committee gave priority to an Apollo site. If the first Surveyor could not land in sunlight at all, it should be put down within the Apollo zone. The committee approved 14 of the 40 sites proposed for the first Surveyor mission, then sent its recommendations to the project offices, which would plan specific missions in more detail for review by the committee and by OSSA's Space Sciences Steering Committee.[21]

OSSA; Oran W. Nicks, lunar and planetary programs dir., OSSA; Samuel C. Phillips, Apollo program dir., OMSF; Lee R. Scherer, Lunar Orbiter project dir., OSSA; and William E. Stoney, Space Environment Div., MSC; and Israel Taback, Lunar Orbiter project office, Langley Research Center.

*Other members were: Cortright, Lee, Stoney, and Christensen, all members of the Surveyor/Orbiter Utilization Committee; Ernst Stuhlinger, Research Projects Office dir., MSFC; and John P. Claybourne, Future Studies Office, Design Engineering Branch, KSC.

Lunar Orbiter project officials set about making detailed plans for the first mission that would satisfy Apollo's requirements. Surveyor managers, however, were reluctant to comply with the committee's recommendations. They pointed out that the ground rules required landing at a less-suitable site even though a better one was available—a significant threat to mission success, JPL believed. They would target the first Surveyor in compliance with the committee's ground rules, but if the spacecraft had any trouble at an inferior site, JPL intended to recommend strongly that the next mission be sent to a better one.[22]

America's first unmanned lunar explorer,* *Surveyor I,* launched on May 30, 1966, landed 63 hours later approximately 90 kilometers north of the crater Flamsteed in Oceanus Procellarum, at the extreme western fringe of the Apollo landing zone.[23] Its first television pictures were encouraging to Apollo engineers: the surface was firm enough to support a lunar landing module and the surrounding area was reasonably level. But beyond the immediate landing area the TV images showed more than a few large boulders and craters. The pictures served to demolish some hypotheses about the lunar surface, notably Gold's idea that the maria were covered with deep layers of dust. The surface was composed of fine particles, firmly packed and cohesive.[24]

Less than three months later, on August 10, *Lunar Orbiter I* roared off the pad at Cape Canaveral; four days later the spacecraft was in orbit around the moon, and by the end of the month it had photographed all nine of its assigned sites in the Apollo landing zone. After completing its photography, *Orbiter* remained in orbit, transmitting electronic data on gravity, meteoroids, and radiation around the moon—another aspect of its service to Apollo and to lunar science.[25]

Malfunctions of the spacecraft and photographic system resulted in fewer good photographs than project officials would have liked; but preliminary photographic analysis began in late August and produced recommendations for the second Surveyor and Lunar Orbiter missions by the end of September. On the basis of crater density and regional slopes, 10 of 23 areas selected for detailed study seemed satisfactory; 8 potential sites were recommended for detailed analysis.[26] On September 29 the Surveyor/Orbiter Utilization Committee reviewed these results and outlined the requirements for the second Orbiter mission.** Again Apollo sites were stressed, particularly those types of terrain shown by *Orbiter I* to be the most promising—landing areas smooth enough to land on, having little enough slope in the approach path to avoid confusing the lunar module's landing radar system. Lighting conditions for photography would be chosen so that slopes of 7 degrees, and surface protuberances 6 feet (2 meters) in diameter and 1.5 feet (0.5 meter) high, could be detected in an area 20 feet (7 meters) square.[27]

Lunar Orbiter II, modified to correct the problems encountered by its predecessor, was launched November 6, 1966, and its results far surpassed those of the

---

*But not the world's first; on Feb. 3, 1966, the Soviet Union, apparently intent on establishing space "firsts," had logged another by landing *Luna IX* on the moon several hundred kilometers northwest of *Surveyor I*'s landing site. Similarly the Russian *Luna X* became the first moon-orbiting satellite on Apr. 3, 1966, four months before the U.S.'s *Lunar Orbiter I. Luna X* did not relay photographs to earth, however.

**A small thruster failed on *Surveyor II* (launched on Sept. 20, 1966) after its midcourse correction maneuver, causing the spacecraft to spin at one revolution per second. All attempts to correct the malfunction failed and the lander crashed 250–300 km southeast of crater Copernicus.

first mission. All of the 30 preplanned sites were photographed at medium and high resolution, giving geologists and cartographers excellent material with which to work.*[28] The mapmakers believed that with more stereo photographs of the quality returned by *Lunar Orbiter II* they could produce terrain maps with two- to three-meter contours. MSC's Apollo project office representatives were pleased with the second mission's results and believed that one more flight would satisfy their requirements for the first one or two lunar landings. Apart from some additional site photography, Apollo wanted one or more Orbiter spacecraft to stay in lunar orbit long enough to allow precision tracking by earth-based radar. All navigational calculations for the lunar module were to be made by ground-based computers using data from the Manned Space Flight Network, Apollo's communication and tracking system, and early certification of this system was essential.[29]

In mid-December 1966 the Apollo Site Selection Board met to review the status of site selection. In general the process was going well. Geologic interpretation of the Orbiter photographs had already yielded some general conclusions about the nature of the lunar surface and where the best landing sites were likely to be found. Cartographers could not accurately determine slopes in the landing area without steroscopic photography, but MSC gave higher priority to finding a landing site with a minimum of rocks and craters; the slope of the terrain leading to a suitable site could be evaluated some other way. No serious problems in site selection were apparent, and the board could only await the results of Lunar Orbiter III and more detailed analysis of the available data.[30]

As 1967 began Apollo planners did not want much more from the unmanned lunar investigation projects. They would use whatever additional data became available, but priorities were changing. Three more Lunar Orbiters and four Surveyors would be sent to the moon during the next 13 months, but after the third orbiter they would do more for lunar science generally than for Apollo specifically.[31]

Apollo Lunar Surface Experiments

MSC's original specifications for the lunar landing module included a generalized "scientific instrumentation system" to provide for selenological research. Specific experiments could not be listed in 1962, but investigations of the lunar atmosphere, surface, and interior were contemplated. The contractor would include weight, volume, and power allowances for the instruments in the module design; specific instruments would be defined later.[32]

After the lunar module contractor, Grumman Aircraft Engineering Corporation, was selected in later 1962, MSC forged ahead with spacecraft design. Definition of the lunar surface experiments, which was the responsibility of the Office of Space Sciences and Applications, proceeded at a much more deliberate pace,

---

*A famous by-product of the *Lunar Orbiter II* mission resulted from the need to advance film periodically even if no picture was taken. To avoid wasting that film, mission controllers were given a limited choice of targets for these exposures. One was used to take an oblique picture of crater Copernicus under almost ideal lighting conditions, showing hitherto invisible surface details. News media called this "one of the greatest pictures of the century."

Surveyor III bounced twice before finally landing, as this Apollo 12 photograph of its landing pad shows. Imprints were as clear as they were when the unmanned spacecraft landed three years before. The surface sampler, partially visible at upper left, was controlled from Earth. It scooped out several trenches, seen just beyond the landing pad, and deposited a scoopful of lunar soil on the landing pad.

however (see Chapter 3). In early 1963 Houston expected that the first experiments would be selected by the end of the year;[33] but months went by with little progress. By September, spacecraft engineers urgently needed data on the experiments—weight, volume, and power requirements—to feed into lunar module design, but since nothing was available, MSC awarded a 10-month study contract to Texas Instruments to investigate instrumentation requirements for manned lunar exploration.[34] Headquarters was somewhat unhappy with this unilateral action, but lunar module designers had to have the information and the scientific community seemed to be in no hurry to supply it.[35] The study provided an indication of the type of instrumentation likely to be useful, and in the next few months the spacecraft office worked out preliminary weight and volume allotments for the experiments: 250 pounds (113 kilograms) and 15 cubic feet (0.42 cubic meters) in the descent stage, which would be left on the moon; 80 pounds (36 kilograms) and 3 cubic feet (0.085 cubic meters) in the ascent stage and in the command module for a sample-return container and film. Requirements for electrical power and a source to supply it were yet to be determined.[36]

Throughout 1964 and the early part of 1965, Headquarters was busy studying the requirements for lunar surface instruments and experiments, working with outside scientific planning groups and with MSC (see Chapter 3). By May 1965 a tentative list of experiments had been devised and MSC had prepared a procurement plan for a "Lunar Surface Experiments Package" (LSEP). On George Mueller's instructions MSC divided the procurement into two phases, a program definition phase to be conducted by several contractors and an implementation phase in which one of the contractors would build the experiments.[37] The request for proposals was sent out in June and three contractors for Phase I were picked in August.[38]

As defined in Houston's request for proposals, the LSEP would be a self-powered scientific station capable of collecting data for a year, returning information to earth by telemetry. It comprised a passive seismometer to record natural seismic events ("moonquakes"); an active seismic experiment, which would record the effects of small explosive charges detonated on the lunar surface; a lunar gravimeter, which was expected to show tidal effects useful in deducing

Technicians prepare a test version of the Apollo command and service module in the space environment simulation chamber at MSC. The lights mounted at left simulate solar irradiation. The chamber walls can be cooled by liquid nitrogen to -193° C, simulating the radiation-absorbing void of space. After the huge door is closed, the entire chamber, 55 feet in diameter and 90 feet high, can be pumped down to less than one ten-millionth of atmospheric pressure.

the internal structure of the moon; an instrument to measure heat flow from the moon's interior; radiation and meteoroid detectors; and a lunar atmospheric analyzer. A 70-watt power module converted heat from a radioisotope fuel capsule into electricity by means of thermocouples. The instruments, their power supply, and their data-transmitting equipment were limited to 150 pounds (68 kilograms) and 12 cubic feet (0.34 cubic meters) and were to be housed in the lunar module's descent stage. The specifications called for three units which could function on the moon at the same time without interfering with each other.[39] The program directive assigning management responsibility to MSC specified three packages, one to be flown on each of the first three lunar landing missions, plus a flight-qualified spare. The first was to be delivered by July 1, 1967.[40]

Newell's office, meanwhile, had been evaluating proposals for the lunar surface experiments, and on October 1 transmitted authority to Houston to begin negotiations with principal investigators for the lunar gravimeter and the active seismic experiment.[41] A third experiment, investigation of the lunar magnetic field, was tentatively approved on December 15.[42] The science complement for the first

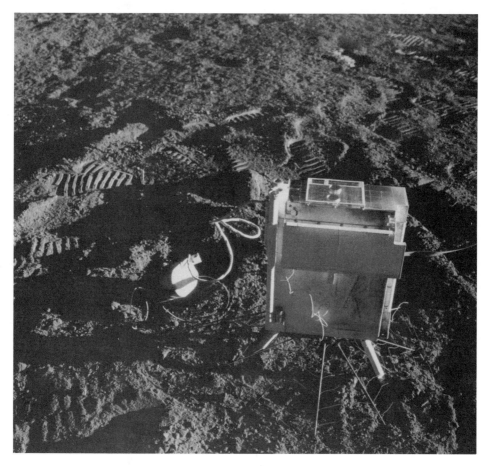

Two lunar surface experiments, the superthermal ion detector (standing on tripod) and the cold-cathode ion gauge (attached by cable, left), measured components of the moon's extremely tenuous atmosphere.

few missions was completed early in 1966 with the public announcement of seven instruments and investigator teams.* Newell noted that the experiments fulfilled the basic recommendations made the previous summer by the science teams at the Falmouth conference and had been approved for flight by the Space Sciences Steering Committee. Since the design of the instruments was not yet fixed it was not certain what combination would be flown on each mission, but modular design would allow each package to carry a group of instruments tailored to the constraints of its mission. The instrument collection was christened "Apollo Lunar

The "Swiss flag" (solar-wind collector) stands on the lunar surface outside the spacecraft Antares. The aluminum foil collector trapped particles from the solar wind, which were then identified in the laboratory of Dr. Johannes Geiss of the University of Berne, Switzerland. One of the simplest of experiments, it was carried on all missions except Apollo 17. Note the almost feature-less lunar landscape at the Fra Mauro landing site.

---

*Passive lunar seismic experiment, Dr. Frank Press (MIT) and Dr. George Sutton (Lamont Geological Observatory); lunar triaxis magnetometer, Dr. C. P. Sonett (NASA Ames Research Center) and Jerry Modisette (MSC); medium energy solar wind experiment, Dr. C. W. Snyder and Dr. M. N. Neugebauer (JPL); suprathermal ion detector, Dr. J. W. Freeman, Jr. (Rice Univ.) and Dr. F. Curtis Michel (MSC scientist-astronaut); lunar heat flow measurements, Dr. Marcus G. Langseth (Lamont Observatory) and Dr. Sydney Clark (Yale); low-energy solar wind, Dr. Brian J. O'Brien (Rice); and active lunar seismic experiment, Dr. Robert L. Kovach (Stanford Univ.) and Dr. Joel S. Watkins (USGS).

Surface Experiments Package," or "ALSEP."[43]

In authorizing MSC to develop the lunar surface science package, Newell assigned specific experiment combinations to each of the first four lunar landing missions and classified each as primary or backup. The first two lunar landers would carry the magnetometer, the passive seismic experiment, the suprathermal ion detector, the medium-energy solar wind experiment, and the heat flow instrument. Subsequent missions might carry different combinations, subject to Newell's approval of any changes. MSC was authorized to spend $5.1 million to develop flight-qualified prototype instruments and provide for operational and support software and data analysis.[44]

All that remained to get instrument development under way was to select a contractor. This was accomplished a month later when Headquarters picked the Bendix Systems Division of Bendix Corporation, Ann Arbor, Michigan, for negotiation of a contract to build four ALSEP packages. Under a cost-plus-incentive-fee contract NASA anticipated a total cost of about $17 million.[45] Bendix was not inexperienced in lunar surface exploration, having worked with JPL from 1963 to 1965, and had made a major corporate commitment to that phase of the space program.[46] Bendix's activity was actually twofold: it would build the "central station" that transmitted data to earth, and integrate the entire experiment package. Some of the instruments were to be built by Bendix's subcontractors to principal investigators' specifications; other principal investigators chose to build their own.

MSC Develops a Science Organization

Mercury project engineers at MSC had reluctantly allowed a few scientific exercises to ride their spacecraft, but this patched-on effort was of small importance. To pave the way for worthwhile experiments in manned space flight, Homer Newell persisted in urging scientists to devise experiments that would take advantage of man's presence. By the time Gemini was fully operational in early 1965 a fair number of scientific exercises had been proposed and accepted for flight.[47] After Mercury, the Manned Spacecraft Center established an Experiments Coordinating Office to ensure that science plans were compatible with the spacecraft and the operational constraints of the mission.[48] Within the Gemini Program Office a Gemini Experiments Office supervised the science exercises undertaken in Gemini.

As the Apollo science program evolved in 1964-65, the horizon of manned space science expanded to include lunar surface experiments as well as in-flight (earth- and lunar-orbital) science. In view of its new responsibility to oversee the development and integration of the Apollo experiments as well as integrating the Gemini science, the Houston center decided to centralize the management of all manned space flight experiments in a single office. In June 1965 MSC Director Robert R. Gilruth appointed Robert O. Piland to head a new Experiments Program Office within the Engineering and Development Directorate.[49] Piland, formerly a research scientist at Langley, had served briefly on the staff of James R. Killian, President Eisenhower's science advisor, before joining the Manned Spacecraft Center in

1959. After contributing to the early planning and study efforts that led to the Apollo spacecraft program, he was appointed deputy manager of MSC's Apollo Spacecraft Program Office in 1962. His new office absorbed the staff and functions of both the Gemini Experiments Office and the Experiments Coordination Office and immediately took over work on the ALSEP contracts.

Piland's job was to keep track of the complex requirements of the Apollo spacecraft and mission plans and see that the experimenters understood those requirements and adhered to them. His responsibilities extended from the conception of the experiment to the management and distribution of data. The Experiments Program Office worked with spacecraft engineers, flight planners, scientific investigators, and contractors in developing, testing, and integrating the experiments into the missions.[50]

The in-flight experiments were an important part of Apollo lunar science, but MSC's involvement with science was growing in other phases of the program as well. During 1965 the Houston center was developing the concept of the lunar receiving laboratory, which went into NASA's budget proposal for fiscal 1966 (see Chapter 4). When Apollo began to return samples of lunar material to the earth, MSC's relations with the outside scientific community would expand considerably. Those samples were of incalculable scientific value, and scientists would assuredly demand a say in how they were handled from the time they were collected on the moon to the time they were parceled out to investigators. Apollo was about to create a relationship between MSC and the scientific world that was new to both groups and would require careful handling.

There is not much room for doubt that MSC considered itself perfectly competent to manage the lunar samples with minimal help from the outside; the science community could simply lay out its requirements and MSC, if it concurred, would do the rest. Scientists, however, would never agree to stand in line at MSC's dispensing window to receive their designated allotments of lunar material. The deliberations of the various advisory groups and ad hoc committees convened to define the lunar receiving laboratory make it plain that scientists saw the proper staffing of the LRL as one of the most important questions of the entire project. At the very least they would insist that a scientist of considerable repute be appointed to head the laboratory and take charge of the sample analysis program, with the advice and consent of the scientific community.

Besides the anticipated lunar science program, MSC had to recognize George Mueller's increasing interest in a science-oriented post-Apollo program. Having established an Apollo Applications Program (AAP) office in August 1965, Mueller went to Congress the following spring to seek funding for it.[51] If he should get what he was asking for, AAP would bring science to the forefront of manned space flight; although MSC lacked the staff to support it (concurrently with its Gemini and Apollo commitments) at the time, science clearly had to have a place in Houston—otherwise MSC might find itself playing a support role to some other center. At the end of March Faget announced the establishment of a Space Science Office within his Engineering and Development Directorate, described as an "interim arrangement pending development of a permanent scientific organization." For this purpose MSC regrouped a number of scattered center activities around Piland's Experiments Program Office and under his direction. While most of those activities were conducted in support of manned space flight operations and lunar exploration, the Space Science Office was also charged with developing, monitoring, and coordinating experiments for all manned space missions involving science.[52]

The following week Gilruth sent MSC's plan for a more extensive reorganization to George Mueller. A new Space Medicine Directorate would consolidate all center medical activities in a single organization. For science, a new Space Science Division was to be established—not a science directorate on the same level as Space Medicine or Engineering and Development, but an upgraded version of the Space Science Office just established, still under Faget's jurisdiction. The proposal more specifically included management of the lunar receiving laboratory, providing MSC's point of contact with the external scientific community, and giving MSC scientists the opportunity to generate their own experiments. Some 76 people from other offices would comprise the new division, and a scientist would be recruited to head it as soon as possible. Agreeing with Newell's view that the division should concentrate primarily on one scientific field, Gilruth suggested that lunar and earth sciences would be its most appropriate disciplines.[53] For the next several months Headquarters and MSC discussed the proposed reorganization and the question of a director for science at MSC.[54].

While those discussions were going on, Headquarters was working to clarify agency-wide management responsibilities for future manned flight activities. Apollo Applications, emerging as the most likely successor to Apollo, embraced a much greater variety of scientific projects than Apollo and appeared to require more interlocking of effort among the field centers (chiefly MSC and Marshall Space Flight Center) as well as Headquarters program offices. Accordingly, on July 26 Deputy Administrator Robert C. Seamans divided responsibility for prospective programs among the various entities—in effect ratifying the arrangement Homer Newell and George Mueller had been working under since 1963 (see Chapter 3). Mueller's Office of Manned Space Flight was to be responsible for the conduct of Apollo and AAP missions, developing and funding the experiments that were selected by Newell's Office of Space Science and Applications. Seamans went one step further, assigning to each center primary responsibility for specific areas: Marshall to develop the Apollo telescope mount (a major component of Apollo Applications), Goddard to handle atmospheric science, meteorology, and astronomical experiments, and MSC to manage the Apollo lunar surface experiments package, lunar science, earth resources, and life-support systems. Future assignments would depend on center capabilities and NASA's long-range plans.[55]

In November, "in response to the growing significance and responsibilities of the Center in the area of science and applications," Houston informed Headquarters that it proposed to create a Science and Applications Directorate, on the same organizational level as those for Engineering and Development and Medical Research and Operations. (At long last Homer Newell's view of the importance of science at MSC prevailed.) It would subsume the functions of Space Science Division and would collect all space- and lunar-science-related functions of the center, along with the many people scattered throughout the center who were then engaged in scientific work in support of Apollo. The directorate would be responsible for planning and conducting all MSC programs in space science and applications and would be the center's point of contact with the scientific world outside. Pending appointment of a permanent director, Bob Piland, named as deputy director, would run the operation.[56] Administrator James Webb approved the new MSC organization on December 23, 1966.[57]

After several months of searching, Gilruth announced on February 17, 1967, selection of a director of science and applications: Wilmot N. Hess, chief of the Laboratory for Theoretical Studies at Goddard Space Flight Center. Hess was a nationally recognized scientist whose major scientific interest was high-energy nuclear physics and space radiation studies.[58] A space physicist seemed a curious choice in view of the scientific responsibilities foreseen for the center, but Hess was considered to be a competent administrator, and he had the scientific stature to give credibility to Houston's scientific efforts.

Apollo at the End of 1966

By the time 1966 drew to a close the Apollo spacecraft and Saturn launch vehicle projects had been through some rough times. The lunar landing craft, under contract to Grumman Aircraft Engineering Company of Bethpage, New York, was slow to reach design maturity and Grumman had considerable difficulty with the main propulsion engines, landing gear, and radar systems. Across the country in southern California, North American Aviation's Space & Information Division had its own set of problems with the command and service module.[59]

North American had other problems as well, notably with the S-II second stage of the mammoth Saturn V launch vehicle. The S-II was similar to Saturn V's third stage (the S-IVB) in that it used cryogenic propellants (liquid hydrogen and liquid oxygen at extremely low temperatures), but it was much larger and presented more difficult manufacturing problems.[60] Besides, North American lacked the experience that the S-IVB stage contractor, Douglas Aircraft Company, could draw on from its prior development of the smaller S-IV stage of Saturn I. In 1965 North American's troubles in managing its two programs—especially the S-II—were the most serious obstacle to achieving the end-of-the-decade goal for Apollo. During 1965 the company's handling of its Apollo and Saturn contracts drew extraordinary attention from Headquarters and from Marshall Space Flight Center, culminating in a top-to-bottom evaluation of NAA's program by teams of NASA experts. In December Maj. Gen. Samuel C. Phillips, Apollo program manager at Headquarters, sent a devastating critique of NAA's program management to his bosses and to the company's executives.[61]

The overall picture for manned space flight was not bleak, however. During 1965-66 the Gemini program had built a solid foundation for operations, sending missions into earth orbit at an average of one every two months. Important questions about the human ability to function in zero gravity were settled. Rendezvous was demonstrated in so many ways it seemed strange that anyone had ever doubted it was feasible. Saturn and Apollo enjoyed some successes as well. Marshall Space Flight Center proved out the Saturn I and IB launch vehicles—important junior partners to Saturn V—and put up earth-orbiting satellites to dispel worries about the hazard from micrometeoroids in space. By the end of 1965 all three stages of the Saturn V had been successfully (but separately) test-fired, and the Manned Spacecraft Center had proved that the Apollo launch escape system worked, easing some concerns about aborts on the launch pad.[62]

Since the status of science in the Apollo program had been nebulous in 1963, the next three years saw substantial progress. Widespread debate over the goals of manned space flight and the validity of the Apollo commitment during the summer of 1963 had given the Office of Manned Space Flight the incentive to accommodate science to some degree in its programs. The Office of Space Science and Applications took the lead in creating an office to coordinate the efforts of the two program offices. The major decisions concerning mission mode and space-craft had been made when OMSF's new director, George Mueller, took over in late 1963; he could therefore direct a good deal of his attention to other matters. Mueller, unlike his predecessor, was not perceived as hostile to scientific investigations in manned space flight, although it took some time for the scientists to decide that he was basically supportive of their efforts.[63]

Advice from the scientific community was brought to a clearer focus between 1962, when the Iowa summer study said little about manned lunar exploration, and 1965, when the Woods Hole and Falmouth conferences defined the objectives of lunar science in more specific terms. After the Woods Hole study posed 15 scientific questions about the moon that bounded Apollo science, study teams defined the essential experimental measurements and the instruments with which they would be made. Additional refinements to these preliminary definitions led to specific studies for the lunar surface experiments package, which was put under contract in 1966.

Houston's Manned Spacecraft Center was finally brought around to at least an acceptance of science by the end of 1966. In the Gemini program it demonstrated a willingness to incorporate scientific exercises into its operational missions and worked out a system for assessing the compatibility of experiments with manned programs. Whether or not these exercises were scientifically important, they brought the scientists and the flight planners together so that they might better understand each other's problems. On its own initiative the Houston center proposed and undertook to develop a laboratory in which to receive, catalog, and conduct preliminary scientific examination of the returned lunar samples. After getting its two spacecraft projects more or less in hand, MSC assumed the responsibility for directing development of the lunar surface experiments and then created a directorate in which (it might be hoped) the center could ultimately develop its own science program. And finally it yielded to the clamor of the scientific community in picking its first class of astronauts from the ranks of promising young scientists. By the end of 1966, in fact, MSC was preparing to select a second group.

On another front, MSC was cooperating with Headquarters in selecting the sites where the lunar missions would land. MSC worked with the Apollo Site Selection Board and Langley's Lunar Orbiter project, supplying its criteria for landing sites and evaluating the information gleaned from Surveyor and Lunar Orbiter. As 1966 ended, the list of candidate sites for the last Lunar Orbiter missions had been considerably narrowed.

At the beginning of 1967 those who wanted to see science become an integral part of manned space flight could feel that some progress had been made in three years. Spacecraft engineers and mission planners could feel that many of their big problems were behind them. In spite of many problems with the command module, the first manned earth-orbital mission was only weeks away from launch. On August 26, 1966, spacecraft 012 arrived at Kennedy Space Center to begin a first-article checkout that would last through the end of the year.[64]

7

SETBACK AND RECOVERY: 1967

Early in 1967 Project Apollo suffered its worst setback when all three members of the first crew to fly in an Apollo spacecraft died in a fire. The tragedy forced a reexamination of the project, especially NASA's supervision of its prime contractors, and delayed the first lunar landing by some unknown length of time.

The fire had no direct effect on the lunar science program other than to provide vitally needed time to catch up to the launch schedule. The experiments package, the selection of landing sites, and the lunar-surface geology program all put the time to good use.

Death at the Cape

On the afternoon of January 27, 1967, Virgil I. Grissom, Edward H. White II, and Roger B. Chaffee, prime crew of Apollo mission AS-204,* were reclining in spacecraft 012 atop their Saturn IB launch vehicle at Kennedy Space Center's launch complex 34. Flight and launch crews were conducting a ''plugs-out'' simulation to determine that the spacecraft would function properly on internal power. The test had been frequently delayed by problems with communications and the environmental-control system; these were exasperating but not abnormal and certainly not a portent of the day's climactic event. Just after 6:31 p.m. horrified ground crews heard a cry of alarm over the communications circuits and saw a bright glow through the spacecraft window. Seconds later the command module ruptured, filling the ''white room'' at the end of the access arm on the service structure with thick clouds of smoke. Technicians worked frantically to pry open the hatch but were repeatedly driven back by the smoke and heat. By the time they got the hatch open Grissom, White, and Chaffee were dead. A few minutes later medical help arrived, only to find that nothing could be done.** Officials quickly secured the launch pad and began the grim task of removing the bodies. NASA Administrator James Webb immediately appointed Floyd L. Thompson,

---

*Unofficially called ''Apollo 1,'' the flight was more commonly referred to as ''AS-204''—the fourth flight of a Saturn IB vehicle (Saturn IB flights were numbered in the 200s, Saturn V missions in the 500s). When flights were resumed, mission numbers started with Apollo 4, for reasons that were never completely clear.

**Postmortem examination disclosed that the crew had died of asphyxiation by toxic fumes produced by incomplete combustion of the synthetic materials in the spacecraft. All had been burned, but not severely enough to cause death.

director of Langley Research Center, chairman of an investigating board* to determine the cause of the tragedy.[1]

Manned space flight unquestionably entailed hazards, and it can be argued that much of the public's fascination with the early manned space flight programs grew out of the perception that it was an exceptionally dangerous business.[2] This perception somewhat exaggerated the true situation. Space flight *was* dangerous, but NASA engineers at every level clearly realized the hazards and went to considerable lengths to minimize them. All the astronauts were aware of the risks and considered them acceptable. Not a man among them would have stayed in the program if he had believed his life was being wilfully risked for the sake of an ephemeral propaganda triumph. Emphasis on producing safe, reliable hardware permeated the program. The astronauts were active participants in design and development; frequent astronaut visits to contractor plants helped to ensure a workable design. Crewmen who had been assigned to flights followed their own spacecraft through assembly and testing. Not less important, the visits impressed on contractor employees the fact that the lives of real people—not anonymous consumers, but people they knew—depended on every component of the system.

One result was that in 5 years 19 Americans had flown 16 earth-orbital missions without serious mishap. Although disaster had flirted with both Mercury and Gemini,** flying in space seemed to be no more dangerous than piloting high-performance aircraft—perhaps less so, for the only astronauts to die before the fire were killed in airplane accidents.

Not surprisingly, most attention had been focused on the dangers in space. Immediately after the fire, Administrator James Webb expressed a widely held view when he said, "Although everyone realized that some day space pilots would die, *who would have thought the first tragedy would be on the ground* [emphasis added]?"[3] Perhaps the AS-204 fire was more traumatic because it did occur on the ground. Whatever the reason, public reaction was vigorous. After a brief period of shock, the nation's press began to ask questions, not only about the cause of the fire but about the wisdom of the manned lunar exploration program.[4] While Thompson's investigating board probed the charred spacecraft and traced its history from California to the Cape, NASA clamped a tight lid on speculation as to possible causes, issuing only brief interim reports. In spite of calls for an independent congressional investigation, both the Senate and House space committee chairmen agreed to defer public hearings until NASA could complete its own probe.[5]

---

*Other members were astronaut Frank Borman, MSC; Maxime A. Paget, MSC; E. Barton Geer Langley; Col. Charles F. Strang, USAF; Robert W. Van Dolah, U.S. Bureau of Mines (replacing Franklin A. Long of Cornell University, who had represented the President's Science Advisory Committee); George C. White, Jr., NASA Headquarters; John J. Williams, Kennedy Space Center; and George T. Malley, Langley, counsel. George W. Jeffs of North American Aviation was asked to serve as a consultant.

**Grissom's Mercury capsule flooded and sank when its hatch accidentally blew off as he awaited recovery from the second suborbital Mercury flight. On Mercury's second earth-orbital mission Scott Carpenter overshot his landing area by nearly 300 kilometers, and for a while his fate was in question. Gemini had produced two incidents: *Gemini VI's* Titan vehicle shut down immediately after ignition, leading to a few anxious minutes, and a malfunctioning thruster set the *Gemini VIII* spacecraft spinning wildly, requiring premature termination of the mission.

In early April the investigating board submitted its report concluding that the precise point of origin of the fire could not be positively identified. Investigators had found physical evidence of electric arcing from wires with damaged insulation. Sometime during manufacture or testing, apparently, an unnoticed incidental contact had scraped the insulation from a wire, thus providing the path for a spark—exactly where, the investigators could not say, but the evidence pointed to a spot near Grissom's couch where components of the environmental control system had repeatedly been removed and replaced during testing.[6] The arc had ignited flammable material and in the pure oxygen atmosphere* the resulting fire had spread with astonishing rapidity.

Contributing to the disaster were an appalling number of factors that could only be called oversights, to put the best possible face on it. The simulation had not been considered hazardous because neither the launch vehicle nor the spacecraft contained any fuel, nor were the Saturn's pyrotechnics installed; consequently no emergency equipment or personnel were at the launch pad. Wiring carrying electrical power was not properly protected against accidental impact. Far too much flammable material—some 70 pounds (32 kilograms) of nylon netting, polyurethane foam, and Velcro fastening—had been haphazardly spread around the command module, creating unobstructed paths for flames. No provision had been made for the crew to get out of the spacecraft quickly in case of emergency. The hatch could not be opened in less than 90 seconds. Neither the board's report nor the congressional hearings that followed could explain why so many technical experts had failed to notice that spacecraft 012, as it sat on the launch pad on January 27, was simply a bomb that needed only a trigger to set it off. Confident in Apollo's design approach, which emphasized eliminating the possibility of ignition by electrical components, and unaware of the intensity and speed of fires fed by pure oxygen, both NASA and contractor engineers had grossly underestimated the consequences of a flaw in their hardware or procedures. Many other problems had demanded attention throughout the fabrication of this prototype command module, and—perhaps lulled by success in past programs—everyone had overlooked the hazards that were accumulating in the spacecraft.

In the months following NASA's investigation, responsibility for these conditions was liberally distributed among contractors and NASA managers alike. Charges of sloppy workmanship and poor quality control by the spacecraft contractor—which NASA should have corrected—seemed justified. Critics asked again whether the "end-of-the-decade" goal was, for no good reason, pushing Apollo out of control and whether there really was a "space race" that justified such haste. James Webb and George Mueller, who took most of the heat of Congress's investigation, doggedly and successfully defended the program's objectives as well as its schedule, aided by generally sympathetic congressional committees.[7]

The Apollo project survived the fire shaken but undeterred. NASA continued to aim for a lunar landing before 1970, but management (especially contractor supervision) would be tightened, procedures (especially safety precautions) would

---

*During the simulation, as it would be at launch on a real mission, the command module atmosphere was pure oxygen at 16.4 pounds per square inch (113 kilonewtons per square meter) pressure—10 percent above normal atmospheric pressure.

be thoroughly investigated, combustible materials in the spacecraft would be rigorously controlled, and new and less flammable materials (particularly fabrics) would be sought.[8]

Perhaps the greatest damage was to NASA's standing with Congress. The space agency no longer seemed larger than life, especially to members who had never been strongly committed to either side of the manned space flight debate.[9] Webb left an atypically bad impression in his appearances before the Congressional committees. He responded testily to suggestions that the Thompson board, made up of NASA's own people, was unlikely to get to the bottom of the accident, and was not cooperative when committee members asked for a report on the performance of the spacecraft contractor.[10]

The Thompson board pointed out numerous deficiencies in the design of the spacecraft and recommended changes in management and quality control throughout the program. Even while the board was preparing its report NASA was hard at work evaluating changes. Obviously the command and service module needed the most attention, but the lunar module was equally rigorously scrutinized, and no aspect of the Apollo program was spared detailed examination for hazards.[11]

How much the lunar landing would be set back no one knew.* Throughout 1966 the Office of Manned Space Flight had been working toward two unmanned test flights of the Saturn V starting in 1967, to be followed by three manned flights to check out the launch vehicle, both spacecraft, and the complex support system for the lunar landing. NASA's public position was that "lunar flights," orbital or landing, would begin before the end of 1969. Planning schedules showed several "simulated" lunar missions—which might orbit the moon but not land—the first of which might be flown as early as October 1967 on AS-503 or as late as August 1968 on AS-506. The initial landing might be assigned to AS-506, but the earliest mission unambiguously categorized in OMSF's master schedules as a "lunar mission" and not a "simulation" was AS-507, scheduled for November 1968.[12]

The fire wrecked that timetable, and for four months afterward all monthly OMSF launch schedules were stamped "UNDER REVIEW." At the end of May 1967, a new master schedule showed only four Saturn V flights preceding the first lunar landing: two unmanned, to check out the launch vehicle; one earth-orbital flight to gain experience in simultaneous operation of the command and service module and the lunar module; and one lunar mission simulation. The Saturn V flights were interspersed among eight Saturn IB missions; as many of these would be flown as necessary to discover and correct flaws in the spacecraft and operations. The May schedule still showed the first landing in the last quarter of 1968, but no one was authorized to mention any date more specific than "before the end of 1968" in public.[13]

---

*Two years later Mueller told a congressional subcommittee that the fire had delayed the first manned flight by 18 months, but in the interim progress was made in areas other than the command and service module.

The Fire and the Science Program

In March 1966, when NASA contracted with the Bendix Corporation to build the Apollo lunar surface experiments package (ALSEP), delivery of the first flight-qualified set of instruments was scheduled for July 1967, seven months before AS-504, the first Saturn V mission to which an experiment package was assigned.[14] It was an optimistic schedule, even though preliminary design work for several of the experiments had already been funded by NASA grants.[15] By late fall the package was in schedule trouble. Two instruments were experiencing minor difficulties, the central data-collecting station was in a critical state, and the magnetometer was having serious development problems.[16] In late December 1966 Headquarters was considering shifting certain instruments from the second mission to the first on account of the lagging magnetometer. Scientists were particularly anxious about this, because the data from the magnetometer were essential to interpreting the results of two other experiments. Experimenters urged that a search be started for a simpler magnetometer in case Ames's sophisticated instrument could not be made ready in time.[17]

Besides the experiments themselves, the radioisotope thermoelectric generator (RTG) required work. The RTG consisted of a large "fuel cask," packed with plutonium-238, supplying heat to an array of thermocouples that produced electricity for the instruments. Project engineers were having difficulty assuring that the radioactive fuel would not be dispersed into the atmosphere in case of an abort during launch.[18] At the critical design review the astronauts discovered several hazards to the crew member who had to remove the hot (500°C, 932°F) fuel capsule from its storage space in the LM and insert it into the thermocouple assembly while setting up the instruments. Redesign of the package or revision of procedures was necessary.[19]

As the status of Apollo cleared in the months following the fire, a degree of optimism returned to the experiments schedule. In July 1967 the first lunar mission was AS-506, set for late November 1968, and the experiments for the first four lunar missions were no longer lagging.[20] Even so, problems remained. In late June 1967 Leonard Reiffel of Apollo Program Director Sam Phillips's scientific staff wrote Phillips suggesting that "we do not schedule the ALSEP for the first lunar landing." Reiffel cited the problems of the magnetometer, the many unknowns that could affect the deployment and function of the experiments, and the weight problems that were hindering production of the lunar module. He offered his personal opinion that, except for the seismometer, the scientific experiments would not yield fundamental information about the moon that would be of immediate importance. All in all, Reiffel thought, the program might be better served in the long run by waiting until the second mission to fly the full complement of surface instruments: *"An uncrowded time line on the lunar surface for the first mission would seem to me to be more contributory to the advance of science than trying to do so much on the first mission that we do nothing well* [emphasis in the original]."[21]

Reiffel's misgivings about the astronauts' work load and the time available for surface activities were not off target. Early in development, Jack Schmitt, one of the astronauts providing crew advice to the lunar surface experiments designers, discovered an undesirable legacy from earlier conceptual work on the instruments:

> . . . In the early days, the crews . . . were worried about having enough to do on the moon. . . . In the early design stages they [the ALSEP designers] took to heart the crew input to "give us something to do," and it was a monster. . . . The way they had that thing put together it was going to take forever to deploy.

This design philosophy was intended to give substance to the argument that men were essential in lunar exploration, but to Schmitt it was the wrong approach. Precious time on the moon should not be wasted in the purely mechanical activity of deploying the instrument package. He and other astronauts worked hard to improve the design of the package so that deploying it did not take so much time, but it was slow going.[22]

Even before Reiffel's pessimistic evaluation of the science prospects for early flight, Phillips had been worried about the progress of the first instrument package. In early June he appointed a review team to look into the development of the magnetometer and another to examine the safety problems with the RTG.[23] The magnetometer investigation team found that the technically sophisticated project had encountered severe schedule delays and cost overruns, but concluded that Ames and its contractor had arrested the project's negative trends. Still, the magnetometer clearly could not be ready for the first scheduled lunar landing. A simpler instrument proposed by investigators at Goddard Space Flight Center was briefly considered, but it could not meet the schedule either and was dropped. At the end of August 1967, Phillips recommended to Deputy Administrator Robert Seamans that the magnetometer be taken off the first ALSEP and replaced by a laser reflector, a completely passive experiment which was under development. The prospects seemed good that a complete ALSEP package as originally planned could be flown on the second lunar landing mission.[24] Ames's magnetometer remained in the schedule for the first landing throughout 1968, however.[25]

Without doubt the delay in the first lunar landing caused by the Apollo fire relieved some of the pressure on the ALSEP experimenters and developers. For all its human and economic cost, the 204 accident forced a pause that was put to good use by those segments of the program (such as the science projects) that were less vitally affected by the fire than the spacecraft. The command module was suffering from many problems in early 1967, and it can be argued that sooner or later something as serious as the fire would have halted progress. The tragedy was that the price of straightening out the program was three lives.

PSAC Examines Post-Apollo Science Plans

During 1966, while the lunar landing program still seemed on track for successful completion within the decade, the question of a manned program to follow Apollo* took on considerable importance. Administrator James Webb was not

---

*At the time no milestone clearly marking the end of Apollo had been defined. It could be argued that the first lunar landing would represent completion of Apollo and that subsequent missions would be part of Apollo Applications. Most officials seemed to think of Apollo as comprising the first two or three lunar landings and AAP as including all lunar exploration more extensive than those landings could accomplish.

inclined to propose another ambitious project, apparently preferring to build a strong, versatile organization and wait for the country to tell NASA what to do with it. George Mueller, Associate Administrator for Manned Space Flight, could not wait. Whatever his office was going to do after the first few lunar missions had to be started very soon or the expensive infrastructure built for Apollo would begin to deteriorate. But Lyndon Johnson's administration had begun to feel the fiscal pinch of an expanding war in Southeast Asia, the president's Great Society programs, and congressional concern for the foreseeable budget deficits. As a result, Mueller's post-Apollo plans were not approved by the White House in fiscal 1967. Though nothing better seemed in prospect and the president was reluctant to let manned space flight wither away, a decision on AAP was postponed.[26]

As part of the efforts to define suitable goals for the nation's space program after Apollo was accomplished, the President's Science Advisory Committee (PSAC) undertook to evaluate NASA's post-Apollo plans. Its report, completed in 1966, concentrated on agency-wide plans and the decades following 1970; but it had some advice concerning later Apollo missions as well. About lunar exploration, PSAC warned that

> the repetition of Apollo flights for more than two or three missions will be unjustifiable in terms of anticipated scientific return without the modification of the system to provide for additional mobility on the Moon's surface and the capacity to remain on the surface for a longer period of time.

After a few initial flights, NASA should adapt the remaining spacecraft and launch vehicles to those ends—for example, by converting the lunar module to an unmanned supply vehicle that could be stocked with expendables, scientific equipment, and mobility aids to support explorations of 7 to 14 days. Manned lunar missions following the first few should be conducted at the rate of not more than one or two per year, carefully coordinated with an expanded program of unmanned exploration to reduce the overall cost of lunar scientific exploration and to investigate areas that Apollo could not safely reach.[27]

PSAC's report, published in February 1967, attracted little public notice in the aftermath of the fire. Its recommendations on Apollo exploration, though brief, were not overlooked by officials at the Manned Spacecraft Center, however. In spite of its preoccupation with the first lunar landing, MSC turned attention to the later lunar missions as soon as it could.[28] Shortly after taking over his duties as Director of Science and Applications at MSC, Wilmot Hess sought the help of lunar scientists in planning scientific exploration of the moon.

Lunar Science and Exploration: Santa Cruz, 1967

Following the pattern set by OSSA and the Space Science Board, Hess organized a summer study in 1967 to discuss lunar exploration. On July 31 more than 150 scientists and NASA officials met on the campus of the University of California at Santa Cruz to provide the expert advice NASA would need to conduct an effective lunar exploration program. The stated objectives of the conference were to prepare detailed science plans, establish an order of priority for lunar investigations, and recommend major programs to develop instrumental and technological support for the advancement of lunar exploration. Eight disciplinary working groups* were organized, each focusing on a different field within the range of scientific interest in the moon.

Unlike the Woods Hole and Falmouth conferences of 1965, which dealt with broader questions of lunar science (see Chapter 3), the Santa Cruz conference concentrated on specifics. Hess tasked the working groups to prepare working papers on particular aspects of lunar exploration: the scientific requirements for lunar surface mobility and mission duration; the scientific use of lunar orbital flights; the scientific utility of planned major hardware items; and mission profiles for trips to the craters Alphonsus, Aristarchus, and Copernicus. Working groups were to study both manned and automated systems and to combine both modes of exploration to optimize the scientific return. They were to consider how to use current spacecraft and systems with minimal modification, or major hardware items already under consideration, not devise new systems that would require substantial development.[29]

Hess opened the conference by summarizing the results of the past two years' work. By way of assessing its own progress in attaining the goals laid out by the Falmouth conference, MSC had tabulated the recommendations made at Falmouth and the extent to which they had been implemented in a document called "Falmouth Plus Two Years, or How Much Nearer is the Whale to the Water?"** According to this tabulation, which Hess used in his summary, only 9 percent of the Falmouth recommendations for Apollo had been rejected (after consideration); 55 percent had been implemented and 8 percent tentatively accepted. Only 6 percent had not yet been acted on. The rest were in various stages of planning or implementation. For post-Apollo recommendations, no action had been taken on 29 percent, 9 percent had been implemented, and 62 percent were in various stages of study.[30]

After two weeks of discussion the working groups made individual reports and produced a consolidated set of recommendations (see Appendix 3). On several points all the groups substantially agreed. For manned exploration the most immediate need was to extend the 500-meter (0.3-mile) range of an astronaut on foot to more than 10 kilometers (6.2 miles) by means of some kind of surface mobility aid. Some favored a wheeled vehicle; others preferred a one- or two-man rocket-propelled flying vehicle, which would enable the astronauts to take sam-

---

*The working groups and their chair persons were: Astronomy, L. W. Frederick, Univ. of Virginia; Bioscience, Melvin B. Calvin, Univ. of California, Berkeley; Geochemistry, Paul W. Gast, Columbia Univ.; Geodesy and Cartography, Charles Lundquist, Smithsonian Astrophysical Observatory; Geology, Alfred H. Chidester, U.S. Geological Survey; Geophysics, Frank Press, Mass. Institute of Technology; Lunar Atmospheres, Francis Johnson, Southwest Center for Advanced Studies; Particles and Fields, D. J. Williams, Goddard Space Flight Center.

**A footnote to the title identified the whale as "Neobalaena shoemakerensis." At Falmouth Eugene Shoemaker, impatient with the lack of progress toward defining Apollo's scientific goals, had compared efforts "to make something happen, like getting some science on Apollo," to trying to push a beached whale back into the ocean." If you push very hard you make a dent in the whale, but as soon as you stop pushing, [the dent] comes right back out again," leaving the whale right where it was. E. M. Shoemaker interview, Mar. 17, 1984. When the laughter had died down, an unidentified Headquarters participant topped the joke with the comment, "Gene, you ought to try pushing that whale off the beach *from the inside.*" H. H. Schmitt interview, May 30, 1984.

ples over a wider area and to explore peaks and valleys that a wheeled vehicle could not reach. All the working groups agreed on the need for long unmanned traverses on the lunar surface, for which a dual-mode "local scientific survey module" was necessary. Besides carrying the lunar explorers from place to place, this module would be capable of unmanned operation directed from earth, traveling across the surface from one Apollo landing site to another. Along the way it would deploy several small geophysical instruments and collect samples for return by the next manned mission. An unmanned vehicle somewhat like this was being considered, but the Santa Cruz conferees favored their own version, which was more sophisticated.[31]

To support longer stays on the moon, the scientists recommended that NASA develop the capability to launch two Saturn Vs per mission: one to carry the crew, the other, unmanned, to transport a modified lunar module carrying additional expendable supplies and scientific instruments.* To supplement the more flexible manned missions they proposed, the working groups recommended that a network of geophysical stations be established on the moon using improved Surveyor spacecraft or similar instrumented modules. Other recommendations for improving Apollo's scientific return included increasing the quantity of lunar samples returned to 250 pounds (110 kilograms), modularizing the experiment package so that instruments could be more easily interchanged to meet the scientific requirements of each mission, and developing the capability to deploy instrumented satellites in close orbit around the moon.[32]

In spite of MSC's efforts to be responsive to experimenters' needs, the scientists at Santa Cruz evidently felt that communication between investigators and engineers still left much to be desired. They recommended that a project scientist, preferably someone involved in the experiment, be assigned to each experiment to make sure that scientists and engineers understood each others' needs and limitations.[33]

Turning to the question of astronaut participation in lunar exploration, the Santa Cruz conference conceded the primacy of piloting skills in the choice of Apollo crews, but strongly recommended that the second criterion for crew selection be ability in field geology. On the later, more complicated scientific missions, the conference considered that *"the knowledge and experience of an astronaut who is also a professional field geologist is essential."* As far as astronaut training was concerned, the conferees insisted that *"in the interest of maintaining career proficiency, astronauts should be provided time to engage in some form of research activity within their professional fields* [emphasis in the original in both cases]."[34] The latter point was particularly important. NASA was about to announce the names of a second group of scientist-astronauts and the conference was sending a message to MSC that this group should, from the start, be scientists first and pilots second.

Rounding out the summary recommendations of the conference, the working groups outlined in some detail a sequence of missions to follow the first few Apollo

---

*A logistic support system based on the Saturn V had been in NASA's plans ever since the lunar-orbit rendezvous decision in 1962. It was part of the price OMSF paid von Braun for his support of lunar-orbit rendezvous, and was to be designed and built by Marshall Space Flight Center, which had no major development responsibilities after Saturn V. See Brooks, Grimwood, and Swenson, *Chariots for Apollo,* p. 81.

landings: (1) manned orbital flights around the moon to obtain better photographs of proposed landing sites and study the surface with remote sensors, (2) single-launch manned missions, and (3) dual-launch manned missions. Three exploration missions were sketched out. One would send two men to the crater Copernicus for three days, exploring and sampling the crater floor and its central peaks with the lunar flying unit. The second, a six-day voyage to the Aristarchus region, required a dual Saturn V launch carrying an unmanned lunar-traversing vehicle. This mission would explore the "Cobra's Head," a curious crater lying at the head of the sinuous rille known as Schroeter's Valley. Finally, plans were outlined for a seven-day mission to crater Alphonsus employing most of the exploration aids recommended by the working groups. Scientific objectives were spelled out in some detail for each site, and the conference report recommended that all these mission proposals be given immediate detailed analysis to test their practicability.[35]

As important as its scientific recommendations was the mechanism the conference set up for effecting them. Near the end of the second week Hess established a Group for Lunar Exploration Planning (GLEP)* to work continuously with NASA mission planners to incorporate as many of the Santa Cruz recommendations as possible into the remaining Apollo and Apollo Applications missions.[36] For the next few years GLEP and MSC planners would meet periodically to examine and refine mission plans.

Santa Cruz gave the Manned Spacecraft Center as much material for study as it could have wanted, and Houston's planning groups began studying its recommendations within weeks of its conclusion. Toward the end of September Elbert King summarized the Santa Cruz proceedings for MSC's lunar missions planning board. The board's reaction to the conference's primary recommendations was not especially sanguine. Neither the lunar flying unit—considered vital by the scientists—nor the local science survey module was much more than a concept at the time, and both would take considerable time to reach operational maturity. Max Faget, MSC's director of Engineering and Development, agreed that both vehicles were desirable but was not optimistic about either cost or schedule. Since the Santa Cruz conferees had suggested that lunar exploration be deferred until the flying unit was developed, the whole program could be held up if it ran into difficulty. In the discussion it was pointed out that Marshall Space Flight Center had studied similar mobility aids and was not convinced that a remotely controlled surface vehicle was feasible. Marshall was currently exploring a smaller lunar roving vehicle that could carry little more than the two astronauts and some scientific equipment. The lunar missions planning board did not settle the question of surface mobility at this meeting. Although it agreed on the necessity to extend the range of the astronauts' exploration, the board was not too keen on starting to work on either of the vehicles suggested at Santa

*Members were chairman Wilmot N. Hess, Maxime A. Faget, Harold Gartrell, Elbert A. King, Jr., Harrison H. Schmitt, and William T. Stoney, all of MSC; Richard J. Allenby, OSSA; James R. Arnold, Univ. of California, San Diego; Melvin B. Calvin, Univ. of California, Berkeley; Philip E. Culbertson, OMSF; Paul W. Gast, Columbia Univ.; Richard Jahns, Stanford Univ.; Francis Johnson, Southwest Center for Advanced Studies; Charles Lundquist, Smithsonian Geophysical Observatory; Frank Press, Mass. Institute of Technology; Nancy Roman, OSSA; Eugene M. Shoemaker, U.S. Geological Survey; and Donald J. Williams, Goddard Space Flight Center.

Cruz. Faget guessed that the long-range vehicle could cost as much as $250 million to develop.[37]

On November 16 and 17 OSSA's Lunar and Planetary Missions Board, which was responsible for advising NASA management on the overall balance between lunar and planetary programs, met at Houston to review lunar exploration plans. Hess presented the status of MSC's plan, the elements it should contain, and how it was to evolve. Board members were clearly uneasy, for they asked Hess to review the plan again before submitting it to Headquarters for approval.[38] Three days after the Houston meeting, on request of the Lunar and Planetary Missions Board, Associate Administrator* Homer Newell sent Robert Gilruth a motion passed by the board to guide MSC's discussions with the Group for Lunar Exploration Planning. While the board found most of the Santa Cruz report acceptable in principle, it requested that the Group for Lunar Exploration Planning review that report "in light of the severe fiscal constraints currently in effect." The board suggested that to reduce the cost of lunar exploration, some of the manned landings could be replaced by unmanned, automated, and mobile exploration systems launched by Saturn Vs, and asked GLEP to submit a few examples of programs at different budget levels.[39]

The Purse Strings Draw Tighter

The "severe fiscal constraints" cited by the Lunar and Planetary Missions Board had been building throughout 1967, and for the unmanned space programs, at least a year before that. NASA's appropriations peaked in fiscal 1965 at $5.25 billion and declined by $75 million and then $207 million in the following two fiscal years. In 1966 Lyndon Johnson had concentrated on establishing his Great Society programs; in 1967 the American military presence in southeast Asia grew substantially. Apollo suffered little by comparison with some other programs, but post-Apollo programs, including George Mueller's grandiose Apollo Applications Program, had been postponed.[40] In the spring of 1967 NASA submitted a request for fiscal 1968 of $5.1 billion, which Congress cut by $517 million. On signing the space agency's authorization bill in August, the President indicated that he would not oppose the reductions, noting that the federal deficit might run as high as $29 billion instead of the $8.1 billion forecast early in the year. The country faced hard choices, he said, and would have "to distinguish between the necessary and the desirable."[41]

Three months later Administrator James Webb presented NASA's proposed operating plan for fiscal 1968 to the Senate space committee. His figures showed that Congress had cut only $50.5 million (2 percent) out of the $2,546.5 million request for Apollo. Apollo Applications had been slashed by 31 percent, and advanced mission studies had been completely eliminated. The Office of Space

---

*In August Newell had been appointed Associate Administrator, NASA's third-ranking officer, succeeding Robert C. Seamans, Jr., who had moved up to the post of Deputy Administrator after the death of Hugh L. Dryden in December 1965. John E. Naugle succeeded Newell as Associate Administrator for Space Science and Applications.

Science and Applications had been hit relatively much harder. Congress had cut $146.6 million (22 percent) out of its total request of $674.6 million for all space science programs. OSSA's lunar and planetary missions were reduced by 12 percent and Voyager, a major planetary program, was eliminated entirely.[42] Such were the "severe fiscal constraints" that motivated the Lunar and Planetary Missions Board to ask for reexamination of MSC's lunar exploration plans.

The fiscal crunch of 1967 was the beginning of a long period of comparative austerity for the American space program. Domestic programs and the nation's growing participation in the Vietnam war created severe pressure on all government programs, and space was no exception. At first Apollo suffered less than other space projects, but eventually it too would shrink as a result of changing national priorities.

Lunar Exploration Program Office Established

As long as Apollo was a goal yet to be attained, the nation and the Congress seemed willing to support it at substantially the level NASA considered essential. Less exciting programs and less tangible space objectives fared less well when the federal deficit grew. This trend in congressional support of space programs from 1965 was not lost on NASA's managers, least of all on George Mueller. In 1963 he had begun looking for a viable post-Apollo program; since 1965 he had worked for a specific (though ill-defined, in the opinion of some) program to use the Apollo spacecraft and organization to produce useful results. His proposals, deferred in fiscal 1966 and again in fiscal 1967, only achieved concrete status as the Apollo Applications Program (AAP) in the fiscal 1968 budget—and then with pitifully small financial support.

In early 1967, Apollo Applications still included (at least on paper) all manned programs after the first few lunar landings—specifically, extended lunar exploration using improved Apollo spacecraft. From mid-1966 onward, however, Mueller's concept of Apollo Applications seemed to be narrowing. Congressional committees had not warmed to his plans for a wide-ranging, multifunctional program whose major goal seemed to be to use up the hardware already developed and exercise the organization. Debate within his own manned space flight organization and discussions with aerospace executives led Mueller to the conclusion that AAP should serve more specifically as a bridge to the next major manned space program—whatever that might be.[43]

Thus as 1967 wore on and the consequences of the AS-204 fire were dealt with, Mueller saw merit in separating lunar exploration from unrelated activities in AAP. Lack of enthusiasm for AAP outside his own office—particularly at higher levels of NASA management—undoubtedly contributed to this changing view. In May, management discussions led to a number of decisions that subordinated AAP to the main objectives of Apollo: the lunar landing had priority over AAP; no hardware would be modified (a key feature of AAP) unless authorized by the deputy administrator; and all AAP launch schedules and hardware assignments would remain tentative until progress in Apollo could be assessed and AAP payloads could be better defined.[44] Under these conditions, continued exploration of the moon stood a better chance of support under the Apollo banner. Since planning for manned lunar exploration was more or less independent of which program office actually oversaw it, site selection, experiment development, and mission planning continued without regard to organizational lines.

In December 1967 Mueller established an Apollo Lunar Exploration Office under the Apollo Program Office in Headquarters. As director of the new office he named Lee R. Scherer, a retired Navy captain who had directed the successful Lunar Orbiter program (completed four months previously). Scherer's office would assume responsibility at Headquarters for directing all activities connected with lunar exploration. Two divisions, Flight Systems Development and Lunar Science, would oversee spacecraft modifications and science plans, respectively. The Lunar Science Division would look to the Office of Space Science and Applications for all science planning.[45] The major functions of the old Manned Space Science Division were distributed between the Lunar Exploration Office and the Apollo Applications Program Office, and its director, Willis Foster, became assistant to the associate administrator of OSSA in charge of manned space flight experiments.

Lunar Receiving Laboratory:
Management and Sample Handling

At the end of 1966 construction of the Lunar Receiving Laboratory (LRL) at MSC was well under way. On January 11, 1967, Gilruth advised Homer Newell that the laboratory's design was essentially fixed and that any new requirements arising out of proposals by scientific investigators could only be accommodated by design changes, which would cost money and time.[46] Construction proceeded without significant delay throughout the spring, and in mid-April program manager Joseph V. Piland reported to Gilruth that he expected to close down his office by June 30 and turn the laboratory over to its operating staff.[47]

The principal unsettled question in early 1967 was that of a laboratory staff. For several months the LRL Working Group, appointed by the Planetology Subcommittee of OSSA's Space Sciences Steering Committee, had insisted that the laboratory must be a permanent scientific research organization headed by a nationally respected scientist who would report to the director of MSC (see Chapter 4). MSC's announcement creating the new Science and Applications Directorate did not mention the Lunar Receiving Laboratory, confirming the "worst fears" (as the group's chairman, Clark Goodman, put it) of the working group, which was informed of the action only after the fact. Headquarters and MSC officials spent considerable time convincing Goodman that those fears were groundless: the laboratory would be a part of MSC's science directorate and MSC was working closely with OSSA in selecting an outstanding scientist to direct it; the director would be appointed only with the concurrence of the associate administrator for space science and applications; and he would be given a free hand in revising the organization of the laboratory and selecting its staff.[48] Three days after announcing the new science directorate, MSC issued an interim plan for the management and operation of the Lunar Receiving Laboratory, placing it in the Lunar and Earth Sciences Division, whose head would report to the Director of Science and Applications.[49]

In the weeks following the establishment of the science directorate at MSC, it seemed that every group having a claim on the functions of the receiving laboratory wanted immediate action on its primary problems. After the LRL Working Group it was the Planetary Biology Subcommittee of the Space Sciences Steering Committee, which met in Houston in mid-January to look after the life sciences' interests in lunar samples and the biological training of astronauts. Some investigators suspected that MSC intended to enlarge the receiving laboratory's activities to the point of usurping some of the responsibilities of outside scientists; but Houston's presentation to the subcommittee evidently laid that fear to rest.[50] The following week it was the Public Health Service, which was concerned with back-contamination and quarantine. MSC Deputy Director George Low and PHS officials met at Atlanta and agreed that the chief of the Biomedical Branch, one of five branches under the Lunar and Earth Sciences Division of the receiving laboratory, would oversee quarantine. When the lab was completed and a formal organization was in place, MSC would recommend appointment of Dr. G. Briggs Phillips, the PHS's liaison officer at MSC since mid-1965, to that post.[51]

Scientific activity in the receiving laboratory came to the fore in the spring of 1967 when the scientists who would investigate the first material returned to earth from the moon were named. On March 16, Headquarters announced that 110 scientists, including 27 working in laboratories outside the United States, had been selected to receive lunar samples.[52] The Manned Spacecraft Center then faced the task of allocating the lunar material to these investigators, considering the type and quantity of sample each one wanted. Before the selection was announced, MSC had asked Headquarters for a list of each experimenter's special requirements. OSSA returned the question to Houston to be settled in face-to-face discussions between MSC and the scientists. Newell suggested that the principal investigators be called together to consider these problems and that they themselves should take much of the responsibility for allocating samples. Headquarters would cooperate in every possible way, but basically it was leaving the management of the lunar investigations to MSC.[53]

In September MSC convened a three-day meeting of the investigators to allow them to update their proposals, revise their budget estimates, and devise an equitable plan for sample allocation. Wilmot Hess and his staff briefed the scientists on the general guidelines expected to govern the distribution of samples and the publication of results, and reviewed the general requirements for managing contracts and grants. In revising their proposals and funding requests, the scientists were not allowed to make any major changes in experiment plans but were encouraged to make small ones if that would significantly improve their investigations. In November MSC forwarded to Headquarters its revised budget estimates for support of lunar investigations: $3.556 million for fiscal 1968, down from the previously estimated $4.135 million, and $3.759 million for fiscal 1969, reduced from $4.592 million.[54]

From the scientists' point of view an equally important result of the September conference was the creation of two teams of scientists to participate in the early phases of sample examination. With the help of the investigators, Hess organized a lunar sample analysis planning team* to assist him in formulating detailed guidelines for the selection and allocation of samples and the necessary scientific functions in the receiving laboratory. A second group, the lunar sample preliminary examination team,** would participate in the initial examination of samples in the LRL. Besides investigators approved to perform experiments in the laboratory during quarantine, the latter group included a mineralogy-petrology team who would help obtain information that would be used in allocating samples to other investigators. These appointments did much to increase the confidence of the scientists in MSC's plans for management of the samples.[55]

The last gap in LRL management was closed with the appointment of a director for the laboratory. On August 1, 1967, P. R. Bell, senior physicist at the Oak Ridge National Laboratory, assumed duties as Chief of MSC's Lunar and Earth Sciences Division and director of the LRL. Bell's career extended from pre-World War II work with the National Defense Research Committee at the University of Chicago through wartime research at MIT's Radiation Laboratory to postwar work at Oak Ridge. An expert in instrumentation, he had conceived the vacuum system that was being built by Union Carbide at Oak Ridge to be installed in the receiving laboratory.[56]

Complexities of Quarantine

Quarantine took up much of the time to the LRL during 1967. An Interagency Committee on Back Contamination (ICBC), established in 1966 (see Chapter 4), served as the policy-making board on questions of quarantine.† As the receiving lab moved toward operational readiness, some possible complexities of operations became apparent. Theoretically the approval of all the concerned regulatory agencies would have to be secured before the samples could be released to lunar scientists. In January 1967, Robert Gilruth sent his deputy, George Low, to Atlanta to discuss matters with ICBC chairman David Sencer. Low and Sencer readily

---

*Members initially appointed were Wilmot N. Hess, MSC, chairman; Elbert A. King, MSC, secretary; Edward Anders, Univ. of Chicago; James R. Arnold, Univ. of California, San Diego; P. R. Bell, LRL manager, MSC; Clifford Frondel, Harvard Univ.; Paul W. Gast, Lamont Geol. Observatory, Columbia Univ.; Harry H. Hess, Princeton Univ.; J. Hoover Mackin, Univ. of Texas; Eugene M. Shoemaker, U.S. Geological Survey; M. Gene Simmons, Mass. Inst. of Technology; Brian J. Skinner, Yale Univ.; Wolf Vishniac, Univ. of Rochester; and Gerald J. Wasserburg, Calif. Inst. of Technology.

**Members initially appointed were Wilmot N. Hess, MSC, chairman; Klaus Biemann, Mass. Inst. of Technology; Almo L. Burlingame, Univ. òf California, Berkeley; Edward C. T. Chao, U.S. Geological Survey; Clifford Frondel, Harvard Univ.; Elbert A. King, MSC; J. Hoover Mackin, Univ. of Texas; G. Davis O'Kelley, Oak Ridge National Laboratory; Oliver A. Schaeffer, State Univ. of New York, Stony Brook; and M. Gene Simmons, Mass. Inst. of Technology.

†Original members of the ICBC in 1966 were David J. Sencer, USPHS, chairman; John R. Bagby, Jr., USPHS; Charles A. Berry, MSC; Aleck C. Bond, MSC; John Buckley, Dept. of Interior; Harold P. Klein, Ames Research Center; G. Briggs Phillips, USPHS; John E. Pickering, OMSF; Leonard Reiffel, OMSF; Ernest Saulmon, Dept. of Agriculture; and Wolf Vishniac, Univ. of Rochester, National Academy of Sciences representative. In January 1967 Robert O. Piland of MSC replaced Bond; later in the year Wilmot N. Hess replaced Piland, James H. Turnock, Jr., OMSF, replaced Reiffel, George L. Mehren, Dept. of Agriculture, replaced Saulmon, and an OSSA member, Lawrence B. Hall, was added.

agreed that a single official at MSC should be given the authority of the ICBC to approve the quarantine protocol and to release the lunar samples when the protocol had been satisfied. (NASA Headquarters had already indicated that NASA's authority to release the samples could be delegated to Gilruth.) The ICBC would develop quarantine procedures, its representative in Houston would certify to Gilruth that those procedures had been satisfied, and Gilruth would order quarantine terminated.[57] If a long chain of approvals—from each of the regulatory agencies through Headquarters to MSC—could be avoided, the academic investigators could have their lunar material much sooner.

Once in the lunar receiving laboratory, crews and lunar samples were safely sealed off from the earth, but between the command module floating on the Pacific Ocean and the crew quarters in Houston was a gap in the containment of extraterrestrial material. By late August 1966 it had been agreed that a "mobile quarantine facility" would be used to transport the lunar explorers from the recovery ship to the receiving lab. Essentially a travel trailer that could accommodate six people for four days, the isolation van would be modified to prevent the escape of infectious agents. It would be sealed aboard the recovery ship, carried from the splashdown point to Hawaii, flown in a C-141 cargo aircraft to Ellington Air Force Base near MSC, and hauled by truck to the LRL, where the astronauts would enter quarantine through a plastic tunnel extending from the van to the LRL door.[58]

A similar method might be used aboard ship to transfer the astronauts from the spacecraft to the mobile quarantine facility. If the command module was assumed to be contaminated it should not be opened; hence the sealed spacecraft with its human contents should be hoisted onto the deck of the recovery ship, where the astronauts could pass through a plastic tunnel from spacecraft to isolation van. Landing operations managers were unwilling to risk this, however, because of the possibility of injuring the crewmen if the spacecraft were dropped. Yet there seemed no feasible way to open the command module while it floated on the ocean and still keep it isolated from the earth's biosphere. After examining the spacecraft's environmental control system, the ICBC was satisfied that it would effectively filter out airborne bacteria during the long return trip from the moon.[59] Thus only the astronauts—not the atmosphere of the command module—were likely to harbor biological contaminants, and some means had to be devised to keep them from infecting the world on their short trip from the spacecraft to the isolation van. The solution was to bag them in plastic: a "biological isolation garment," a zippered plastic coverall equipped with a respirator. The recovery crew would toss these into the spacecraft; the astronauts would don them before entering the recovery raft and remove them after they were sealed in their temporary quarters aboard ship.[60]

Another key problem was the selection of biological tests—the "protocol"—that would give maximum assurance that the samples harbored no dangerous organisms while requiring minimum time and facilities in the receiving lab. Early in 1966 MSC had contracted with Baylor College of Medicine in Houston to develop the test protocol, which the ICBC reviewed at its December meeting.[61] The objective of the protocol was to permit a biomedical assessment of lunar material so that samples could be released within 30 days of their arrival at the LRL,

An Apollo mobile quarantine facility (MQF) on its pallet, in storage at MSC. The unit is a considerably modified version of a commercial Airstream travel trailer. On the roof are two air-conditioning units, front and back, and two filtered air inlets, center. The main door and viewing window are in the rear (right). A side door (center) allowed transfer of rock boxes from the command module through a plastic tunnel (not shown), after which they would be sterilized and passed out through the transfer lock just forward of the side door. Auxiliary power units, visible on the front end of the pallet, provided electricity during transfers from ship to shore and onto the transport aircraft; at other times the unit took electricity from the recovery ship or the transport aircraft. The MQF was 35 feet long, 9 feet wide, 8 feet 7 inches high and weighed 18,700 pounds when fully stocked with consumable supplies and support equipment.

unless some evidence of extraterrestrial organisms appeared. The samples would be examined microscopically for evidence of living organisms, and then a variety of plants and animals would be exposed to lunar material to determine whether they were affected by it. After some discussion at the meeting, committee members were asked to review the plan, considering the limitations of time, facilities, and laboratory personnel, and recommend a collection of tests that would provide the best statistical validity for the protocol.[62]

After looking over the proposed quarantine protocol, Wolf Vishniac, biologist at the University of Rochester representing the National Academy of Sciences, raised a question that the committee would discuss for the rest of the year. Baylor's proposed battery of biological tests was quite comprehensive; as a result, Vishniac noted, it would require a very large sample. He then asked what would be done if, as "an extreme case, . . . no samples of lunar material [were] collected and brought back [other than] dust brought from the lunar surface into the space capsule and inhaled by the astronauts." Between this and the several kilograms of lunar soil that was the nominal sample lay a range of possibilities in which only grams or fractions of grams of material might be available. If the crew should have to leave the moon with only a small sample, how could a satisfactory quarantine examination be carried out and still leave material for other scientists to work with? Vishniac took the position that if extraterrestrial organisms existed they were bound to be introduced into the earth's biosphere eventually, no matter how rigorous the precautions, if planetary exploration continued. Therefore the Apollo quarantine protocol should search not for living organisms in general but only for infectious organisms that constituted a clear danger to the earth. Thus the quarantine protocol could be considerably simplified.[63] Vishniac's suggestion was privately welcomed at MSC, where some project managers felt that quarantine procedures were on the verge of becoming unworkable.[64]

Evidently other members of the committee were coming to a similar conclusion, because at the next ICBC meeting they agreed to keep the quarantine protocol under continuing review. They also agreed that samples would be tested on a few well understood living systems rather than a larger number of less common ones. Systems of greatest sensitivity would be selected so that results could be quickly assessed, minimizing the amount of sample required and the time needed to certify the lunar material as harmless.[65]

At its June meeting the ICBC reviewed the revised quarantine protocol and considered a statistical approach to determining the size of sample required to give the desired reliability. It appeared that 1.2 kilograms (about 2.5 pounds) of material would be needed to conduct an acceptable quarantine protocol. If, as the Santa Cruz conference had recommended, no more than 5 percent of the total lunar material were used for quarantine testing, 24 kilograms (53 pounds)—roughly the nominal lunar sample—would be enough to provide a sample for quarantine testing. The committee would continue to work toward a minimum test protocol in case less than 24 kilograms were returned.[66]

By the end of the summer the Interagency Committee had agreed on an outline of the procedures for releasing lunar material and astronauts from quarantine. If, as expected, the biological tests showed no exotic organisms and the astronauts developed no symptoms of infection within 21 days, the ICBC would review the data and certify that the crewmen could be released. Any change in the astronauts' general health would call for diagnosis. If the change was noninfectious or could be attributed to familiar terrestrial organisms, quarantine could be ended. But if no cause for the change was readily apparent, the question of release would be passed to a NASA medical team. A similar plan was prescribed for testing and releasing the lunar samples. If there were any doubt about living organisms being in the lunar material, release might be conditional, requiring sterilization or stipulation that the sample could be examined only behind a biological barrier. In all cases, doubtful results could lead to retention of the samples in the receiving laboratory until further testing satisfied all concerned agencies that no hazard existed.[67]

While the ICBC worked toward a final quarantine protocol, others were looking ahead to premission testing of the receiving laboratory and its facilities. Early in the summer George Mueller sketched out for MSC Director Robert Gilruth a tentative timeline for LRL tests and simulations. According to Mueller's concept, laboratory personnel could be trained in the operation of the equipment and conduct of the quarantine tests before the laboratory was actually occupied. By the end of 1967 some partial simulations should be conducted to confirm the suitability of the lab's systems and the ability of the technical support team to handle the anticipated work load. In the following three months some actual sample-handling simulations should be conducted. By mid-1968, Mueller said, a full-scale dress rehearsal of a mission should be conducted, from splashdown through the end of quarantine. Samples and test subjects would be treated exactly as they would be during an actual mission, with data being accumulated, handled, and reduced in "real time," to iron out any deficiencies remaining in procedures and equipment.[68] Houston estimated that preparation for an end-to-end simulation of a lunar mission would require 11 months.[69]

1967: A Critical Year

The six months after the Apollo fire were probably the busiest of the entire moon-landing program. At the Manned Spacecraft Center and its spacecraft contractor, North American Aviation, procedures were tightened up to give managers a better grip on details of the program. Both MSC and NAA brought in new managers for the spacecraft project. George Low, picked by James Webb to run Houston's spacecraft program, appointed a tough configuration control board at MSC which met every week to review proposed design changes. Every system and subsystem in the command module was examined for hazards. New materials and test procedures assured that the risk of fire was reduced to the absolute minimum.[70]

The fire bought time for the Saturn project as well. While the first and third stages had few major problems, the second (S-II) stage—also a North American project—had been having serious problems since 1965. S-II was the largest rocket stage ever built to use liquid oxygen and liquid hydrogen, and its builders ran into unique technological problems and needed all the help they could get to meet launch schedules. At one point, in fact, S-II was the single most troublesome part of the Saturn program.[71]

The toil, tears, and sweat expended by NASA and contractors during the post-fire months produced striking results before the year ended. On November 9, 1967, the first complete Saturn V to be flight-tested, AS-501, lifted off from the brand new launch complex 39 at Kennedy Space Center, carrying a boilerplate command and service module and a mock-up of the lunar module into earth orbit. Apollo 4, as the flight was designated, was certainly the most complex mission launched up to that time. Its long list of "firsts" included checking out all systems of the Saturn V, including the first in-orbit restarting of the S-IVB third stage. As far as the spacecraft was concerned, Apollo 4 proved the soundness of the command module's heat shield and redesigned hatch during simulated reentry from a lunar mission. George Mueller's 1963 decision to test the Saturn V "all-up" was apparently vindicated when AS-501 was officially evaluated as a success in all respects.[72]

With that success it was easy to feel that Apollo was "on its way to the moon," as Program Director Sam Phillips put it in a postlaunch press conference. Apollo

4 was the first of six steps that MSC mission planners had set down as essential precursors to the lunar landing. These flights, lettered "A" through "F," would progressively test the systems of the command and service module and the lunar module and verify flight operations procedures, first in earth orbit and then in deep space (lunar orbit), before mission "G" landed on the moon. Four Saturn Vs and two Saturn IBs (or three and three, depending on how the Saturn V and the spacecraft systems performed) would be used for the prelanding missions. This classification of missions, though unofficial, was the framework on which subsequent planning was built.[73]

The year also saw a considerable evolution of scientific activity at the Manned Spacecraft Center. Much of the cause of scientists' complaints about the Houston center was removed by the creation of the Science and Applications Directorate and the appointment of a research scientist of recognized stature as its head. When that directorate became operational, Headquarters's Office of Space Science and Applications delegated to MSC most of the responsibility for management of the lunar samples, including direct contacts with the participating scientists. The lunar receiving laboratory, although it was not accorded the bureaucratic stature at MSC that the lunar scientists had insisted on, was placed sufficiently high to assure adequate autonomy, and its first manager was a research scientist who had been a major contributor to the design of the LRL's sample-handling equipment.

NASA Administrator James Webb spent considerable time in 1967 exploring a more prominent role in planetary science for the lunar receiving laboratory. In view of the historic significance of the lunar samples and the possibility that probes might eventually return samples from the planets, Webb wanted the LRL to become the world's premier site for scientific research on the moon and the planets. In discussions with the National Academy of Sciences during 1967, Webb worked for an arrangement to bring this prospective "Lunar Science Institute" under academic management, which would be essential if the institute was to achieve the stature Webb wanted for it. By the end of the year NASA and the Academy had defined the role of the institute and were working to form a consortium of universities to manage it, but ideas had not yet crystallized and quite a bit of negotiating was still to be done.[74]

As 1968 began, Apollo program officials could look back on a year of accomplishment and ahead to more hard work. It would be a busy year at the Cape: the official manned space flight schedule projected six developmental missions for 1968. Three were Saturn IB missions, two of them unmanned tests of the lunar module, and the third a manned test of the much-reworked command and service module in earth orbit. Three Saturn V flights—two unmanned, verifying spacecraft development, and one manned—would precede five manned Saturn V missions in 1969. "The first lunar landing," the schedule stated for public consumption, "is possible in [the] last half of 1969."[75] Manned space flight operations were about to go into high gear.

When mission G was ready to fly, lunar scientists could expect the facilities for handling its history-making cargo to be ready. During 1967 the lunar receiving laboratory had been completed and much of its specialized equipment installed. Perhaps more important, MSC and the community of lunar investigators had set up the administrative mechanism by which the lunar material would be parceled out for scientific examination. Within the constraints of quarantine and of MSC's responsibility to safeguard and account for the samples, the scientists, through their chosen representatives, would allot the lunar rocks and soil to the approved investigators. As far as the official records reveal it, MSC and the academic science community were developing a cooperative relationship along lines that the scientists found acceptable, if not ideal.

8

FINAL PREPARATIONS: 1968

Apollo took longer strides in 1968, perhaps, than in any previous year. In April, fifteen months after the fire, a Saturn V carried an unmanned command module into earth orbit to test launch vehicle and spacecraft systems and to simulate re-entry from a lunar voyage. Only months before Apollo 7 tested the redesigned command module with a crew in October, officials at the Manned Spacecraft Center began to think about a much more ambitious flight. After selling their proposal to Headquarters, they began making plans to send a crew to the vicinity of the moon. For Christmas 1968 NASA gave the world a present to remember: live television pictures and oral commentary from the crew of the Apollo 8 spacecraft in lunar orbit.

On the science side, progress was less spectacular. Technical problems with the lunar surface experiments were largely overcome, but doubts about the astronauts' ability to unload and deploy the instruments in the time available eventually caused program officials to substitute a smaller science package on the first lunar mission. Several of the systems for handling lunar samples in the lunar receiving laboratory caused major delays, but when the year ended the staff was gearing up for a full-scale simulation.

Spacecraft and Mission Progress

Most of MSC's activity during 1967 and early 1968 was directed toward reducing the fire hazard in the command module. The spacecraft hatch was redesigned so that the crew could open it unaided in five seconds. Electrical systems and the plumbing that carried oxygen and combustible coolants were thoroughly examined and modified. New materials to replace flammable nylon were investigated. The advantages and disadvantages of atmospheres other than pure oxygen were weighed. Repeated testing gathered data to support recommended changes. A Senior Flammability Board, a Materials Selection Review Board, and a Crew Safety Review Board were established to oversee specific parts of the process, and a tough Configuration Control Board was set up to pass on all proposed changes in the command module. The spacecraft builder, North American Aviation, was prodded to supervise its subcontractors and vendors more closely to ensure on-time delivery of critical subsystems.[1] The lunar landing module was less affected by the fire, but it had its own problems. As had happened two years earlier, its weight continued to creep upward, putting a squeeze on the scientific payload that could be landed on the moon. Stress corrosion—cracks in aluminum structural members—and fragile wiring caused delays and concern for reliability. Perhaps the most serious problem was combustion instability in the lander's ascent engine, which would blast the astronauts in the upper section back into lunar orbit after they had completed their lunar exploration. These were worrisome problems at this late stage, but concentrated effort by the lunar module contractor and two engine contractors produced enough progress by mid-1968 to give NASA managers reason for guarded optimism.[2]

Late in January 1968 flight tests resumed when MSC put an unmanned lunar lander through its paces in earth orbit. Designated Apollo 5 and launched on a Saturn I-B, this flight verified operation of the lunar module's propulsion and attitude-control systems and checked out the performance of the instrument unit on the launch vehicle's uppermost stage, the S-IVB. The success of this mission was more encouraging than the one that followed. On April 4 Apollo 6, the second "all-up" unmanned test of the Saturn V, tested launch-vehicle systems and the emergency detection system and allowed launch crews and vehicle engineers to rehearse their tasks once more. Along for the ride, more or less, was a Block I command module with some Block II* modifications; no mission objectives were to be satisfied by this spacecraft except verification of its performance during reentry from a lunar mission. Apollo 6 was trouble from the start. The first-stage burn produced intolerable "pogo" effects (longitudinal oscillations due to irregular fuel feed), and a section of the adapter that mated the spacecraft to the booster blew off during ascent. Two of the five engines on the second stage (S-II) shut down prematurely. Perhaps the most troublesome fault was the failure of the third stage (the S-IVB) to reignite in orbit—the maneuver that, on a lunar mission, would send the Apollo craft on its way to the moon. Marshall Space Flight Center immediately got to work on the problems; yet another unmanned Saturn V test might be needed unless Marshall's engineers could correct them.[3]

In spite of the difficulties with Saturn V and the lunar module, in mid-1968 it still seemed possible that manned flights could resume before the end of the year. Command and service module number 101, delivered to the Cape at the end of May, had fewer discrepancies on arrival than any previous spacecraft. It would fly on Apollo 7, the first manned earth-orbital test of the second-generation spacecraft, scheduled for the last quarter of the year. Since that mission would use a Saturn I-B rather than a Saturn V and would not carry a lunar module, its prospects seemed good.[4]

While spacecraft and operations engineers worked toward getting Apollo flying again, their counterparts in the science programs were equally busy preparing for the first lunar landing. Of all the science-related efforts, the lunar surface experiments were in the best shape in early 1968. The lunar receiving laboratory with its complex scientific equipment, and the elaborate procedures for back-contamination control, had much farther to go before they would be ready to handle their part of the program.

---

*The two versions of the command module were designed respectively for earth-orbit and lunar-landing missions. They differed chiefly in onboard systems (guidance and navigation, docking, life-support, etc.).

Lunar Surface Experiments

During 1967 the Bendix Corporation and MSC ironed out the development problems of the Apollo lunar surface experiments package (ALSEP), and the project was on schedule for the first lunar landing mission (see Chapter 6). But as the year wore on, doubts about the ability of the astronauts to set up the instruments in the time available,* expressed in July by a Headquarters official[5] (see Chapter 7), arose in other quarters as well. Planning for lunar surface operations called for one period of activity in which the astronauts would collect a contingency sample, inspect the lunar module, and familiarize themselves with the low-gravity environment. During this excursion—whose length was limited by the capacity of the portable life-support systems—they would scoop up 10 kilograms (22 pounds) of surface material and load it into the sample-return container. No one was sure how much time these activities would require. MSC tended to be very conservative in estimating how much work astronauts could do under lunar surface conditions, and since it was certain that setting up the experiments would take a good deal of time and effort, deployment was not scheduled during the first extravehicular period. "For planning purposes," preliminary plans called for a second moon walk during which the crew would unload, lay out, and connect the science instruments.[6]

As planning went on, however, and the limitations of time and life-support systems became clearer, the flight of the planned group of instruments on the first lunar landing mission became more doubtful. Simulations showed that a suited astronaut had serious difficulty in unloading the package from the lunar module. To complicate the situation further, weight was a growing problem with the lunar module and difficulties were developing with the radioisotope thermoelectric generator that provided power for the experiments. By mid-1968 planners were already discussing the need to develop a smaller, less complex package for that flight, so that some minimum scientific return would be realized.[7]

MSC's normal conservatism in matters of crew safety and mission success grew even stronger during the summer. The more engineers and mission planners looked at alternatives the more attractive a simplified plan for lunar surface operations became. The first lunar landing would, after all, be the most hazardous (potentially catastrophic) space mission yet, and it would be done with a spacecraft that had not been (and could not be) tested under conditions exactly duplicating the mission. Houston's concern had its effect in Washington. In June George Mueller asked Gilruth to schedule a simulation of instrument deployment so that he could make his own evaluation.[8] After this was completed in mid-August 1968, Sam Phillips, Apollo program director at Headquarters, conferred with the chairman of the Science and Technology Advisory Committee** on proposed changes in plans for the first lunar landing. When they reached agreement, Phillips noti-

---

*An additional question, raised by MSC's Flight Operations Directorate, was whether any lunar module systems should be turned off during the lunar stay. FOD favored keeping critical systems running so that the astronauts could depart immediately in case an emergency developed; this would require additional consumables which would cut into LM weight margins. Minutes, Lunar Missions Planning Board, May 19, 1967.

**Principal external advisory group to the Office of Manned Space Flight, established by George Mueller in 1964.

Commander Jim Lovell practices the "barbell carry," transporting a mockup of the lunar surface experiments package in a walkthrough of the first traverse planned for Apollo 13. The black object on the right with fins, is the radioisotope thermal generator (power unit).

fied committee members of the new plans. Instead of a 26-hour stay on the moon and two-person excursions, he said, "it now seems prudent to limit the lunar surface staytime to about 20 hours and the EVA [extravehicular activity] to a single one-person excursion of 2 to 2½ hours duration." This decision was based on experience in Gemini, where MSC had been unpleasantly surprised by the difficulty of working in null gravity. Whereas previous plans had called for the astronauts to set up a high-gain (strongly directional) antenna for television transmission, conduct preliminary geological exploration, collect samples, and emplace the lunar surface experiments package, mission planners now proposed to curtail the geological exploration, to depend on the 64-meter antenna at Goldstone, California, to pick up TV signals from a less directional antenna that did not have to be deployed by the astronauts, and to carry no scientific instruments at all. Phillips regretfully acknowledged that much scientific information would be lost but noted that subsequent missions would make up for it. Offsetting that loss would be operational advantages that looked extremely attractive: an increase in the safety margins for the lunar module's propulsion systems; maintenance of the lunar module in a state of readiness for quick departure in an emergency; and simplification of the training program, which was becoming undesirably complex.[9]

Phillips's action was welcomed by most of the MSC officials directly involved, but not by Wilmot Hess, director of science and applications. Hess, who had been fighting the battle for Apollo science on many fronts for the past two years, was dismayed by Phillips's proposals. In a vigorous remonstrance he deplored the

Apollo 13 commander Jim Lovell (left) and lunar module pilot Fred Haise load up the tool carrier in a training session at Kennedy Space Center.

severe loss to lunar science and the loss of credibility among the scientists that MSC would suffer if the proposed changes were adopted:

> What can I answer to the critics of the manned program? . . . People in NASA and outside . . . have repeatedly told me that no useful science had been done on Gemini and that none would be done on Apollo. My answer has been that it . . . would start with the lunar landings. This is not the case now. A person who says now that the scientific program of Apollo could be carried out as well using the Surveyor Block III spacecraft has a very good story. I don't know how to answer him.

Hess strongly urged that the proposed single EVA be open-ended, lasting up to three hours if all should go well, and carrying all the scientific instruments:

> . . . if there is a 50–50 chance of getting one experiment deployed . . . it would be better to carry ALSEP and take the chance of not deploying it rather than not carry [it] and . . . lose any chance to do this important experiment.

He closed with a strong recommendation to carry several small, easily deployed experiments on the first landing and to make a firm announcement that the second lunar landing would include a 35-hour stay, three EVAs, and the conduct of the entire science program: collection of samples, deployment of the full package of surface experiments, and carrying out the field geology program.[10]

In response to Hess's plea Phillips asked MSC to propose a contingency science program for the first landing; this was prepared and reviewed at MSC early in October. In discussions with Phillips and George Low, MSC Apollo spacecraft program manager, Hess, won back some important points for science. The new experiments package would contain three simple instruments: a laser retroreflector, requiring no electrical power, with which variations in the earth-moon distance could be measured with great accuracy; a passive seismometer powered by solar panels; and a solar-wind composition experiment, also passive and requiring very little astronaut attention to set up. The astronaut on the lunar surface would not spend time simply testing his mobility and agility, but would carry out some productive scientific tasks while acclimating himself, such as collecting and packaging samples and taking documentary photographs of samples as they were collected. Hess forwarded his proposals to Robert Gilruth and prepared a procurement plan for the new set of experiments.[11]

On November 5, Headquarters approved the new package and authorized MSC to modify Bendix's contract to build it. Manufacture could start before the final terms were negotiated, as time was critical. Some difficulty was expected in fabricating the laser reflector (an array of a hundred individual "corner reflectors," internal corners of cubes made of silica, polished to fine tolerances and precisely aligned) and MSC was authorized to pursue parallel approaches with the University of Maryland and the Air Force Cambridge Research Laboratory, both of which were working on similar projects.[12] The new package would be called the "early Apollo scientific experiments package," or "EASEP."[13] Total funding for the package was $5.3 million, and delivery was to be scheduled for May 15, 1969.[14] Formal assignment of EASEP to the first landing mission was made by direction of the Apollo configuration control board on December 5.[15]

Hess's vigorous advocacy saved a minimal scientific return for the first lunar landing mission, and the original group of experiments—with some changes— would be flown on all subsequent missions. Whether the dire consequences he anticipated in the event of complete cancellation would have materialized is debatable. As early as May 1968 the Planetology Subcommittee of the Space Science and Applications Steering Committee had been warned of such a possibility and had not objected; it merely urged that a set of backup experiments be developed in parallel, to be ready in case the complete package could not be flown.[16] The scientists who were to conduct the experiments agreed—after the fact—that the changes were acceptable under the circumstances, but expressed understandable vexation because they were not consulted before the decision.[17] Still, the mood of many scientists outside NASA was such that removing the lunar surface experiments from the first mission might well have provoked a new storm of criticism of the Apollo program.

Lunar Receiving Laboratory

Construction of the lunar receiving laboratory was virtually complete by the middle of 1967; Hess held a press conference to open the new facility on June 29.[18] For the rest of the year and much of 1968, NASA and Brown & Root-Northrop, support contractor responsible for operating the laboratory, were occupied with installation and testing of the specialized equipment required for quarantine testing and sample handling. Brown & Root-Northrop set up training programs for the technicians who would do much of the work in the laboratory. In April 1968, to assure coordination of effort between quarantine studies and scientific investigations, the two MSC directorates principally involved—medical research and operations and science and applications—began holding monthly meetings on the status and problems of the receiving laboratory.[19] MSC officials planned an operational readiness inspection for the last quarter of 1968, and contemplated partial and complete simulations as soon as they could begin, to exercise the laboratory's functions and uncover flaws in equipment and procedures.[20]

As could be expected for a facility of such complexity, many technical problems arose in the lunar receiving laboratory during the installation and testing of its equipment. Both in the sample-handling area and in the biological laboratories, difficulties slowed progress until late summer.[21] By mid-September, however, problems with autoclaves (pressure vessels for sterilizing items to be transferred out of the biological containment area) were the only serious concern. Laboratory managers prepared and sent to the MSC director a request to appoint an Operational Readiness Inspection Board for the receiving laboratory.[22]

While awaiting completion of the laboratory, scientists on the preliminary examination team and the lunar sample analysis planning team were busy defining their procedures and preparing for simulations.[23] Both teams met at Houston frequently during 1968 to discuss issues of importance in operation of the laboratory and to maintain liaison with the principal investigators. For the outside members of these teams, many of whom were university researchers, preparations for the laboratory work to follow the first lunar landing entailed a considerable sacrifice of time from their normal duties.[24] Simulations would require from 10 to 30 consecutive days of work in the laboratory.

Late in October the science teams gathered in Houston to conduct a training session and simulation of operations in the sample-receiving and -processing sections of the receiving laboratory.[25] The 10-day exercise uncovered 82 major and minor faults in equipment. A substantial number of these impaired effective operation of the vacuum system in which the sample return containers were opened. The vacuum chamber, like most of the other cabinets that comprised the primary biological barrier, was a "glove box," designed so that various tools, stored inside, could be manually manipulated through a pair of impermeable gloves built into the chamber wall. The gloves had to withstand a pressure difference of around 100 kilopascals (15 pounds per square inch—a high vacuum inside and normal atmospheric pressure outside); consequently they were stiff, making it difficult for the operator to use the small hand tools with any dexterity and sensitivity. The simulation also revealed that the viewing ports left blind spots for the operator in some corners of the chamber. These and other problems necessitated more

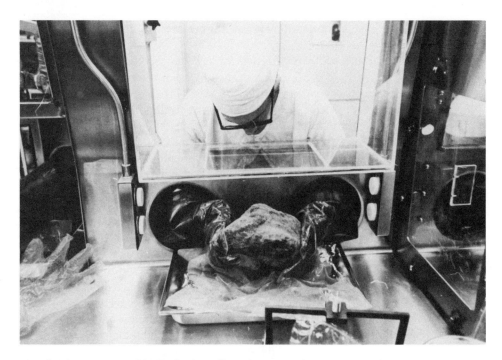

A technician examines "Big Bertha," Apollo 14 lunar sample no. 14321 in the non-sterile nitrogen atmosphere of a processing cabinet in the lunar receiving laboratory.

than 80 major and minor changes to the system and procedures before a full mission simulation could be conducted some time in early 1969.[26] MSC established a configuration control board for the receiving laboratory to pass on proposed changes and keep nonessential ones from proliferating.[27]

By October 1968 sufficient progress had been made to conduct an operational readiness inspection—a mandatory procedure for all MSC facilities. MSC Director Robert R. Gilruth appointed a 10-person committee to review the facilities, staffing, and operational plans.[28] After an initial meeting in early November and a complete briefing two weeks later, committee members spent a month scrutinizing physical facilities, staffing and personnel training, and operational procedures.[29] The committee's recommendations, submitted in mid-December, included 72 mandatory and 91 desirable changes necessary to render the laboratory acceptable for operation.[30]

Preparations for a 30-day simulation of receiving laboratory operations, scheduled for March and April, occupied most of the early months of 1969. In the interim between the operational readiness inspection and the simulation, laboratory staff and contractor employees worked to iron out the remaining problems.

Problems with Back-Contamination Control

The receiving laboratory was only part of the scheme for preventing contamination of the earth by alien organisms. Between the spacecraft floating on the

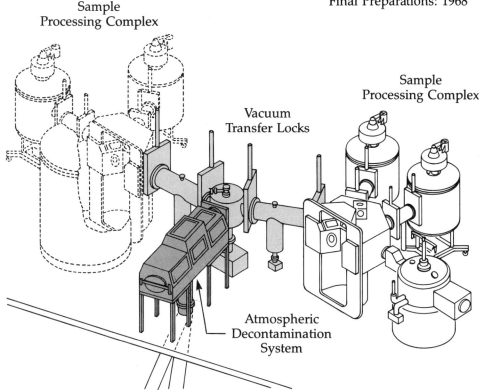

Sample
Processing Complex

Sample
Processing Complex

Vacuum
Transfer Locks

Atmospheric
Decontamination
System

Sketch of the layout of the vacuum processing complex in the lunar receiving laboratory. Phantom units at left provided for expansion but were never actually built. Samples containers were passed into the atmospheric decontamination system, through vacuum transfer locks, into the sample-processing complex, where they were opened and their contents were examined and processed.

Pacific Ocean and the laboratory in Houston was a long chain of events that offered several chances to contaminate the environment. Early in 1968, spurred by expressions of concern by scientists outside the government, the Interagency Committee on Back Contamination (ICBC) revived the question of whether lunar contaminants could be completely prevented from escaping into the biosphere during recovery operations, particularly between the floating command module and the mobile quarantine facility aboard the recovery ship. For two years the committee had been uneasy about this problem, and at its February meeting the chairman opened the discussion once more. The committee asked MSC's landing and recovery division to provide

> a detailed discussion on the return lunar mission [focusing on] containment countermeasures on the lunar surface, in the Lunar Module (LM) ascent stage, during LM-CM transfer, during CM earth return, splashdown, retrieval, operations onboard the recovery vessel, transfer into the mobile isolation unit, and delivery and transfer into the LRL.

Committee members also wanted details of MSC's contingency plans for biological containment in case the spacecraft came down outside the primary recovery zone.[31]

These requests surprised Apollo officials at Houston, who thought those questions had long been settled. Nonetheless, MSC engineers reviewed the subject again at the June meeting of the ICBC, and the committee kept the containment question on its agenda for further study.[32] At its October meeting the ICBC reiterated its considered opinion that the recovery crew should attach some type of biological filter to the spacecraft's post-landing ventilation valve, which circulated air through the command module and exhausted it to the outside. When this question had been raised two years before, MSC had demurred, warning that an effective filter would reduce the efficiency of the ventilation system to an unacceptable level and that attaching it to a floating spacecraft was too hazardous to the recovery crew, especially in a rough sea. The committee insisted, however, that development of a filter attachment with a supplemental fan should be "energetically pursued to avoid what could be compromised and untoward decisions."[33]

When these discussions came to the attention of MSC Director Robert R. Gilruth, he asked for a detailed briefing on the situation and ordered another review of the question.[34] Further discussions at MSC reaffirmed the center's basic position that biological isolation garments for the crew (see Chapter 7) constituted the best available containment in view of the difficulties presented by more secure measures. MSC officials did, however, propose to take more stringent precautions against bringing lunar dust into the command module.[35] Preliminary discussions of these procedures with an ad hoc committee of the ICBC produced tentative agreement.[36]

At the full committee meeting, however, agreement could not be reached. MSC presented its plans for minimizing the transfer of lunar dust into the lunar module and thence into the command module, which apparently mollified the ICBC somewhat. But after reviewing the results of a simulation of Apollo 9, which had included the use of the biological isolation garments and other phases of recovery, the committee was convinced that recovery procedures allowed a breach of containment that must be corrected.[37] Dr. David Sencer, the ICBC chairman, then wrote a strongly worded letter to Thomas O. Paine, newly appointed NASA administrator,* expressing dismay over MSC's apparent failure to appreciate the problem and make serious efforts to correct it. He pointed out that if recovery procedures allowed lunar organisms to contaminate the earth, the rest of the elaborate (and expensive) chain of quarantine was rendered useless and the space agency would lay itself open to severe criticism by the public and the scientific community. Sencer told Paine that the ICBC had considered recommending that the first mission be delayed so the situation could be corrected, but had decided instead to recommend that a filtration system for the command module be positively required on all future lunar missions—on the first, if possible, but only if it would not delay the mission. The risk of contamination was acceptable only if strict housekeeping procedures were enforced to minimize the amount of lunar dust brought into the command module and if recovery teams were thoroughly trained and required to observe contamination-control measures.[38]

Sencer's letter produced immediate results. The questions were aired at a Manned Space Flight Management Council meeting on April 9, after which Sam Phillips directed Houston's spacecraft project manager, George Low, to begin

---

*Paine had been deputy administrator under James Webb since early 1968. When Webb retired in October, Paine became acting administrator; he was appointed and confirmed as administrator in March 1969.

immediate action on the ICBC's concerns.[39] Low assigned a variety of tasks to various MSC offices: reexamination of the feasibility of adding a biological filter to the postlanding ventilation system, a look at the possibility of maintaining air flow from the command module to the lunar module during crew transfer in lunar orbit, and a study of reinforcing the command module's hoisting ring to allow the spacecraft to be safely lifted onto the recovery ship with the crew still inside.[40] Houston reported the early results of its studies to Headquarters two weeks later, then continued to work the problems with a view to submitting concrete proposals to the next interagency committee meeting in early May.[41]

But time was running short. Plans called for the first lunar landing mission to be launched in July; back-contamination procedures were still not approved, and the receiving laboratory, critical to the program, was still far from ready. Bob Gilruth became uneasy. In early April he put his special assistant, Richard S. Johnston, in charge of all management activities relating to the receiving laboratory and back-contamination procedures, with the full authority of the director's office.[42] Johnston had been chief of the Crew Systems Division at MSC for many years before moving up to the post of Special Assistant to the Director, and had built a reputation as a forceful administrator who could get things done.

Johnston's first task was to present a summary of MSC's position on back-contamination problems at the May meeting of the ICBC. Stressing that in MSC's opinion the questions of controlling back-contamination during recovery had been settled earlier, Johnston went ahead to outline MSC's plans in detail. Flight plans called for the astronauts to contain as much lunar dust as possible in the lunar module by vacuuming it off their space suits, then doffing the suits and sealing them in storage bags before returning to the command module.* During the transfer the LM pressure relief valve would be opened and the oxygen flow in the command module would be increased, to maintain a current of gas flowing from the command module to the lunar module, thus minimizing the amount of dust carried into the command module. Tests convinced spacecraft engineers that the command module's environmental control system—specifically the canisters of solid lithium hydroxide that absorbed carbon dioxide—would effectively filter particles (including microorganisms) out of the spacecraft atmosphere during the 63-hour trip back from the moon. Thus only the astronauts would be a potential source of biological contamination, and they could be isolated from the environment by donning the special garments provided for the purpose. Johnston reiterated that adding a biological filter to the command module ventilation system was not necessary and was undesirable. However, MSC would continue to study the implications of adding such a filter to later missions. As for modifying recovery procedures to minimize possible contamination, MSC was unwilling to compromise safety. Hoisting the closed spacecraft to the deck of the recovery ship, as the committee preferred, was not acceptable because of the many hazards

---

*The crew of *Apollo 10*, after testing the lunar module in lunar orbit, recommended against this procedure. They believed it was unacceptably hazardous because in the cramped lunar module the body movements involved could too easily result in contact with fragile areas, such as the windows. Their recommendation was later accepted by the ICBC. Donald K. Slayton to Special Asst. to the Dir. and Mgr., Apollo Spacecraft Program, "Suit Doffing in LM," June 2, 1969; Richard S. Johnston to Dir., Flight [Crew] Operations, same subj., June 11, 1969.

involved. Houston officials preferred to leave recovery operations unchanged and to indoctrinate the recovery crews in techniques for minimizing contact between the spacecraft interior and the earth's atmosphere.[43]

The interagency committee found MSC's proposed procedures generally acceptable, at least for the first mission, but wanted to see the results of more tests before certifying them as effective. Members still felt a need to add the post-landing ventilation filter for the second and subsequent flights. They also thought MSC should pay more attention to contingency recovery procedures, since there was more chance of contamination escaping if the spacecraft had to be picked up by a ship other than the primary recovery vessel.[44] But they seemed satisfied that under Johnston's direction the Manned Spacecraft Center was making a stronger effort than before to comply with their recommendations.

Final Simulations and Certification

The successful flight of *Apollo 8* into lunar orbit during the Christmas season of 1968 emphasized the fact that a lunar landing was fast approaching and that preparations for back-contamination control were becoming critical. Headquarters began to take a more active interest in the receiving laboratory and related operations as 1969 began. Apollo program director Sam Phillips came to Houston in early February for a thorough review of containment from splashdown to release from quarantine.[45] Ten days later George Mueller, whose Office of Manned Space Flight had just been given full responsibility within NASA for back-contamination control, brought a group of advisers from the regulatory agencies to scrutinize Houston's preparations.[46] These advisers found the receiving lab far from ready to handle a mission: equipment problems, a shortage of technicians, incompletely trained personnel, and deficient protocols for biological testing all required work.[47]

These and other deficiencies were expected to surface during a month-long dress rehearsal of laboratory operations scheduled to start March 3. This exercise could not exactly simulate every detail of operations because of equipment limitations, but it would come as close as possible. During the second week a simulation of recovery operations would be run in connection with the flight of Apollo 9, to evaluate the transfer of astronauts from the recovery ship to the mobile quarantine facility and then to the crew reception area of the lunar receiving laboratory.[48]

Expectations were borne out by the results of the simulation. Scores of faults, most of them minor but some of them critical, emerged in both equipment and procedures.[49] Among the more serious problems was contamination of the first work station—the vacuum chamber in which the lunar sample containers were opened. Somehow minute traces of organic matter had found their way into the system, threatening to vitiate the search for indigenous organic material in the samples. Worse yet, the level of contamination was not constant, so the investigators could not correct their findings for it. When it appeared that the contamination came from the vacuum pumping system, investigators began to consider alternatives to opening the sample containers under vacuum to keep the lunar material in pristine condition. The best choice seemed to be to fill the chamber with a gas (sterile nitrogen) that would not interfere with subsequent analyses.[50]

The receiving laboratory staff continued to work on the many details of laboratory operation throughout the spring. Progress continued toward certification of the quarantine facilities and back-contamination procedures.[51] By the time the interagency committee met on June 5, MSC had completed action on most of the outstanding questions and the committee granted certification of the lunar receiving laboratory as a biological containment facility.[52] The launch of Apollo 11 was only five weeks away when another mission simulation got under way in the laboratory, with only minor technical problems remaining to be worked out.[53]

Considering the technical sophistication of the receiving laboratory and the rigorous procedures for handling the lunar samples, it is hardly surprising that problems persisted as long as they did. Management arrangements contributed as well; the outside scientists who were intimately involved in examination and distribution of the samples could not spend full time at MSC, making continuity of activity more difficult. Wilmot Hess was sensitive to the need for close cooperation between MSC and the outside scientific community, and the relationships established in the two years before the first landing went a long way toward alleviating the problem.

Quarantine and back-contamination control added to the overall complexity. Without a doubt, most engineers and lunar scientists at MSC took the back-contamination problem much less seriously than did the interagency committee—which, unfortunately for the engineers, had the authority to impose its requirements on the program. The reemergence in early 1969 of concern with back-contamination during recovery operations can be attributed to the committee's perception that MSC was not cooperating to solve the problem because the engineers considered it unimportant.[54] In the end, the committee yielded at least as much as the engineers on the question of biological containment, but it required considerable effort and tactful interaction with the committee to produce that result.

The start of operations in the receiving laboratory coincided with a period of retrenchment in NASA, and technical difficulties were compounded by personnel problems. Rising budget deficits in the last two years of Lyndon Johnson's administration had alarmed Congress and put a real squeeze on federal programs, not excepting NASA's. Apollo suffered less than other programs,[55] but it was not completely immune to economy measures. Ever since authorization hearings began in early 1968 the field centers had been under pressure to reduce expenditures and cut staff.[56] MSC's inability to add professional staff in the receiving laboratory affected more than just the science. Four months before Apollo 11 flew, the Office of Manned Space Flight's operations forecast for 1969 showed lunar landing missions being launched at two- to three-month intervals after the first. Gilruth took exception, however, advising Headquarters that, with its current complement of professionals, the lunar receiving laboratory could support a lunar mission only once every four months and that only if all went well during each quarantine period.[57] The laboratory had enough professional staff to operate only two fully productive shifts a day; a third "holding shift" was manned principally by technicians, who maintained the laboratory but did not continue processing the samples.[58]

Nonetheless, on June 4, 1969, mission operating conditions were established in the laboratory,[59] and a Task Group was formed to direct operations during the Apollo 11 mission.[60] With the launch date for Apollo 11 inexorably approaching, the laboratory staff continued to refine sample-handling procedures and work on the last remaining technical problems.

Picking Landing Sites

Landing site evaluation had begun as soon as photographs from the first Lunar Orbiters became available (see Chapter 6). By March 1967 MSC's Lunar and Earth Sciences Division had prepared a short list of candidate sites,* which it presented to the Apollo Site Selection Board. Operational considerations predominated at this stage of planning; the sites were all within the "Apollo zone of interest" (5 degrees north to 5 degrees south latitude, 45 degrees east to 45 degrees west longitude) and appeared from the Orbiter photographs to be comparatively level and smooth. MSC recommended that sites for the first landing be chosen from that list by August 1. The board accepted that recommendation.[61] For the rest of the year the Apollo Site Selection Board worked steadily, with input from Bell-

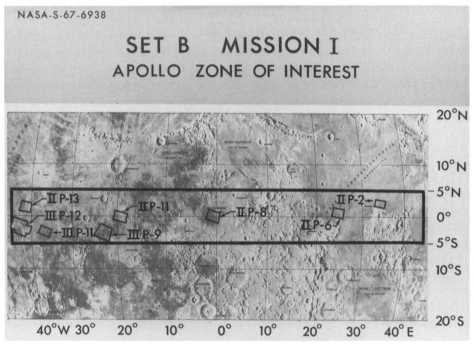

Eight lunar landing sites recommended by MSC in March 1967 for further evaluation. Sites selected from Lunar Orbiter photographs were designated primary (P) or secondary (S) according to apparent merit; "II P-6" was the sixth primary site chosen from the Lunar Orbiter II photographs.

---

*"Landing sites," as somewhat loosely used in the following discussion, corresponds more nearly to the "landing areas" previously mentioned (Chap. 6). However, within the larger areas studied, MSC evaluators did examine several specific landing ellipses—"sites," in the sense that term was used before.

comm, the Jet Propulsion Laboratory, the U.S. Geological Survey, and the Group for Lunar Exploration Planning, evaluating additional photographs from Lunar Orbiters as they became available.* Meanwhile MSC continued to study the operational constraints of its chosen sites; planners were mainly concerned with slopes and irregularities in the terrain along the approaches to the sites, which could interfere with the lunar module's landing radar. The data necessary to make the final choice of sites (principally topographic maps) were slow in coming, however, and MSC could not meet the August 1 date it had proposed.[62] Instead the site selection board met at Houston on December 15 to decide on sites for the first two missions and to lay out a schedule for future activities.

MSC presented the results of its studies on approach paths, which showed that all the "Set B" sites were acceptable. Simulations showed that the lunar module's landing radar could guide the spacecraft to a landing at any of the eight sites. A considerable part of the discussion dealt with the number of sites to be retained in planning for each mission. At this stage no one could be sure that every mission would be launched on time. If a technical malfunction caused a countdown to be stopped, up to 66 hours' delay in launching might be neces-

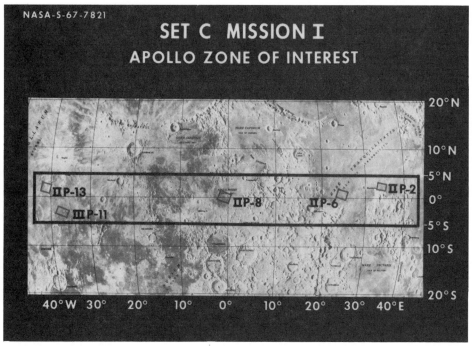

Five lunar landing sites from set B (p. 126) recommended by MSC in December 1967 for the first landing mission.

*Lunar Orbiters III, IV, and V, launched on Feb. 4, May 4, and Aug. 1, 1967, all returned photographs used in Apollo landing site selection. Four more Surveyors were landed by early 1968 as well; although they were mainly equipped for scientific studies, they also provided Apollo with useful data on properties of lunar soil.

sary, in which case a more westerly landing site would have to be substituted for the original. This entailed complications in other aspects of mission preparation—it required the astronauts to become familiar with three different landing approaches and geologic areas, for example, and increased the requirements for maps and other data—but the alternative was to postpone the launch for at least a month.[63]

After summing up the detailed evaluations, John E. Eggleston, chief of MSC's Lunar and Earth Sciences Division, recommended five of the best sites ("Set C") for the first landing mission and another six for the second, which the board approved.[64] Eggleston and Wilmot Hess, chief of MSC's Science and Applications Directorate, then directed the board's attention to the need to begin evaluating sites for the third and subsequent missions. Landing areas topographically different from those already chosen should be examined. Evaluation of highland sites in or near the Apollo zone should be started. Board chairman Sam Phillips agreed in principle that the third mission could be more ambitious, but he advised planners to stay on the conservative side for the time being and look for science targets in the areas covered by the Set B sites. A U.S. Geologic Survey representative suggested that in some cases shifting the landing point only a few kilometers within a selected area would bring the astronaut-explorers within walking distance of some scientifically interesting features. He also pointed out that if the first landing was made in an eastern mare the second should be targeted to a western one, since maria showed different characteristics in the two regions.[65]

From early 1968 onward the site selection teams would get only limited additional lunar photography, since after the fifth* no more Orbiters would be flown. Orbiter photographic coverage, however, was not comprehensive; it did not cover every interesting site with sufficient resolution to permit detailed evaluation. Until the middle of 1967 the Office of Manned Space Flight was planning to fly an advanced lunar mapping and survey system that would furnish more detailed coverage of a greater portion of the lunar surface; but in August its potential value to Apollo was judged to be marginal and its further development was canceled to save money.[66] The following March, however, Sam Phillips noted a continuing need for lunar photography and asked MSC to consider how it might be provided by astronauts in lunar orbit on scheduled Apollo flights.[67] Houston enumerated several types of cameras and film that could be useful in site selection and in lunar cartography, and scientific interpretation as well.[68] Planning was started for a considerable amount of lunar surface photography, including candidate sites for future landings, to be conducted on every manned lunar mission.

Looking Beyond the First Landings

As 1967 ended, the short list of sites for the first manned lunar landing had been approved and a longer list for the second mission had been tentatively

---

*Launched Aug. 1, 1967, *Lunar Orbiter V* completed its photographic mission on Aug. 18 and crashed into the moon Jan 31, 1968.

selected. The site selection board would continue to work with flight planners and scientists to narrow down the choices. Meanwhile, as evidenced by the comments of Eggleston and Hess at the December meeting of the site selection board, attention was shifting to planning for the later missions.

As early as April 1967, the Manned Spacecraft Center's Science and Applications Directorate had put together a plan for lunar scientific exploration, to identify the scientific observations essential to basic understanding of the moon. As guidelines, the review used the fifteen scientific questions about the moon formulated by the Woods Hole summer study of 1965 (see Appendix 3). The plan envisioned eight manned lunar missions, including one orbital mission which would take photographs and gather data with remote sensors, arranged in a logical geological sequence and spaced to allow scientists to digest the results of each mission before executing the next. No more than three landings in the "Apollo zone" were contemplated, and a highland area rather than a mare might be considered as early as the third landing. Both geological exploration by the astronauts and emplacement of instruments for long-term data collection were contemplated. The plan concluded that each mission should stay long enough to accomplish the scientific tasks planned for it; there was no point in staying 14 days unless the extra time could be put to good use. Short-range mobility aids for the astronauts, such as a one-man flying vehicle, would be necessary for later missions, but there was no real need for a vehicle that could cover hundreds of kilometers.[69]

The MSC plan was partly an effort to focus local attention on hardware and mission planning requirements for the later scientific flights, partly a contribution to higher-level agency planning. Throughout 1967 a joint study team from

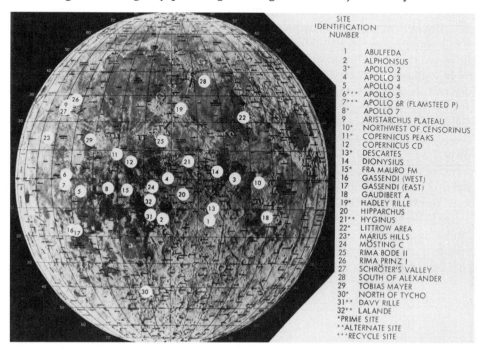

SITE IDENTIFICATION NUMBER	
1	ABULFEDA
2	ALPHONSUS
3*	APOLLO 2
4	APOLLO 3
5	APOLLO 4
6***	APOLLO 5
7***	APOLLO 6R (FLAMSTEED P)
8*	APOLLO 7
9	ARISTARCHUS PLATEAU
10*	NORTHWEST OF CENSORINUS
11*	COPERNICUS PEAKS
12	COPERNICUS CD
13*	DESCARTES
14	DIONYSIUS
15*	FRA MAURO FM
16	GASSENDI (WEST)
17	GASSENDI (EAST)
18	GAUDIBERT A
19*	HADLEY RILLE
20	HIPPARCHUS
21**	HYGINUS
22*	LITTROW AREA
23*	MARIUS HILLS
24	MÖSTING C
25	RIMA BODE II
26	RIMA PRINZ I
27	SCHRÖTER'S VALLEY
28	SOUTH OF ALEXANDER
29	TOBIAS MAYER
30*	NORTH OF TYCHO
31**	DAVY RILLE
32**	LALANDE
*PRIME SITE	
**ALTERNATE SITE	
***RECYCLE SITE	

Landing sites under consideration in late 1969.

the Offices of Manned Space Flight and Space Science and Applications in Headquarters had been working on the same question, in collaboration with Wilmot Hess and the Group for Lunar Exploration Planning (formed after the 1967 Santa Cruz conference; see Chapter 7).[70] In February 1968 Headquarters issued its own lunar exploration plan through the new Apollo Lunar Exploration Office. This plan, more comprehensive than MSC's, outlined a strategy for lunar exploration including and extending beyond currently planned Apollo missions.

Headquarters' strategy called for a combination of orbital and surface missions to photograph and map structural features of interest and collect representative samples, emplace a network of geophysical instruments for long-term monitoring of the moon, and carry out long-range traverses to link local and regional studies together. The lunar exploration program would be evolutionary, stressing initial use of existing equipment, extending its useful life, and introducing new developments as required. Desirable modifications included upgrading the Saturn V to increase its lunar payload by 12 to 15 percent; extending power and life-support systems on the lunar module to support astronauts for three days on the moon; and reducing the lander's conservative propulsion allocations, as experience warranted, to increase scientific payload.[71]

The plan projected several new developments to extend the range of exploration and increase the amount of useful work the astronauts could perform: one-man flying units, a small roving vehicle, a local scientific survey module with longer range that could be operated remotely, and a dual-launch system using two Saturn Vs. The second (unmanned) booster of a dual-launch mission could carry either a lunar payload module providing extra scientific equipment or an extended lunar module with additional expendables for life support, to be landed at the selected exploration site. The proposed schedule called for introducing the extended lunar module, which could support three-day stays on the moon and would carry short-range mobility aids, by mid-1971. Thereafter no more than two missions per year should be flown, new capabilities being added as early as possible within budgetary limitations. Sites for nine missions of special scientific interest requiring systematic increases in mission capability were described in an appendix.*[72]

Estimates of what such a program would cost were necessarily tentative, but the report calculated upper and lower limits based on current assumptions about production of Saturn Vs and spacecraft. Over the next eight fiscal years, new payloads and additional spacecraft and launch vehicles would cost from $4.20 billion to $5.54 billion, depending on which Apollo mission made the first landing. Funding was expected to peak in fiscal 1972 at $1 billion to $1.3 billion.[73]

Lee Scherer, head of the Lunar Exploration Program Office, toured NASA installations in March to brief center officials of the plan. His first stop was in Houston, and afterwards MSC Director Robert Gilruth offered his center's comments to George Mueller, calling the plan "thoughtfully developed, well integrated,

---

*The sites, identified by the names of nearby craters or other physical features, were: Censorinus, Littrow, Abulfeda, Hyginus, Appenine front-Hadley Rille, Tycho, Schroeter's Valley, Marius Hills, and Copernicus. Hadley Rille and a site near Littrow were visited by 3-day Apollo missions.

[and] unified." His sole reservation was that the funds provided for the next year seemed inadequate to make a real start immediately:

> I believe the FY 1969 budget requested should be increased to the amount that can be spent wisely to initiate procurement activities to support the program. Without additional monies early, the proposed flight dates will be very difficult to achieve. More serious, however, the impression may be created that the program can wait another year to really get started.[74]

A month later Gilruth again urged that Mueller start development for the advanced Apollo missions. Reminding Mueller of the President's Science Advisory Committee's public statement* that only two or three lunar landing missions with the basic Apollo systems could be justified, Gilruth emphasized that

> additional stay time, mobility, and scientific payload are considered essential for later missions to produce adequate scientific payload to justify the mission. . . . In order to obtain the increased capabilities . . . a small but significant amount of funds must be committed in FY 1969.

Both the lunar flying unit and the extended lunar module could benefit from additional funds in the current fiscal year. Acknowledging a general reluctance in Headquarters to propose "new starts" (projects not already approved by Congress for Apollo), Gilruth nonetheless urged Mueller to start on an unmanned logistic support system.** Should development be postponed, he said,

> it appears to us that the alternatives to significantly more money for lunar exploration . . . in FY 1969 will all be very embarrassing to NASA in 1971.

If the longer, more productive missions could not be conducted because the systems had not yet been developed, the choices might all be unattractive: dual missions of standard Apollo equipment at excessive expense, single Apollo missions of limited productivity—certain to draw criticism from the scientists—or intervals of a year or more between launches, which would make it difficult to maintain launch capability at the Cape.[75]

Mueller had presented the basic lunar exploration plan to the House Manned Space Flight subcommittee earlier in February, but with no details of schedule or budget. It had provoked little discussion. About a logistic support system, he said only that it continued to be an objective and that studies would continue to determine the best way to carry larger scientific payloads and support longer stays on the lunar surface.[76]

If anyone had reason to be pleased with the lunar exploration plan, the lunar scientists did. To provide scientific advice to Scherer's planners, the Group for Lunar Exploration Planning had met for two days in January 1968 to create a list of recommendations based on the conclusions of the Santa Cruz conference the

---

*In its report, *The Space Program in the Post-Apollo Period*, published in February 1967; see Chapter 7.
**In 1962 Joseph Shea, Deputy Director for Systems in the Office of Manned Space Flight, had helped to secure Wernher von Braun's support for lunar orbit rendezvous by suggesting that Marshall Space Flight Center develop a logistic support system and a lunar roving vehicle; see Courtney G. Brooks, James M. Grimwood, and Loyd S. Swenson, Jr., *Chariots for Apollo*, NASA SP-4205 (Washington, 1979), pp. 80–81.

previous summer. The plan issued in February incorporated virtually every major recommendation they made—the extended lunar module, the lunar flying unit, and the logistic support system—and included the most favored science sites.[77]

Extended lunar exploration would not be delayed by lack of cooperation at the Manned Spacecraft Center. Gilruth's recorded reaction to the plan seems to show that Houston was ready to participate fully in conducting it. But at Headquarters George Mueller seemed less eager to move rapidly. Mueller had many other problems in 1968; the Saturn V had not yet been fully qualified, and both the command module and lunar module had problems yet to be solved.[78] Against that background, Mueller was not in a hurry to seek approval of plans to modify the lunar spacecraft before the principal objective had been achieved. At the same time he was working hard to establish a post-Apollo program that Congress would support.[79]

Mission Planning and Operations, 1968

Site evaluation continued throughout 1968, with considerable attention given to reducing the number of potential landing sites being considered. Keeping five sites as alternatives created problems in providing the necessary maps, photographs, and terrain models for the simulators. On the other hand, until the team at the Cape could be sure of on-time launches, alternate sites had to be kept in the plans to avoid a month's delay in launching a mission.[80]

Experience alone would answer many of the critical questions, and in the last half of 1968 some very important experience was gained. By early fall the reworked command and service module was ready for an earth-orbital test. The 11-day flight of *Apollo 7* (October 11–22) was successful in all respects, demonstrating the performance of spacecraft, crew, and mission support facilities in extended earth-orbital flight. Among other critical activities, the crew simulated rendezvous with the S-IVB stage (which would carry the lunar module on a lunar landing mission) and fired the service module engine eight times without failure. That was especially gratifying, because if that engine failed the spacecraft could neither enter lunar orbit for a lunar landing nor return from lunar orbit after completing a landing mission.[81]

Even before *Apollo 7* flew, project officials made one of the more daring decisions of the entire program: to send a crew to orbit the moon on the first manned flight of the Saturn V. Planners had long felt the need for such a mission to verify communications and navigation at lunar distances. In the alphabetical scheme of flights (see Chapter 7), the "E" (fifth) mission was to approximate these conditions in high earth orbit. In August, MSC Apollo manager George Low set his people to work plans to extend the flight to lunar orbit. In a week of intensive discussions no "show-stoppers" emerged, and neither Marshall nor Kennedy Space Center officials expressed any reservations about it. By mid-month, Houston had clearance from Headquarters to prepare for a "C-prime" mission, contingent on successful qualification of the spacecraft on Apollo 7. The lunar flight, to be designated Apollo 8, would take a manned command and service module into orbit around the moon, check out critical systems at lunar distances, and return.[82]

It had to be done sooner or later, and as more than one official mentioned during the planning discussions, it was essential to assuring a successful lunar landing by the end of 1969. Yet the decision to send men to the moon on the Saturn V's third flight seems almost breathtaking. The flight record of the big booster at the time (two flights, one completely successful, the other plagued by malfunctions) was enough to give managers pause; and the risks to be taken on a lunar mission were far more serious than those of an earth-orbital mission. Still, among NASA officials and Apollo contractor executives, no serious doubts were raised about sending Apollo 8 to the moon at the end of 1968.[83]

The first men ever to see the moon at close range, Frank Borman, James A. Lovell, Jr., and William A. Anders, left Launch Complex 39A at Kennedy Space Center early in the morning of December 21, 1968, aboard Apollo command module 103. Four days later they fired their main propulsion engine and the spacecraft became a captive of the moon's gravity, in an elliptical orbit which they later changed to a near-circular orbit 60 nautical miles (111 kilometers) above the surface. For the next 20 hours they took photographs, relayed their visual observations and descriptions of the topography back to Mission Control, and gave viewers on Earth spectacular television views of the moon and of their home planet. On Christmas day in the early morning, at the end of the 10th trip around the moon, they fired the service module engine once more to return to Earth. The success of this maneuver prompted Lovell to notify Houston, "Please be informed there is a Santa Claus."[84] Had it failed, he and his crewmates would have remained where they were, circling the moon until their oxygen ran out.

The space program was spared that grisly possibility, however, and the propulsion burn was so accurate that only one of three planned course-correction maneuvers had to be conducted on the way back—an adjustment of 4.8 feet per second (1.5 meters per second). After an uneventful return trip, they splashed down before dawn on December 27 in the Pacific Ocean, less than three miles (4.8 kilometers) from the recovery ship.[85]

As the climax of nearly two years of painful recovery from the tragedy of AS-204, Apollo 8 was a triumph the American space program badly needed. It not only restored the prestige of NASA and boosted the morale of the entire Apollo work force, it provided the confidence in spacecraft and ground-support systems that no other, less daring, mission could have provided. About all that remained before the first landing was to prove that the lunar module could carry out its part of the mission; two flights in 1969 would test that last important link in the system.

Besides its operational tasks, Apollo 8 had some photographic assignments that would give flight planners additional information about potential landing sites and settle important questions about the use of landmarks to determine the spacecraft's position. Overlapping vertical and oblique photographs were taken to determine the elevation and geographical position of features on the far side, information needed for accurate determination of the lunar module's position before it descended for a landing (see Chapter 6). The crew was also given a list of photographic targets of opportunity, selected from sites of scientific interest, to supplement the imagery from *Lunar Orbiter IV*. Although the photographic objectives were not completely fulfilled, the pictures returned by Apollo 8 added considerably to the store of knowledge about lunar surface features.[86]

Equally valuable were the crew's visual observations of landmarks and potential landing sites. Their reports on the sharpness of landmarks, the ease of locating them, and the topographic details that could be seen from 60 nautical miles, reduced many of the uncertainties about what a lunar landing crew might be able to do by eye. Among other things, Borman's crew determined that lighting constraints (the allowable sun angle for optimum visibility) were on the conservative side; they found no difficulty in observing surface details at sun angles of 2 to 3 degrees, considerably less than the 6 degrees adopted as a lower limit. They reported that the maps and photographs they carried, together with last-minute instructions from the ground, were adequate for locating surface features.[87]

The Apollo Decade Draws to a Close

The *Apollo 8* crew was still debriefing when 1969 came around, the last year of the decade set by John Kennedy for accomplishing the manned lunar landing and return. Early projections of the 1969 launch schedule called for five missions, spaced just over two months apart.[88] The progress made in 1968 suggested that most preparations for the first landing could be made sometime during the year. Of the major components of the Apollo system, only the lunar module remained to be checked out.

For the most part, science had yielded to operations during the year. To provide some leeway for increases in the weight of the lunar module, and to avoid overtaxing the astronauts on a mission that still contained many unknowns, the lunar surface experiments had been simplified. The original "ALSEP" (Apollo lunar surface experiments package) had been reduced to an "EASEP" (early Apollo scientific experiments package). Enough suitable landing sites had been picked to provide alternatives in case of launch delays, increasing the chances of launching a mission during a given month's opportunity in spite of any technical problems that might halt the countdown before liftoff. But the sites were chosen for maximum probability of successful landing, not for their immediate scientific interest. Even so, no one would argue that whatever samples of lunar material the first astronauts were able to return would not be of extraordinary interest to scientists.

But if science yielded to operational considerations, it received in return the assurance that missions after the second—perhaps after the first—would include as much science as the system could accommodate. The organizational mechanism was in place and functioning to assure that mission plans encompassed scientific objectives as far as possible. If the first landing attempt should be successful, enough launch vehicles and spacecraft were in the pipeline to conduct nine more landings. Project officials at the Manned Spacecraft Center were ready to extend the duration of missions as much as the hardware would allow without major redesign and were urging Headquarters to start work on mobility aids, to extend the range an astronaut could cover on the moon. As soon as they had the operational experience to justify it, MSC managers were willing to do more than merely land a human on the moon and return him safely to earth.

9

PRIMARY MISSION ACCOMPLISHED: 1969

Crew Activities, 1968–1969

Crews for the early Apollo manned missions were named in 1966, but their assignments were canceled after the spacecraft fire in January 1967. Mission-specific training languished for a while, but in early May 1967 the crew of Apollo 7 was selected and began preparing for their earth-orbital mission. Deke Slayton, director of flight crew operations at MSC, expedited training by assigning a third group of astronauts (a support crew) to each flight (see Appendix 6). The support crew assisted the prime and backup crews in training, doing a great deal of necessary but time-consuming work: maintaining the flight data file (flight plan, check lists, and mission ground rules); keeping the prime and backup crews advised of all changes; helping training officers to work out procedures on the simulators before the prime crews used them; and generally taking a load of details off the prime and backup crews.[1] Experience and seniority appeared to be the major factors in Slayton's choice of crews. For the first three missions, prime and backup crews were chosen from the first three groups of pilots selected; all commanders had flight experience in Gemini. Support crewmen were all from the fourth group of pilots.[2] The scientist-astronauts, who ranked fourth in seniority but last in experience, had varied assignments, participating in design reviews and preflight tests and working on plans for Apollo Applications experiments; geologist Jack Schmitt spent much of his time working on site selection.[3]

A Second Group of Scientist-Astronauts

George Mueller was making ambitious projections for the Apollo Applications Program in 1966, and if they developed, more scientists would be needed for its crews. In September 1966 the National Academy of Sciences and NASA announced that applications for a second group of scientist-astronauts would be accepted (see Chapter 5). Nearly a thousand hopefuls applied; the Academy's selection committee forwarded 69 names to NASA in March 1967, and five months later 11 new astronaut trainees, 9 Ph.D.s and 2 M.D.s, were named (see Appendix 6). The group contained not a single qualified pilot; after six months of orientation and basic "ground school" training at MSC, they would all be sent to Air Force flight training.[4]

Addition of this class of candidates brought the astronaut corps to a strength of 56—more than Apollo seemed to need, since no more than 10 lunar missions

were contemplated.* The qualifications of the new group—who were mostly physicists, astronomers, or physicians, not earth scientists—suggests that MSC was
preparing for Apollo Applications missions rather than lunar exploration. When
they were chosen in mid-1967, however, even that prospect was growing dim
and would be dimmer yet by the time they had finished flight training. Headquarters program officials continued to talk of frequent Apollo Applications missions,[5] and Slayton had to be prepared to supply crews for whatever missions
should be assigned and to anticipate some degree of attrition. On the other hand,
if the program did not materialize as planned, the astronaut corps would be over-
staffed and many aspiring space explorers would have small chance of flying in
space. That could be an especially touchy point with scientists, who risked their
careers by enlisting as astronauts.** In interviews with the scientist applicants,
Slayton tried not to raise their expectations about the availability of flight assignments, to the point of frankly admitting that the astronaut corps had no urgent
need for them at the moment. With that understanding, however, they were welcome in Houston, because the astronaut corps needed all kinds of talent to support MSC's missions.[6] Members of the new group soon dubbed themselves "The
Excess Eleven," or "XSXI."

Crew Training and Flight Operations

Crews for both Apollo 9 and Apollo 10 were in training as 1969 began; on January 9 a third was named, for Apollo 11. Since November 1967 the prime crew
of Apollo 9—James A. McDivitt, David R. Scott, and Russell L. Schweickart—
had been following their spacecraft through assembly and rehearsing the complex procedures of orbital rendezvous. Both McDivitt (who came in with the second group of astronauts) and Scott were veterans, each having flown a Gemini
mission that had paved the way for rendezvous and docking operations.
Schweickart, like Scott, was a "Fourteen" (third group); he had not yet flown
a mission but had participated extensively in the development testing of the space
suit. The crew of Apollo 10, named on November 12, 1968, was the first in the
entire manned space flight program composed entirely of experienced astronauts:
Thomas P. Stafford, John W. Young, and Eugene A. Cernan. Stafford, the first
of his class to fly two orbital missions, and Young, who had also flown twice,
were from the second astronaut class; Cernan, from the third group, had a single flight. They were also the first crew to be promoted as a unit, having served
as backup crew on Apollo 7. The crew for Apollo 11, also all veterans, consisted

---

*Given Slayton's system of promoting a backup crew to prime crew three missions later, and assum-
ing that no crewman would be assigned to more than one mission in any capacity, 10 lunar missions
would have required 55 astronauts (if support crews later became backup and prime crews). Slayton's preference for flight experience in crew members, however, makes that assumption questionable; hence fewer astronauts could have filled the positions on prime, backup, and support crews.
On the 7 lunar landing missions flown, 39 astronauts filled 63 available crew positions.

**So did the pilot-astronauts, for that matter, but provision was made for them to keep their profes-
sional skills (flying) well honed as part of the program. The scientists enjoyed no such concession,
as was repeatedly pointed out by NASA's outside scientific consultants.

of Neil A. Armstrong from the second group and Edwin E. ("Buzz") Aldrin, Jr., and Michael Collins from the third. Armstrong and Aldrin had been on the backup crew for Apollo 8; the third member of that crew, Fred W. Haise, was replaced by Collins on the flight crew because Deke Slayton wanted an experienced pilot in the command module for Apollo 11, which, if all went well, would most likely be the first mission to attempt a lunar landing.[7]

The success of *Apollo 8* left only the qualification of the lunar module to be accomplished before a lunar landing could be attempted. In the alphabetical sequence of missions adopted a year before (see Chapter 7), the next mission ("D") was to fly both the command/service module and the lunar module, to check out all LM systems and rendezvous techniques in earth orbit. This, according to 1967 plans, would be followed by missions "E" (combined CSM/LM operations in high earth orbit, later discarded in favor of the "C-prime" mission, Apollo 8, which did not carry a lunar module) and "F" (all operational procedures of a landing mission except the actual landing). On the "F" mission the crew would take the lunar module down to 50,000 feet (15,240 meters) above the moon, get as good a look as possible at the proposed landing site for the first mission, and return to lunar orbit for rendezvous with the command module.[8]

Mission "D" (Apollo 9) would be the most complex yet flown, involving the checkout of two spacecraft at the Cape and simultaneous operations with two vehicles in orbit. An additional objective was operational checkout of the extravehicular space suit and the portable life-support system that lunar explorers would use on the moon's surface.[9]

Apollo 9 was launched on March 3, 1969, for 10 busy days of operations. After extracting their lunar module from the S-IVB stage, McDivitt and his crewmates performed numerous tests of the engines on both modules, assessing the dynamic behavior of the linked spacecraft as well as the performance of the propulsion systems. On the third day they separated the vehicles and began independent operations, testing the LM systems, and carrying out a complete docking maneuver, all successfully. Schweickart's tests of the extravehicular space suit and life-support system were equally gratifying. The most serious anomaly on the flight was the appearance of motion sickness, which had also affected *Apollo 8's* crew. Schweickart vomited twice and Scott reported being on the verge of nausea for considerable time after reaching orbit. Some tasks had to be cut short on this account, but even so, all the primary mission objectives were satisfied. The Apollo 9 command module came down on March 13 within three miles (4.8 kilometers) of the target point in the western Atlantic.[10]

Two months later, launch operations crews at Kennedy Space Center were getting ready for the final countdown for Saturn no. 505, the launch vehicle for Apollo 10. This mission was to "confirm all aspects of the lunar landing mission exactly as it would be performed, except for the actual descent, landing, lunar stay, and ascent from the lunar surface." Besides giving the entire mission support team one more workout, it would also allow the manned space flight network to track the spacecraft as it orbited the moon, to provide a more accurate description of the irregular lunar gravitational field.*[11] On May 18, the big booster sent CSM

---

*Analysis of irregularities in the orbits of *Lunar Orbiter IV* had revealed gravity anomalies attributed to concentrations of mass ("mascons") at certain locations on the moon; they were sufficiently large

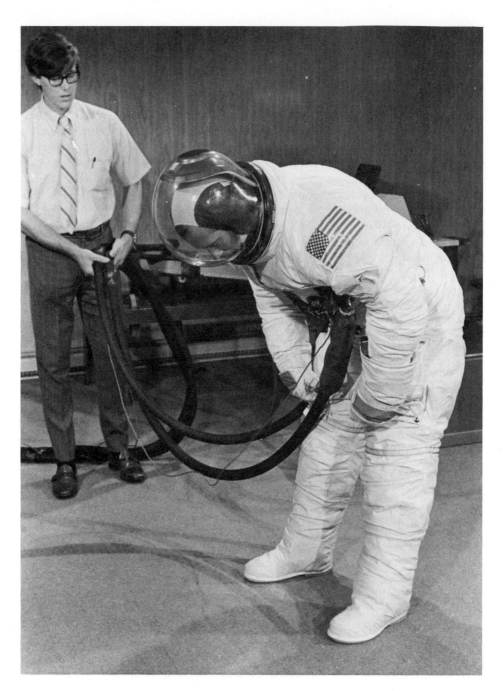

Charles Duke checks the fit of his extravehicular garment.

to deflect the path of a lunar module. Results from Apollo 10 would be helpful in developing the capability to land precisely at a targeted spot, which mission planners considered essential for the later missions.

106 and lunar module LM-4, with Tom Stafford and his crew aboard, toward the moon on a near-perfect trajectory—one designed to duplicate the path to be taken by the first landing mission. Indicative of the pace of the program in early 1969, even before Apollo 10 reached the moon, Cape crews trundled out Saturn 506, the Apollo 11 launch vehicle, to Launch Complex 39A and began preparing it for its epoch-making flight.[12]

Stafford, Cernan, and Young carried out their pathfinding mission with remarkably few problems. On the fifth day the lunar module separated from the command module to photograph the approach path, verify the operation of the landing radar, and survey the first landing site* from low altitude. They reported that the landing radar worked as advertised and that the first astronauts to attempt a lunar landing should have no problems finding a smooth level spot in the eastern end of the site; farther west, however, they might have to maneuver to find a boulder- and crater-free area to touch down. That evaluation completed, Stafford and Cernan took their lunar module back into orbit, checking out the abort guidance system as they left low orbit. After rejoining John Young in the command module they headed back to earth, splashing down on May 26 in the Pacific Ocean some 400 miles (650 kilometers) east of Pago Pago.[13]

Apollo 11 Crew Makes Ready

As backup crew for Apollo 8 since late 1968, the prime crew of Apollo 11 had already rehearsed many of the procedures for a lunar mission; but as the first crew that would attempt a lunar landing, they had to prepare for much more. From January 9, when they were named, until launch day their days were filled with activity: briefings on all the spacecraft systems, fitting of spacesuits, lessons in geology and astronavigation, rehearsal of procedures for rendezvous, practicing unloading the lunar surface experiments, and whatever else the training officers could squeeze into their schedule. For some time they had little access to their primary training devices, the lunar module and command module simulators. Until March, when Apollo 9 flew, they were third in line for those simulators, and only after Apollo 10 was launched in May did they have first priority.[14]

For Neil Armstrong and Buzz Aldrin and their counterparts in the backup crew, practicing the lunar landing itself was among the most important parts of training. The lunar module pilots used a simulator built for the purpose by engineers at Langley Research Center in Virginia. It consisted of a mockup of the lunar module, suspended by cables from a long overhead trestle so that all but one-sixth of its weight was neutralized. Since it did not provide complete freedom of movement in all three directions, the Langley device was inferior as a lunar-module simulator to the vehicle on which the mission commanders trained.[15] This free-flying lunar landing training vehicle (LLTV), developed at the Flight Research Center in California and built by Bell Aerosystems, was a skeleton framework of tubing supporting a control station, fitted in the center with a downward-thrusting jet engine to offset five-sixths of its weight and on the periphery of the

---

*Site II P-6 (see Chapter 8), just north of the equator in the southwestern part of Mare Tranquillitatis.

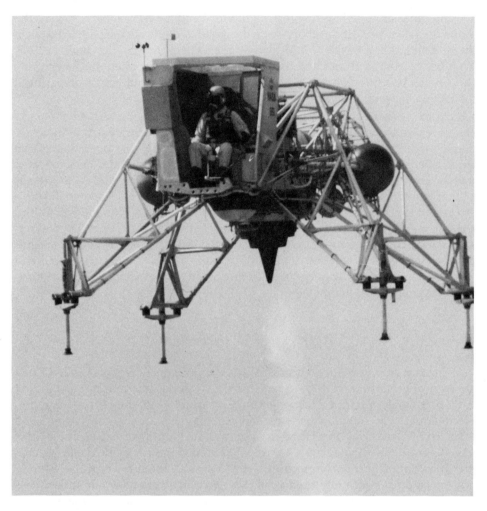

Pete Conrad practices lunar landing in the lunar landing training vehicle (LLTV).

structure with a group of small thrusters to provide attitude and directional control. It was a skittish and somewhat unstable vehicle; on two occasions in 1968 pilots had to eject from it.[16] Nonetheless, it was the only device that could accurately simulate the last few hundred feet of the lunar landing approach and commanders of lunar landing missions and their backups were required to perform as many landings in the LLTV as time permitted.[17]

The crews went through many rehearsals of the lunar surface activity—deploying the experiments and collecting surface samples—during the last few months of training, but they were given comparatively little refresher work in field geology. The last formal geology briefings for the Apollo 11 crew came in mid-April.[18] As launch came closer, Armstrong became concerned about collecting samples and making scientific observations. When he expressed his fear that he might do something wrong, MSC geologist Elbert King emphasized that everything he said and every specimen he collected would be valuable, simply because he would be the first man ever to make scientific observations on the lunar sur-

Apollo 11 commander Neil Armstrong practices collecting lunar suface samples.

face. King's advice was not to worry about making mistakes. If he talked as much as possible about what he saw and collected all the samples he could, no reasonable scientist could fault his efforts.[19]

By the time the Apollo 10 crew left on their mission and the crew of Apollo 11 could get first priority on the simulators, training had become the "pacing item" in the launch schedule.[20] Throughout May and June the spacecraft and its launch vehicle went through preflight preparations with no major problems, staying on schedule for a July 16 launch, while crews spent long hours in the simulators.[21] When the White House proffered an invitation for the Apollo 10 and Apollo 11 crews to dine with the President, Deke Slayton was obliged to notify Headquarters that taking one day out of the training regimen was likely to cost a month's delay in launch.[22] On June 9 the photomosaics of their landing site were delivered to the Cape so that Armstrong and Aldrin could practice the approach to their touchdown point in the lunar module simulator.[23] Armstrong completed his required flights in the lunar landing training vehicle on June 16.[24]

The final days wound down without major interruption, however. The Lunar Receiving Laboratory was certified by the Interagency Committee on Back Contamination, two mobile quarantine facilities—including a spare, just in case—were delivered to the recovery ship, and it appeared that humans would indeed attempt a lunar landing as scheduled.[25]

The First Lunar Explorers

At 9:32 a.m. Eastern daylight time on July 16, 1969, Apollo 11 left Launch Complex 39A at Kennedy Space Center, bound for the moon. Four days later, at 4:18 p.m. EDT on July 20, Neil Armstrong skillfully set the lunar module *Eagle* down in the Sea of Tranquility and reported, "Houston, Tranquility Base here. The Eagle has landed."[26] For the next 10 minutes Armstrong and Aldrin were occupied with several post-landing procedures, reconfiguring switches and systems. Armstrong found time to report to Mission Control what he had been too busy to tell them

"The Eagle *has wings!" said Neil Armstrong as the lunar module* Eagle *pulled away from command module* Columbia, *on its way to the first manned lunar landing. Armstrong turned* Eagle *around so that Mike Collins in* Columbia *could examine it for possible problems. Three long probes signaled contact with the lunar surface just before touchdown.*

during the landing: that he had manually flown the lunar module over the rock-strewn crater where the automatic landing system was taking it. Then he made his first quick-look science report:

> We'll get to the details of what's around here, but it looks like a collection of just about every variety of shape, angularity, granularity, about every variety of rock you could find. . . . There doesn't appear to be too much of a general color at all. However, it looks as though some of the rocks and boulders, of which there are quite a few in the near area, it looks as though they're going to have some interesting colors to them. . . .[27]

After giving Houston as many clues as he could to the location of their module, he added some more description:

> The area out the left-hand window is a relatively level plain cratered with a fairly large number of craters of the 5- to 50-foot variety, and some ridges—small, 20, 30 feet high, I would guess, and literally thousands of little 1- and 2-foot craters around the area. We see some angular blocks out several hundred feet in front of us that are probably 2 feet in size and have angular edges. There is a hill in view, just about on the ground track ahead of us. Difficult to estimate, but might be half a mile or a mile.[28]

Armstrong and Aldrin then started preparing their spacecraft for takeoff, setting up critical systems to be ready in case something happened and they had to leave the lunar surface quickly. A short break in this activity gave Armstrong a chance to pass along more information about the landing site:

> . . . The local surface is very comparable to that we observed from orbit at this sun angle, about 10 degrees sun angle, or that nature. It's pretty much without color. It's . . . a very white, chalky gray, as you look into the zero-phase line [directly toward the sun]; and it's considerably darker gray, more like . . . ashen gray as you look out 90 degrees to the sun. Some of the surface rocks in close here that have been fractured or disturbed by the rocket engine plume are coated with this light gray on the outside; but where they've been broken, they display a dark, very dark gray interior; and it looks like it could be country basalt.[29]

Setting up the spacecraft systems took another hour and a half to complete; then they were ready to get out and explore. The flight plan called for them to eat and then rest for four hours, but Aldrin called Mission Control to recommend starting their surface exploration in about three hours' time. Houston concurred.[30] Although they had been awake almost 11 hours and had gone through some stressful moments during the landing,* it seemed too much to expect the first men on the moon to take a nap before they made history.

While Armstrong and Aldrin tended to their postlanding chores, Mike Collins, orbiting 60 nautical miles (112 kilometers) overhead in the command module *Columbia*, had little to do. Houston enlisted his aid in an attempt to locate *Eagle*,

---

*While Armstrong was maneuvering to avoid a boulder field, alarms sounded in the lunar module indicating that the computer was overloaded. Mission Control quickly told the crew to proceed. Then, as fuel was running low, a dust cloud obscured the surface and Armstrong had to touch down without a good view of his landing spot.

giving him the best map coordinates they could derive from the sketchy information available. With his navigational sextant Collins scanned several spots, without success; *Columbia* passed over the landing site too rapidly to allow him to search the area thoroughly and he never found the lunar module.[31] Determination of its exact location had to wait for postmission analysis of Armstrong's descriptions of the area and examination of the spacecraft's landing trajectory.

Getting ready to leave the lunar module took longer than the crew had anticipated. It was after 9:30 p.m. in Houston, an hour and a half later than they had hoped, when they opened the hatch. Armstrong carefully worked his way out onto the "porch," then climbed down the ladder, pausing on the lowest rung to comment on the texture of the surface and the depth to which the footpads had penetrated. At 9:56 p.m. he stepped onto the moon's surface, proclaiming, "That's one small step for man, one giant leap for mankind"—inadvertently omitting an "a" before "man" and slightly changing the meaning he intended to convey.[32]

Armstrong made a cursory inspection of the lunar module and reported his reactions to the new environment. Aldrin then lowered a camera on the lunar equipment carrier—a clothesline and pulley arrangement that seemed out of place in the high-technology environment of Apollo—which Armstrong immediately began using. Mission Control reminded him to scoop up the contingency sample, which he did. "I'll try to get a rock in here. Just a couple." He noted that the collecting tool met resistance after penetrating a short distance into the surface material. He then stowed the sample in a bag that he tucked into a pocket of his suit. To the scientists on earth he remarked, "Be advised that a lot of the rock samples out here, the hard rock samples, have what appear to be vesicles in the surface. Also, I am looking at one now that appears to have some sort of phenocryst."*[33]

Aldrin then joined Armstrong on the surface, and they spent the next several minutes inspecting the landing craft and reporting on its condition, adjusting to the low lunar gravity and trying various ways of getting around on the surface. After a brief commemorative ceremony (reading the plaque attached to the lunar module) and a short conversation with President Richard Nixon, they began unloading and emplacing the scientific instruments and collecting samples. They supplemented earth's limited television view of their activities with descriptions of what they were seeing and doing. On a couple of occasions they acted like field geologists. Aldrin reported that he saw a rock that sparkled "like some kind of biotite," but he "would leave that to further analysis."** After closely examining some rounded boulders near the spacecraft, Armstrong said they looked "like basalt, and they have probably two percent white minerals in them. . . . And

---

*That is, the rocks had surface pits resembling those caused by the escape of gases from molten material (which could indicate a volcanic origin), and one seemed to have a prominent embedded crystal.

**This comment drew criticism from some scientists on the ground that biotite (a mica-like mineral) could not have formed on the moon. The criticism was unwarranted, because Aldrin had not said that it *was* biotite, but that it *looked like* biotite he had seen on field trips; the criticism was mildly detrimental, because it made the next crew more reluctant to use technical terminology, with which some of the astronauts felt uncomfortable anyway, in describing what they saw. H. H. Schmitt interview, May 30, 1984.

the thing that I reported as vesicular before, I don't believe that any more. . . . they look like little impact craters where BB shot has hit the surface."[34]

The geologists in Houston watching this surface activity on television were quite pleased with the astronauts' performance. At one point Armstrong disappeared from the field of view of the TV camera, causing some momentary anxiety at his apparent departure from the plan. It turned out that some unusual rocks had attracted his attention and he had gone off a few meters to collect them. That was exactly the kind of thing the geologists had hoped people on the moon would do.[35]

By the time the crew had taken two core samples, again experiencing difficulty in driving a sampling tool into the surface, and filled their sample return containers, Houston notified them that it was time to wind up their activity. Just before midnight CapCom* Bruce McCandless told Aldrin to "head on up the ladder," and at 12:11 a.m. Houston time both men and their samples were back in the lunar module and the hatch was sealed.[36] Humanity's first excursion on the surface of another celestial body had lasted 2 hours, 31 minutes, and 40 seconds.

Back inside the lunar module, Armstrong and Aldrin removed their lunar surface suits and portable life-support systems and used up their remaining film. Houston passed up some more instructions in preparation for liftoff and tentatively signed off for the night, but before long CapCom Owen Garriott, who had relieved McCandless, came on the line with some questions from the scientists about the nature of the surface and the problems in driving sampling tools into the surface. Three hours after they returned to the lunar module, the lunar explorers finally were able to turn in for a few hours of fitful sleep.[37]

Next morning Armstrong, Aldrin, and Collins spent most of their time setting up *Eagle* and *Columbia* for liftoff and rendezvous. Before the lunar module left the moon, however, Armstrong gave Mission Control a detailed description of the landing approach path and landing area, in the hope of helping scientists locate their exact landing spot, and summarized the characteristics of the soil and rocks around the area.[38]

Liftoff and rendezvous went smoothly. When the two spacecraft were locked together Collins cracked *Columbia*'s oxygen supply valve and Aldrin opened the lunar module's vent valve, to create a gas flow into the LM when the hatches were opened—part of the procedure to minimize back-contamination—while Aldrin and Armstrong vacuumed the lunar dust from their suits as best they could. Their vacuum cleaner, a brush attached to the exhaust hose of the LM suit system, was not very powerful and the tenacious dust came off only with difficulty. There was not nearly as much loose dust in the lunar module as they had expected when they returned from the surface; evidently it stuck tightly to whatever it touched.[39] They passed the rock boxes and other items over to Collins and then clambered into the command module, where they removed their suits and stowed them in the bags provided. After jettisoning the lunar module and straightening

---

*The "Capsule Communicator"—a term left over from the Mercury days when spacecraft were called "capsules"—normally was the only person who talked to crews in space over the radio circuit. All CapComs were astronauts, most often astronauts who had not yet flown. Appendix 6 lists CapCom assignments for all Apollo flights.

Technicians unload the first box of lunar samples from Apollo 11 at the entrance to the lunar receiving laboratory, July 25, 1969.

up the command module, the three astronauts settled in for an uneventful trip back to earth.[40]

In the early morning hours of July 24, 8 days, 3 hours, 18 minutes, and 18 seconds after leaving Kennedy Space Center, *Columbia* plopped down into the Pacific Ocean about 200 nautical miles (515 kilometers) south of Johnston Island. Recovery crews from the U.S.S. *Hornet* arrived quickly and tossed the biological isolation garments into the spacecraft. After the cocooned astronauts emerged from the spacecraft the swimmers swabbed the hatch down with Betadine (an organic iodine solution); then astronauts and recovery personnel decontaminated each other's protective garments with sodium hypochlorite solution. The biological isolation garments were not uncomfortable in the recovery raft, but aboard the helicopter they began accumulating heat. Both Collins and Armstrong felt that they were approaching the limit of their tolerance by the time they reached the ship.[41] An hour after splashdown they were inside the mobile quarantine facility. As soon as they had changed into clean flight suits, the astronauts went to

The Apollo 11 mobile quarantine facility is unloaded from its C-141 transport.

the large window at the rear end of the mobile quarantine facility to accept the nation's congratulations from President Nixon, who had flown out to the *Hornet* to meet them.[42]

Meanwhile, recovery crews brought *Columbia* on board and connected it to the astronauts' temporary home by means of a plastic tunnel. Through this, the film magazines and sample return containers were taken into the quarantine trailer, then passed out through a decontamination lock. Sample return container no. 2, holding the documented sample, was packed in a shipping container along with film magazines and tape recorders and flown to Johnston Island, where it was immediately loaded aboard a C-141 aircraft and dispatched to Ellington Air Force Base near MSC. Six and a half hours later the other sample return container was flown to Hickam Air Force Base, Hawaii, and thence to Houston.[43]

Scientific Work Begins

The first sample container arrived in Houston late in the morning on July 25 and was delivered to the receiving laboratory shortly before noon.[44] The second arrived from Hawaii that afternoon.[45] For the next few weeks the laboratory was the center of more public attention, probably, than had been focused on any scientific installation in this century. As they had done since the Ranger project and would continue to do throughout the space science program, reporters pressed the scientists for interpretations of their preliminary observations. Obviously any scientific conclusions would have to await detailed study; nonetheless, the scientists answered the journalists as best they could. One was reported to have charac-

The world's first lunar explorers debark from their recovery helicopter aboard the U.S.S. Hornet and walk toward the mobile quarantine facility, encumbered by their biological isolation garments. These were intended to protect the earth from any harmful microorganisms that the astronauts might have picked up on the moon. They were abandoned on the next mission in favor of clean flight suits and respirators.

terized the result as ''instant science,'' noting that ''We have to look at the stuff and then walk into a news conference, where we should be fairly conservative

The contents of the second sample return container from Apollo 11.

and reserved . . . and we talk off the top of our heads, and we're bound to goof some."[46]

After initial inspection the sealed sample return container was transferred to the vacuum laboratory and weighed (33 pounds, 6 ounces; 15.14 kilograms). Film magazines were placed in an autoclave, to be exposed to a gaseous sterilant for several hours before being sent to the photographic laboratory for processing.[47]

Technicians in the vacuum laboratory then started the container through the process of preliminary examination. Before admitting it to the vacuum system, they passed it through a two-step sterilization, first with ultraviolet light, then with peracetic acid (a liquid biocide), after which it was rinsed with sterile water and dried under nitrogen. Next it was passed through a vacuum lock into the main vacuum chamber. Opening the box was delayed for some hours when the pressure in the vacuum system could not be stabilized. Operators suspected a leak in one of the gloves through which the samples were handled, which was torn in several places.[48]

By midafternoon the next day, however, the leak had not worsened and the scientists decided to proceed. At 3:45 p.m. on July 26, as geologists crowded around the observation port at the work station, the cover was removed from the sample return container.[49] The first glimpse of moon rocks was frustrating because they were covered with fine dust that obscured their surface characteristics. About all that visual observation revealed was that they were irregular in shape and angular, with slightly rounded edges.[50]

In addition to the rocks and dust, the sample return container held a solar-wind collector* and two core tubes, with which Aldrin had obtained subsurface sam-

---

*This was a sheet of aluminum foil that was set out at the start of the astronauts' lunar surface activity to trap particles of the solar wind. Dr. Johannes Geiss of the University of Berne, Switzerland, devised the experiment, which went on every mission; crews called it "the Swiss flag" because it was unfurled along with the Stars and Stripes.

ples. These had been least exposed to contaminants from earth and would provide the "bioprime" sample, a 100-gram (3.5-ounce) portion of lunar soil that would be tested for its effects on a variety of living organisms. Core tubes and solar-wind collector were sealed in stainless-steel cans, sterilized, and transferred out of the vacuum laboratory.[51]

The following week more preliminary data trickled out of the receiving laboratory as two teams of scientists (the lunar sample preliminary examination team, or "LSPET," and the lunar sample analysis planning team, or "LSAPT") got down to work. These groups would examine, characterize, photograph, and catalog the specimens and allocate samples to the 142 principal investigators. Working two shifts a day for a while, they determined how the larger rocks would be divided and which ones should be sectioned, basing their judgment on the requirements of the investigators and the availability of the different types of samples. While many members of these teams were later involved in specific investigations, their purpose at this stage was to match samples with research projects to make most effective use of the samples.[52]

On July 27 technicians removed a rock from the box for low-level radiation counting, dislodging much of the dust from its surface. Geologists immediately classified the specimen as igneous (formed by melting).[53] It might be a fragment of lava from a lunar volcanic flow or a rock produced by localized melting, such as might result from a large meteorite crashing into the moon. Careful visual examination disclosed numerous small pits on the surface, as Armstrong had noted during his excursion. The scientists tentatively identified three common minerals, feldspar, pyroxene, and olivine, as the major constituents of the sample.[54]

Spectrographic analysis of a pinch of lunar dust indicated that the surface material from Tranquility Base was much like that analyzed two years earlier by *Surveyor V.** Noteworthy characteristics of the sample included an abnormally high proportion of titanium and relatively small quantities of some volatile elements, which would be depleted in materials aggregated at high temperatures. Microscopic examination of the powdery material ("fines") showed that more than half of it was composed of tiny particles of glass—shiny spheres, rare dumbbell-shaped beads, and sharp, angular needles and fragments.[55]

A few more tidbits emerged from the early work in the receiving laboratory, among them the facts that the samples contained very little carbon (no more than 12 parts per million) and that some of the rocks had lain undisturbed on the lunar surface for a million years or more.[56]

Preliminary examination of samples and preparation of material for the investigators continued throughout August, aiming at distribution of experimental material as soon as the quarantine protocols were completed. For the first few days work in the laboratory went on as planned; some minor malfunctions in the waste disposal system briefly threatened to break the biological containment but were corrected without incident.[57] About the only potential equipment problem was the torn glove in the vacuum chamber. Since replacement was a time-

---

*The inactive *Surveyor V* sat some 25 kilometers (15.5 miles) north-northwest of Tranquility Base. The last three Surveyors (V, VI, VII) carried instruments capable of identifying the major chemical elements making up the surface material,; see *Surveyor: Program Results*, NASA SP-184 (Washington, 1969), pp. v-vi, 7, 271–350.

consuming process, technicians jury-rigged a repair by slipping another glove over the damaged one and taping the two together.[58] But before the first week was out the gloves ruptured, exposing most of the samples in the system to the atmosphere and sending two technicians into quarantine.[59] Work in the vacuum laboratory was suspended while scientists and laboratory managers decided what to do. Problems with the vacuum system had been experienced during premission simulations (see Chapter 8) and provision had been made to fill the chamber with sterile nitrogen gas; that alternative was now adopted.[60]

The Astronauts in Quarantine

After the President and other celebrities departed, the U.S.S. *Hornet* continued steaming back to Hawaii, arriving at Pearl Harbor on the afternoon of July 26. The mobile quarantine facility with its five passengers* was hoisted off the ship onto a truck for transfer to Hickam Air Force Base a few miles away, pausing briefly to acknowledge the greetings of the mayor and several thousand citizens of Honolulu. At Hickam the trailer was loaded into a C-141 cargo aircraft, which departed immediately for Houston.[61] Just after midnight the big plane touched down at Ellington Air Force Base, where a large crowd awaited a glimpse of the astronauts. Three hours later the crew and their companions entered their living quarters at the lunar receiving laboratory, which would be their home for at least the next three weeks.[62] On hand to greet them were the support personnel who had entered the living quarters the week before: a clinical pathologist, five laboratory technicians, three stewards, photography specialist, Brown & Root-Northrop's logistic operations officer, and a representative of MSC's public affairs office.[63]

After a day off to recuperate from the stresses of the preceding two weeks—since July 16 they had been cooped up in very close quarters—the crew began a week of intensive technical and medical debriefings. Periodic examinations and blood tests monitored the physiological effects of their flight and recovery, while the doctors kept a close watch for any signs of exotic infection. The living quarters were equipped for routine clinical tests, or even minor surgery; but if a life-threatening medical emergency arose, whether from recognizable or unfamiliar causes, the victim would be transferred out to a hospital regardless of any concern for back-contamination.[64]

Although all three astronauts were veterans of prior missions, none had ever spent so much time in space, and post-mission activities had never been conducted under strict confinement. Thus quarantine quickly became oppressive, the more so because only meager provision had been made for recreation. An exercise room was available, as well as a Ping-Pong table, and they could read or watch television or talk by telephone to their families, but it was not like being at home. At the conclusion of the technical debriefing sessions, less than a week after they returned, they were called upon to comment on operations in the receiving labora-

---

*A physician, Dr. William R. Carpentier, and a technician, John Hirasaki, joined the crew in the quarantine trailer aboard the recovery ship and remained in quarantine with them.

tory. Armstrong was noncommittal, saying that so far it had been going "about as well as you can expect." Collins's less tolerant response was, "I want out."[65]

Since no press conference with the astronauts was scheduled before quarantine was lifted, reporters' only source of information was a daily briefing by John McLeaish, the public affairs officer confined in the crew quarters. Once or twice a day McLeaish briefed the press pool through a glass wall in the conference room where the crew debriefings took place. Most of the time he had little to report: everyone inside was healthy, the astronauts were busy with debriefings or writing their pilot's reports, and otherwise nothing much was going on.[66] And so it remained for the duration of the three-week quarantine.

Public Interest in Lunar Samples

While the laboratory teams worked two shifts a day to prepare the samples for outside researchers, project managers were considering how to deal with a growing public curiosity about the first rocks returned from the moon. Even before Apollo 11 left the launch pad, requests for lunar specimens for display had begun to arrive at the Manned Spacecraft Center. When MSC scientists expressed concern at what these requests might lead to, Apollo project manager George Low asked Headquarters for help in developing a policy to deal with them. Low's principal concern was that NASA "would arouse the animosity of the scientists in this program by wholesale or capricious distribution of valuable lunar material for nonscientific purposes." Low considered it appropriate to prepare a few lunar specimens for touring exhibits at major museums in the United States and abroad, provided they were eventually returned to the lunar receiving laboratory for scientific use. He also expected to be asked for samples to be presented to various VIPs; this MSC was willing to do, provided the number and size of such samples could be kept quite small. MSC scientists were alarmed, however, by a White House proposal to present lunar samples to the heads of state of 120 countries.[67]

Apollo program director Sam Phillips replied that his office was accountable for the spacecraft and all hardware items that had gone to the moon; presumably that responsibility extended to the lunar samples as well (although it appears that this point had not been explicitly settled), and Phillips asked Houston for suggestions as to procedures. Concerning the lending of samples to museums and presentation of souvenirs to VIPs, however, he noted that "The Administrator will make the final determination of the overall allocation of lunar material for scientific and other purposes."[68] A mid-August general management review at Headquarters reaffirmed this position, stating that the Public Affairs Office had primary responsibility for arranging exhibition, presentation, or other uses of the samples, and it was directed to develop plans for presentation, "if possible, of small amounts of lunar material to Chiefs of State as desired by the President." The Offices of Manned Space Flight and Space Science and Applications were asked to indicate how much material, if any, and what kind (i.e., rocks or dust) could be made available for such purposes.[69]

When Headquarters called for that information, MSC's response reflected the view of lunar scientists. They were not inclined to be liberal in giving away lunar rocks, which they rightly regarded as a priceless scientific resource. Director Robert R. Gilruth replied that he could make available 150 to 200 presentation samples, each consisting of 100 milligrams (less than the volume of an aspirin tablet) of fines from the bulk or contingency samples. Concerning VIP souvenirs, Gilruth said that "we could make available if required about 10 samples of 1 gm each . . . to be given away *if needed* [emphasis in the original]," but the Center would prefer not to part with samples that big. For museums, ten 50-gram (1.75-ounce) rock specimens might be made available on loan, provided they could be recalled on one month's notice by Houston's curator of lunar samples.[70]

Not surprisingly, public interest in the lunar samples was tremendous, and in September the first of several displays of moon rocks was opened in the Smithsonian Institution in Washington. Thousands queued up to get a glimpse of a moon rock; many found it disappointingly ordinary.[71] Presentation samples— tiny portions of lunar dust, packaged in plastic vials and attractively mounted in plaques—were prepared for the world's heads of state and presented by astronauts or government officials on world tours. So far as can be determined, no lunar rocks were cut and polished as paperweights for the desks of VIPs.

Astronauts Given Clean Bill of Health

Premission plans called for the astronauts to be kept in isolation for 21 days after their exposure to lunar material. If one of them had developed signs of infection by an exotic organism, or if some adverse effects had appeared in the living systems being tested, their confinement could have been extended. Samples were to be kept until results were in from all the biological protocols, which would be completed within 50 to 80 days after the samples entered the receiving laboratory, depending on test results.[72]

From the medical standpoint the quarantine of the astronauts was absolutely uneventful. Not the slightest sign of ill effects appeared in any of the astronauts, the support personnel, or the technicians who were quarantined after two breaches of containment in the vacuum laboratory.* So, after reviewing the results of initial biological testing and finding no evidence of infectious agents, the Interagency Committee on Back Contamination agreed to release the astronauts and their companions in confinement at 1 a.m. on August 11, one day earlier than originally planned. The committee recommended, however, that they be kept under medical surveillance until biological testing was complete and the samples were released.[73]

The early morning hour for release was quite likely chosen in the hope of avoiding a wild scramble with the news media at the door of the crew quarters. But after a month of close confinement Armstrong, Aldrin, and Collins were not inclined to stay an hour longer than necessary; so at 9 p.m. on August 10, after the last medical examinations had been completed, they walked out of the living quarters, made some brief remarks to the few reporters present, and were whisked away to their homes.[74] After a press conference on the morning of the 12th, the astronauts were scheduled to leave on a worldwide personal appearance tour.

---

*The first, a rupture of a glove in the vacuum system on Aug. 1 (see above), sent two technicians into quarantine; on Aug. 5 a leak in an autoclave exposed four more to lunar materials. The most direct exposure to lunar soil occurred on July 25 when a photo technician picked up a film magazine that Buzz Aldrin had dropped on the lunar surface and found his hand covered with the tenacious black dust. He was already in quarantine but had to decontaminate himself by showering for five minutes. K. L. Suit, "Apollo 11 LRL Daily Summary Report, 1200 July 25 to 1200 July 26."

Releasing the Samples

Activity in the receiving laboratory remained at a high level during August, aimed at releasing the samples in mid-September. Two batches of material were prepared for biological examination. A "bioprime" sample, taken from the two core tubes, went to the biological section on July 27, to be examined for evidence of living organisms or their relics. A "biopool" sample, comprising several hundred grams of the "fines" plus chips taken from the lunar rocks in the bulk sample container, provided the material that would be tested in numerous living systems to determine its toxicity or pathogenicity.[75] The pooled sample was prepared the following week, and the extensive biological test protocols got under way.[76]

Throughout August the daily LRL summary reports indicated no observable effects in the biological tests. Lunar material was injected into germ-free mice, cultured to detect growth of microorganisms and viruses, and otherwise introduced into both plant and animal species. In no case was any effect noted that indicated a hazard for earth organisms. Gross and microscopic investigation of exposed systems showed only minor and localized abnormalities, if any. No exotic microorganisms appeared in the cultures. One interesting observation was that the lunar samples stimulated growth in some of the plants tested.[77]

The Interagency Committee on Back Contamination reviewed the evidence from the biological tests and concluded that the material returned by Apollo 11 was biologically harmless. The committee notified MSC Director Robert Gilruth that he could release the samples at noon on September 12.[78] Principal investigators began picking up their allotted samples in person at the lunar receiving laboratory, as required by their contracts, and the detailed investigation of lunar material began.[79] Many investigators, however, needed specimens (such as thin sections of rock) that required time to prepare, and sample distribution was only completed several weeks after initial release.[80]

Scientific Side Issues: Security and Publication

With the return of the unique, expensive, and widely ballyhooed moon rocks, NASA and its scientific advisory bodies anticipated some potential problems that could embarrass both the space agency and the community of lunar scientists. One was the theft or unauthorized use of the samples; another was an unseemly scramble to publish results. To minimize the chances of such undesirable events, the contracts covering the scientific investigation of lunar material stipulated that the contracting institution would establish strict security and accounting procedures for the samples, and that the investigator would reserve initial presentation of all results for a symposium to be held as soon as possible after each mission.[81] Both provisions violated established scientific practice to some extent, particularly the agreement to withhold publication; but in view of the unique significance of the lunar samples and the benefits to be gained by discussion at a lunar science symposium, scientists agreed to them.

Not surprisingly, since the samples were released before any of the samples were publicly displayed, requests for permission to display the research samples came in as soon as they were received.[82] These were considered case by case at

Headquarters and usually were approved.[83] This undoubtedly contributed to the public-relations dividends NASA was accumulating from the first lunar landing.

To disseminate the results of the initial lunar investigations to the widest possible audience, NASA took the unusual step of entering into a contract with a single journal[84]: *Science*, published weekly by the American Association for the Advancement of Science and circulated in 120 countries. *Science* undertook to devote a single issue to the preliminary results of the Apollo 11 research and to conduct all the normal work of reviewing and editing the manuscripts on an unusually tight schedule; NASA would cover any excess costs entailed by this accelerated process.[85]

A week after the samples were released, the lunar sample preliminary examination team published the results of its work—the only results of the first lunar landing to appear in the scientific literature before the symposium scheduled for early January 1970. Inasmuch as preliminary examination was not intended to produce answers to the important scientific questions about the moon, the paper was mostly confined to descriptions of the lunar material and the procedures followed in the lunar receiving laboratory.[86]

Nonetheless, this early work had produced data that permitted drawing some tentative conclusions. Two generic groups of samples could be distinguished: fine- and medium-grained crystalline rocks of igneous origin, which were probably originally deposited as lava flows, and breccias (heterogeneous crystalline rocks compacted from smaller particles without extensive alteration) of complex history. There was no water at the Tranquility site and probably never had been any since the samples were exposed. Neither was there any significant amount of organic material (considerably less than one part per million), which might have indicated something about life on the moon. The rocks and dust were chemically similar to each other and contained the same chemical elements as igneous rocks on earth. But in their mineral content the igneous rocks were different from terrestrial rocks and meteorites. Evidence was found that the lunar material had been chemically fractionated, some volatile elements being depleted and some refractory (high-melting) elements enriched. One of the more striking conclusions of the preliminary examination was that the igneous rocks on the moon had crystallized between three and four billion years ago—as old as any found on earth, perhaps even older, given the uncertainty in the measurements. Furthermore, the samples collected at Tranquility Base had been within 1 meter (39 inches) of the surface for 20 to 160 million years. Many of the rocks and fines showed signs of having been subjected to severe shock, such as might result from the impact of meteorites. The rounded, eroded surfaces found on many of the rocks suggested continuous bombardment by micrometeorites. Glass-lined surface pits suggested the same thing.[87]

Interest in the moon rocks tended to overshadow the results of the surface experiments that the astronauts had left behind. The passive seismometer had recorded the footsteps of the lunar explorers as soon as it was activated, and ground-based scientists noted a response when Armstrong and Aldrin tossed out excess equipment before departing. Two fairly large seismic events in the first week of operation excited geophysicists at first, but when they were not repeated the scientists concluded that they could have been false signals originating in the instrument itself. Even so, the very lack of seismic activity was indicative that the moon's interior was not like the earth's.[88] The other instrument, the laser retroreflector, was difficult to locate from earth because it was in strong sunlight, which made detection of the reflected light pulse difficult. Only after lunar sunset would astronomers be able to use the instrument in the manner for which it had been designed.[89]

Sketchy as these results were, they clearly showed that Apollo would revolutionize scientific thought about the moon and its relation to the earth. The detailed studies would show how extensive this revolution might be (see Chapter 14).[90]

The End of the Beginning

While editorial writers and columnists pondered the significance of Apollo 11 and where the space program should go, NASA managers and engineers were preparing for the next lunar flight. In the same manned space flight weekly report that documented the lunar landing, George Mueller noted that Apollo 12 was scheduled for launch on November 14 and Apollo 13 would follow on March 9, 1970. For Administrator Thomas O. Paine's information, Mueller listed the landing sites tentatively assigned to the nine remaining missions. They included one western mare, several sites in the lunar highlands, and two large craters, Tycho and Copernicus—all selected for their scientific interest, some presenting real operational challenges. Missions were scheduled at four- to five-month intervals through December 1972.[91]

After Apollo 11 such a schedule seemed reasonable. The necessary launch vehicles and spacecraft were either completed or on order. Having finally worked through all phases of a lunar landing mission, operations officers were satisfied that it could be done safely. Mission planners could see many ways to improve operations. Both lunar scientists and NASA engineers would have been happier with longer intervals between launches, but Mueller felt that a tight schedule would keep the skills of the launch teams well honed—and that it would keep costs down.

Cost was important, for NASA's appropriations had been steadily declining since fiscal 1965, when they peaked at just over $5 billion. For fiscal 1969, Congress had given the space agency just less than $4 billion. To some degree the reductions resulted from declining expenditures in Apollo; but they also reflected a change in the nation's attitudes during the decade. Heightened public sensitivity to the condition of the cities, the plight of the poor, and the deterioration of the environment put the space program in a different light. When Apollo was proposed, it seemed a way to get the country moving again; in the summer of its achievement, the space program seemed to many an unwelcome drain on resources needed for other purposes.

But it was post-Apollo programs that were most affected by the changing climate of public opinion. The lunar landing project, although it had accomplished its major objective, still offered enough promise for adding to the store of knowledge to retain a constituency and preserve its momentum. After Apollo 11, engineers and scientists alike looked ahead to improving their ability to uncover more of the secrets that lay under the barely scratched surface.

10

LUNAR EXPLORATION BEGINS

For a while during the summer of 1969, NASA basked in the afterglow of two major successes. Within a few days of the return of the first men from the moon, two unmanned spacecraft, *Mariner 6* and *7*, flew by Mars and transmitted hundreds of closeup photographs—the first since *Mariner 4* in 1965 and the most detailed ever.[1] Some continuity in manned space flight was assured when several years of difficult planning culminated in the decision—made while Apollo 11 was on its way back to earth—on a configuration and mission plans for the first post-Apollo project, Skylab.[2]

Successes notwithstanding, the future of manned space flight was far from settled. The Nixon administration, just eight months in office, had not adopted a policy on space and was waiting for recommendations from a special Space Task Group appointed in January 1969. Early indications were that the new administration was more strongly committed to reductions in government spending than to ambitious space programs.[3]

The national enthusiasm generated by Apollo 11 was soon spent. In the words of a journalist who had followed the program from its early days,

> The people who in 1961 said, 'yessir, let's go to the moon and beat the Russians' had become a different people by 1969. . . . There was the feeling: 'we won the war, now bring the boys home.' . . . no one wanted a big space program any more.*[4]

"No one" is an exaggeration—there were many who believed in continuing manned space flight at an ambitious pace, particularly NASA Administrator Thomas O. Paine—but certainly the public's enthusiasm for space flight waned rapidly after Apollo 11. Apollo would not suffer the full effects of this change in public opinion, but neither would it escape them entirely.

Until the first lunar landing was accomplished, George Mueller intended to launch missions as frequently as possible. At the end of June 1969 three Saturn/Apollo vehicles were in preparation at Kennedy Space Center: Apollo 11 on the launch pad, Apollo 12 in the Vehicle Assembly Building, and Apollo 13 components being readied for stacking. Headquarters' launch forecast issued in June projected flights only two months apart, in July, September, and November—essentially the same schedule that had existed for more than a year.[5] After the

---

*"Bring the boys home" was a cry raised throughout the country as soon as World War II ended in August 1945. As a result, the U.S. hastily—some thought unwisely—discharged millions of draftees from the armed services and dismantled its military forces around the world.

first successful landing the interval between missions was extended to four months: Apollo 12 was rescheduled for November 14 and Apollo 13 was targeted for earliest launch readiness by March 9, 1970.[6]

On With Lunar Exploration

Pleased by the results of Apollo 11, scientists called for scientific goals to take priority on subsequent missions. Those goals, stated by the Santa Cruz summer study in 1967, included extending the Apollo landing zone to cover more of the moon's surface, increasing the time the astronauts could remain on the lunar surface, and providing mobility aids to enable the astronauts to cover more ground in the time available (see Chapter 7 and Appendix 3). The scientists also noted the desirability of gathering data from lunar orbit, either by deploying an independent lunar satellite from the command and service module or by using sensors mounted in the CSM. Four working groups had found good reasons to use the moon-orbiting CSM for scientific purposes.[*7]

As soon as Apollo 11 was safely down in the Pacific, Mueller directed the manned space flight centers to shift their efforts to lunar exploration. The manned space flight organization had long agreed in a general way that the scientists' goals were the primary justification for continuing the lunar landing program and had, in fact, begun studying the changes implied in those goals as soon as the Santa Cruz conference was over. Improved mobility was the most urgent need for the scientific missions. The conferees at Santa Cruz had called for both a lunar flying unit and a surface-traversing vehicle. Since both vehicles were likely to require extensive development, the Manned Spacecraft Center's Lunar Exploration Project Office had begun technical discussions with several contractors in the fall of 1967.[8] After a year spent in defining requirements, MSC awarded two seven-month preliminary definition contracts for a flying unit in January 1969. Marshall Space Flight Center let study contracts in April for a dual-mode surface vehicle, which could be remotely controlled from earth after the astronauts left the moon.[9]

Equally high on the priority list for exploration missions was extending the time astronauts could stay on the moon. Grumman Aircraft Engineering Company, the lunar module contractor, had made some preliminary studies on modification of the lander for that purpose in 1967.[10] Life-support systems and electrical power were enough for only 36 hours on the surface; longer stays would require hardware changes. Propulsion systems would have to be upgraded to carry the scientific equipment and mobility module to be taken on exploration missions. In March 1969, MSC directed Grumman to define the changes necessary to allow

---

*This option had been available since 1964, when engineers defining the Block II service module had arranged its systems so that one sector was left empty to accommodate scientific instruments. Courtney G. Brooks, James M. Grimwood, and Loyd S. Swenson, Jr., *Chariots for Apollo: A History of Manned Lunar Spacecraft*, NASA SP-4205 (Washington, 1979), p. 140. The Santa Cruz planners had envisioned one or two lunar-orbit missions devoted entirely to photography and remote sensing, as then-current plans for Apollo Applications provided.

the lunar module to stay three to six days on the moon, with their cost and schedule impacts.[11]

Planning for lunar-orbital science began in May of 1968, when Wilmot Hess, acting on a request from the Lunar Exploration Office in Headquarters, asked the Apollo spacecraft program office to look into the question of placing a scientific payload in the service module.[12] MSC commissioned North American Rockwell, the spacecraft contractor, to study the effects on cost and schedule of adding scientific instrumentation to the spacecraft as early as Apollo 14.[13] As that study drew toward a close without discovering any major difficulties, MSC established a panel to review the operational implications of making scientific observations in lunar orbit and to identify the scientific activities that could be conducted during a lunar exploration mission.[14]

By early 1969 MSC had concluded that a package could be developed in time to fly on Apollo 14 in the last half of 1970 and had compiled a tentative list of instruments for evaluation by the Space Science and Applications Steering Committee. Director Robert Gilruth sent Headquarters the center's procurement plan, which provided for MSC to procure the instruments and deliver them as government-furnished equipment to North American to be integrated into the service module. Early approval and provision of the necessary funds by Headquarters were essential to meeting the projected schedule.[15]

In early April the Office of Space Science and Applications approved a group of experiments for the first phase of the lunar-orbit science project.[16] Apollo Program Manager Sam Phillips wanted more specific information before granting final approval and releasing funds, however.[17] The proposal was discussed in detail by Mueller's Management Council on May 7, after which Mueller gave MSC the authority to proceed—but on Apollo 16 rather than Apollo 14. Details of integrating the experiments into the service module remained to be settled, as did the exact mode of operation of each experiment, and this would take time.[18]

On June 30 Headquarters sent Houston a list of lunar-orbital experiments for Apollo 16 through 20, with authority to continue North American's integration effort and begin procurement of experiments. Assignment of experiments to specific flights could not yet be made, because the list was subject to revision during the summer. Total cost of the project was not to exceed $55 million.[19]

The manned space flight organization worked steadily from 1967 onward to lay the foundations for scientific exploitation of the Apollo systems, but the effort was often overshadowed by preparations for the first lunar landing. With NASA's encouragement, the Santa Cruz conference that summer had called for a maximum effort. The scientists' recommendations had been carefully considered, and by mid-1969 many of the most important ones were on their way to realization. It seems clear that if manned space flight had enjoyed the assurance of a high level of support, the lunar scientists would have gotten more. As it was, lunar exploration had to proceed with little more than had been projected by mid-1969: limited extension of the duration of exploration missions, a limited increase in the range of operations on the lunar surface, and a few lunar-orbital sensors.

Selecting Sites for Exploration

Since its primary objective was to land on the moon and return, Apollo 11 had been targeted for the least hazardous site. When the emphasis shifted to exploration, however, scientific considerations carried much greater weight in the choice of a landing site.[20] Even so, every landing was as risky as the first, and if MSC vetoed a site or expressed strong reservations about its feasibility on operational grounds, Headquarters and the Apollo Site Selection Board were reluctant to override the center's recommendations for the sake of enhanced scientific return.[21]

During 1968 and early 1969 the Apollo Site Selection Board necessarily concentrated on choosing landing areas for the first two missions. Five prime candidates had been chosen by December 1967, from which three—an eastern, a central, and a western, to allow for possible delays in the launch—were picked for the first landing (see Chapter 8). It was more or less taken for granted that if the first landing mission should succeed, then the second would be sent to another of those five sites, since much of the necessary planning would already have been done. If the first mission landed in an eastern mare, the second would be sent to a western one, and *vice versa.*

The Board's advisory groups continued to evaluate Lunar Orbiter photographs and by June of 1969 had produced a list of 22 sites for lunar exploration missions (Table 1). These were chosen for their apparent value in contributing to answers to one or more of the 15 questions in lunar exploration (Appendix 3). For most of these sites, changes in operational philosophy would be required. Only one site, not three, would be available at each launch opportunity; point landings (within 1 kilometer, 0.62 miles) would be necessary, to place the landing module as near the features of interest as possible; approach paths might be rough or undulating rather than smooth; free-return trajectories could not be used; and the high-resolution photography required to certify a site was generally not available.[22]

With more than twice as many interesting locations to visit as there were missions planned, site selection would be a complex process at best. Scientists' priorities might change as the results of early missions became known and as NASA developed more precise landing techniques and extended the area where the spacecraft could land. Reconciling the goals of science with the constraints of mission operations required an early start and continuing tradeoffs as the project progressed.

At the June 3 meeting of the Apollo Site Selection Board, chairman Sam Phillips, anticipating heavy work loads if the board was to accomplish its task within tight schedules, requested that the board meet monthly if possible. He directed Lee Scherer to prepare a thorough briefing on the scientific objectives of lunar exploration and suggested that the Group for Lunar Exploration Planning propose a sequence of missions that would accomplish those objectives. Board members agreed a that to make sensible choices between sites they needed a better understanding of the rationale of lunar exploration and the operational improvements being planned.[23]

The meeting then turned to the question of a site for the second mission. Scientists reiterated their preference for a western (younger) mare if the first mission landed safely at an eastern (older) site. Two western sites were on the short list

Table 1. **Lunar Landing Sites Recommended for Consideration as of June 1969.***

Site	Latitude	Longitude
Censorinus	0°17′ S	32°39′ E
Rima Littrow	21°25′ N	28°56′ E
Abulfeda	14°50′ S	14°00′ E
Rima Hyginus	7°52′ N	6° 7′ E
Tycho	41° 8′ S	11°35′ W
Copernicus Peak	9°36′ N	19°53′ W
Copernicus Wall	10°22′ N	19°59′ W
Schröter's Valley	24°36′ N	49° 3′ W
Marius F	15°10′ N	56°31′ W
Fra Mauro	3°45′ S	17°36′ W
Mösting C	1°55′ S	8° 3′ W
Hipparchus	4°36′ S	3°40′ E
Prinz	25°57′ N	43°40′ W
Gassendi	17°50′ S	40°20′ W
Dionysius	2°31′ N	17°49′ E
Alexander	37°46′ N	14° 6′ E
Alphonsus	13°35′ S	4°11′ W
Rima Bode II	12°47′ N	3°49′ W
Copernicus CD	6°32′ N	14°58′ W
Tobias Mayer P	13°18′ N	31°11′ W
Aristarchus	24°24′ N	47°50′ W

*From minutes of the Apollo Site Selection Board meeting, June 3, 1969.

of preferred sites compiled in 1968: one just below the equator some 450 kilometers (280 miles) south and slightly east of crater Kepler and the other about 250 kilometers (155 miles) northwest of the first. Benjamin Milwitzky of the Lunar Exploration Office then suggested that Apollo 12 land near a Surveyor spacecraft. As early as January 1969 Milwitzky, formerly the Headquarters program manager for Surveyor, had suggested visiting a landed Surveyor and returning some spacecraft parts and nearby surface samples to earth for study. This could yield valuable engineering information on the effects of the space environment on materials, besides allowing postmission verification of Surveyor's scientific results.[24]

MSC representatives then presented a rationale for considering two other western sites. Although these had been eliminated in selecting the final five sites, they met MSC's criteria for operational suitability and offered certain advantages over the first two. Both sites were near Surveyor spacecraft.[25] The Board reacted unfavorably to these suggestions, pointing out that the site where *Surveyor III* was located was in a younger mare that was not much different from those in the eastern sites, whereas the scientists' first two choices were in typical older regions. Examining the Surveyor would detract from the other objectives of the mission. Furthermore, if returning Surveyor parts were set as a goal for the mission, failure to accomplish the necessary precision in landing could be interpreted as failure—which would not, in fact, be the case. Chairman Sam Phillips was reluctant to add any more sites to the list for the second mission. He did not favor either of MSC's choices and instead directed Houston to examine two sites considered highly desirable by the scientists, Hipparchus and Fra Mauro, and report on their suitability.[26]

MSC analyzed the data available for these two sites and found them unacceptable for the second landing mission. Hipparchus had only about half as much good landing area as the average Apollo 11 site and Fra Mauro was worse. Photographic coverage in both cases was marginal. Houston recommended that the site selection board give no further consideration to these two locations, but that it reexamine the *Surveyor III* site, which met all the criteria for the first landing and was in some respects better than the two western sites under consideration.[27] Phillips concurred and directed the Group for Lunar Exploration Planning to assess the scientific merit of the site.[28]

On June 17 the Site Selection Subgroup of the Group for Lunar Exploration Planning met at Houston to try to reduce the complexity of lunar exploration planning. MSC's operations planners needed definite recommendations as to scientific objectives and priorities rather than the unstructured group of sites currently being considered. A short list of high-priority sites was desired, which would not subsequently be changed except through formal change procedures. MSC engineers briefed the subgroup on the increased capabilities that might be expected for the exploration missions. After Apollo 11, four "H" missions were planned, each of which would be able to carry a complete Apollo lunar surface experiments package (ALSEP), could support two periods of surface activity by the astronauts, and would be targeted for a smaller landing zone than the first mission.* On the later "H" missions engineers expected to be able to land within a 1-kilometer (0.62-mile) circle. After the "H" missions, six "J" missions would be flown. These could land with considerably improved accuracy, stay on the surface for three days and allow four excursions to the surface, and carry scientific equipment in the service module for lunar-orbital experiments. Starting with the second or third "J" mission, a powered surface vehicle would extend the astronauts' radius of operations to about 5 kilometers (3 miles).[29]

With these developments in mind, the subgroup reduced the list of candidate sites to 10, ordering them in a sequence that would produce the best scientific

---

*The landing zone for Apollo 11 was about 19 kilometers long and 5 kilometers wide (12 by 3 miles) as a result of uncertainties in the determination of the spacecraft's position and velocity in lunar orbit before landing.

return, and added five more representing lunar features of scientific interest not covered by the existing list. It recommended systematic photography from the orbiting command and service module to provide the necessary planning data for later missions. Finally, the subgroup recommended that *Surveyor III* be deleted from further consideration for the second landing because its location was not expected to yield data significantly different from the two eastern sites already picked for Apollo 11.[30]

The Site Selection Board met again on July 10 for a briefing on the aims of lunar exploration. Donald Wise of the Lunar Exploration Office discussed the types of information the scientists hoped to get from the lunar exploration program: the ages of lunar materials, their chemical composition, clues to the processes that have created lunar landforms, the interior structure of the moon, and the rate of flow of heat from its interior. Farouk El-Baz of Bellcomm described the general areas that should be sampled in the first phase (10 missions) of lunar exploration: two types of mare material ("older" or "eastern" and "younger" or "western"); regional stratigraphic units such as deposits around mare basins; impact craters in both maria and highlands; morphological manifestations of volcanic activity in both maria and highlands; and areas that might give clues as to the nature and extent of processes other than impact and volcanism which may have acted on the lunar surface. He then enumerated the characteristics of each of the 10 landing sites proposed by the Group for Lunar Exploration Planning, relating each to the scientific goals of the program and tying the sequence to expected improvements in spacecraft capabilities and flight operations planning.

Chairman Phillips remarked that the list seemed well thought out and that a short list of desirable science sites must soon be stabilized. After considerable discussion, the Board approved the 10 candidate sites for planning purposes. Phillips directed MSC to study these sites and report on their suitability.

Houston had already taken a quick look at the sites and determined that all would require additional photographs before they could be certified under existing criteria. Photography from the Apollo 10 command module—conducted specifically to evaluate its usefulness for filling gaps in Lunar Orbiter coverage—had proved adequate for site analysis, and MSC's data indicated that by proper choice of sites for early missions, photographs of many of the later ones could be obtained. According to MSC's studies, the *Surveyor III* site offered better opportunities for this "bootstrap" photography than the other western locations on the list.[31]

Two months' work by the Apollo Site Selection Board did not finally determine where each Apollo exploration mission would go. It reduced the scope of the debate somewhat, established the principle that changes were to be made only for good scientific reasons, and provided the means for accommodating changes as the program developed. The list of 10 sites approved in July provided specific targets for mission planners; it would change as operational problems arose and as improved equipment and techniques became available.

Preparations for the Second Mission:
Crew Selection and Training

On April 10, 1969, following the success of Apollo 9, MSC announced the names of the crews for Apollo 12. The prime crew was commanded by Charles (''Pete'') Conrad, Jr., astronaut since 1962 and veteran of two missions, Gemini V and Gemini XI. His lunar module pilot was Alan L. Bean, who joined the program with the third group but had yet to fly a mission. Rounding out the crew as command module pilot was Richard F. Gordon, Jr., a member of the third class of astronauts, who had flown with Conrad on Gemini XI. The three, all Navy aviators, had trained together as backup crew on Apollo 9. Named to the backup crew were commander David R. Scott, who had flown with Neil Armstrong on the prematurely terminated Gemini VIII mission, and two astronauts from the fourth group of pilots, lunar module pilot James B. Irwin and command module pilot Alfred M. Worden.[32] Irwin's and Worden's only prior experience had been on support crews, Irwin on Apollo 10 and Worden on Apollo 9. As support crew for Apollo 12, Deke Slayton picked Gerald P. Carr and Paul J. Weitz, two pilots from the fourth astronaut class, and Edward G. Gibson, the first scientist-astronaut named to any Apollo crew position.[33] For Gibson and his colleagues this was recognition of a sort. No one yet knew how support crews would fare in later competition for prime crew slots*—so far, only one had gotten as far as a backup crew—but his appointment was a sign that they had at least not been completely forgotten.

In April 1969 no one could be certain that Apollo 11 would make the first lunar landing, so training for Conrad, Bean, and Gordon was very much like that for Armstrong, Aldrin, and Collins. Nominally both crews were preparing to land at any of three sites; actually, their site-specific training concentrated on only one, simply because there was not time to prepare adequately for three.[34] Apollo 11's primary target was site 2, in the Sea of Tranquility; 12 focused on site 5, in Oceanus Procellarum.

As was normal when two crews were in training simultaneously, the crew assigned to the earlier mission had priority in use of the command module and lunar module simulators until just before launch. Conrad and his group spent their training time on other phases: briefings on systems, thermal-vacuum tests, design reviews, geology classes and field trips, and the thousand and one other details that went into preparation for a lunar mission.[35] When Apollo 11 returned successful in late July, the launch date for Apollo 12 was moved to November and two months were gained for preparation.

Target: *Surveyor III*

Some of that preparation time would be needed, for after Apollo 11 the objectives for Apollo 12 began to change. Shortly after the site selection board

---

*As it turned out, 8 of the first 13 astronauts named to support crews eventually made lunar flights.

had settled on its list of 10 landing sites for lunar exploration, Sam Phillips selected the *Surveyor III* site as the target for the next mission.[36] Even though the scientists had unanimously rejected its location as unsuitable, the inert Surveyor offered some opportunities that could not be passed up. A demonstration of point landing (i.e., within 1 kilometer [0.62 mile] of a preselected spot) needed to be made as soon as feasible. For that purpose, something like a Surveyor was more appealing to mission planners than picking some specific crater. Scientists might have good reasons for landing next to one crater rather than another, but to nonscientists (including the public) all craters were pretty much alike.[37] Besides, valuable engineering and scientific information could come out of examination of a spacecraft that had spent more than two years in the space environment,* as Milwitzky had pointed out to the site selection board (see above). Houston had already determined that certain components could be removed from a Surveyor with no special difficulty and was working with Hughes Aircraft Company, builder of the Surveyors, on procedures the astronauts could use.[38]

As soon as that decision was made, changes in mission techniques became a major effort for Apollo 12 planners. Howard W. Tindall, Jr., MSC's chief of Apollo data priority coordination, wrote as the planning began, "It is clear that *lunar point landing capability is absolutely necessary if we are to support the exploration program the scientists want* [emphasis added]. That is, mission success intrinsically depends upon it."[39] A preliminary look at the problems of accomplishing it, however, showed that it could not be done as early as Apollo 12. Too many procedural changes were involved, and there was not enough time to incorporate them and prove their suitability. Accordingly, mission planning teams at MSC set out to refine existing techniques, in the hope that they could reduce the unavoidable errors to the point of landing within a mile (1,600 meters) of a preselected spot.[40]

Trajectory planners first needed to know the precise location of *Surveyor III*, which, they discovered, had been determined not long after the spacecraft had touched down.** Geologists had located the spacecraft inside and just below the eastern rim of a subdued crater about 200 meters (650 feet) across and 15 meters (50 feet) deep, some 480 miles (775 kilometers) west-southwest of the moon's center, at lunar latitude 2°57' south, longitude 23°20' west.[41]

Given the precise location of their target, trajectory planners next went to work improving mission techniques. The fundamental difficulty in making a precision landing was determining the lunar module's orbital position and velocity in lunar orbit (its "state vector") with sufficient accuracy. Earth-based radar acquired this information and the communications network transmitted it to the lunar module's primary guidance computer just before the landing craft began its powered descent. No further change was made until the spacecraft reached low altitude and the landing point was in sight. Then the mission commander could assume manual control and adjust the landing point to avoid hazards, as Neil Armstrong had done on Apollo 11, or land at a desired spot.

Surveyor III landed on April 20, 1967, and transmitted 6,326 television pictures and large amounts of scientific data for 14 days.

**Prominent features appearing in *Surveyor III's* television photographs had also been located in high-resolution pictures of the same area taken by *Lunar Orbiter III*.

After the state vector was determined and before descent began, the spacecraft's actual path could deviate from its predicted path for several reasons. Perturbations caused by the irregular shape and gravity field of the moon, unwanted changes in velocity during required spacecraft maneuvers, and the guidance system's tendency to drift slightly from its inertial alignment, all contributed to inaccuracy in the state vector.*[42]

A month of considering the sources of these errors and ways of minimizing them boosted MSC's confidence considerably. Still, after a "three-day mission techniques free-for-all" at the end of July, Tindall concluded that "if we land within walking distance, it is my feeling we have to give most of the credit to 'lady luck.'"[43] Four weeks later, numerous productive suggestions had been explored. Among the more significant changes were separating the lunar module from the command module earlier in the mission and eliminating several attitude changes normally made in the following two orbits. These changes gave more accurate tracking of the LM, could significantly reduce unwanted velocity changes, and might enable Mission Control's computers to measure drift in the inertial navigation system and compensate for it. Another proposed refinement was to give the spacecraft computer a revised landing point, based on radar tracking just before the lunar module started down, while the lander was making its descent, thus taking out all the error accumulated to that point in one step.[44] All these changes got careful study in the next two months, with encouraging results. Two weeks before launch Tindall was more sanguine. "For whatever it's worth," he reported, "my feeling now is that as long as the systems work as well as they have in the past, we have a pretty good chance of landing near the Surveyor." Beyond that he would not make any predictions about the accuracy of the landing.[45]

Firming Up Plans for Apollo 12

On July 18, 1969, MSC issued the mission requirements document for Apollo 12, listing the primary purposes of the mission as investigating the lunar surface environment, emplacing the first Apollo lunar surface experiments package (ALSEP), obtaining samples from a second lunar mare, and enhancing the capability for manned lunar exploration. Major changes from Apollo 11 included the possible use of a hybrid trajectory** rather than a free-return trajectory and

---

*These random (indeterminate) errors led engineers to use statistical methods in predicting landing points. The most commonly used designation was the "3-sigma ellipse," centered on the desired landing point, within which there was 99 percent confidence that the spacecraft would land. For *Apollo 11* the 3-sigma ellipse was roughly 11.7 miles long by 3 miles wide (18.8 by 4.8 kilometers); for Apollo 12, planners were aiming for 2.6 by 1.6 miles (4.2 by 2.6 kilometers).

**Hybrid trajectories were fuel-saving flight paths which, unlike free-return trajectories, would not return the spacecraft to earth if the service module's main propulsion system failed to put it into lunar orbit. They were designed so that in case of such a failure the lunar module's descent engine could correct the resulting flight path (which might put the spacecraft with its three occupants into solar orbit) for return to earth.

scheduling two periods of lunar surface exploration by both crewmen. Five possible landing sites were specified, including site 5 (the western mare site preferred by the scientists) and *Surveyor III*.[46]

After Sam Phillips designated the Surveyor as the spot for Apollo 12 to land, mission planning focused on the problems of precision landing (see above), deployment of the ALSEP, geological observations and sample collection, and examination of the Surveyor and its surroundings.[47] Although some at MSC believed that returning some components of the Surveyor was of considerable importance, other surface activities were given higher priority [48]—probably in deference to the scientists. (The site had been unanimously rejected by the Group for Lunar Exploration Planning, who considered the inert spacecraft to be an ''attractive nuisance'' that would likely divert the astronauts from more important work.[49]) Deployment of the first ALSEP was high on the priority list, since scientists had been disgruntled by the decision to fly a simplified package of surface instruments on Apollo 11 (see Chapter 8). Geologists wanted the Apollo 12 astronauts to be somewhat more selective than their predecessors in collecting samples and stressed the importance of documenting (photographing and describing) them. They also preferred more rocks and less dust, if possible. To determine what the astronauts should do on and around *Surveyor III*, MSC had already begun discussions with Hughes Aircraft Company.

Interlude: Alarums and Excursions in the Scientific Community

The samples returned by Apollo 11 were just becoming available to experimenters when Apollo 12 was launched, and while those scientists eagerly awaited them, many in the scientific community expressed discontent with NASA's management of the lunar exploration program. Some of their dissatisfaction stemmed from specific actions of the manned space flight organization and the scientists' perception of their significance, some of it from disagreement with the priorities of the program. Much of it, however, seems to have had a much more elusive origin.

Scientists had for years bewailed the priority given to test pilots in selecting crews for lunar exploration. Their distress was temporarily alleviated by the selection of two groups of scientists for training as astronauts, but flight assignments remained a sore point. Those scientists most closely associated with planning the missions understood that the capability to land on the moon and return had to be developed first and accepted MSC's contention that test pilots were best suited by experience to deal with the uncertainties of developmental flights. Even they, however, expected that scientists would be assigned to missions very soon after the first successful lunar landing.

Crews for the first two landing missions had been named in January and April of 1969 without producing an outcry from the scientists. Early in August, however, the announcement of selection of the next two crews for lunar exploration did provoke a response. The Apollo 13 crew included James A. Lovell, Jr., commander, Fred W. Haise, Jr., lunar module pilot, and Thomas Mattingly II, command module pilot. Their backups were John W. Young, John L. Swigert, Jr., and Charles M. Duke. For Apollo 14, Alan B. Shepard, Jr., was named commander, Edgar D. Mitchell, lunar module pilot, and Stuart A. Roosa, command module pilot, backed up by Eugene A. Cernan, Ronald E. Evans, and Joe H. Engle.[50] Not one of the 13 scientist-astronauts was included.* Worse yet from the scientists' point of view, seniority—which hitherto had seemed to be one of Deke Slayton's primary criteria for crew appointments—appeared to count for nothing when the first class of scientist-astronauts acquired it: of the 12 crewmen named, only Lovell, Young, Shepard, and Cernan had been in the program longer than they.**

This alone might not have been cause for alarm on the part of scientists had it not been followed shortly by the resignation of several men occupying positions of some prominence in NASA's science program. At Headquarters, Dr. Donald U. Wise, Lee Scherer's deputy in the Lunar Exploration Office, left to take an academic appointment. Dr. Elbert A. King, Jr., a prime mover in establishing the lunar receiving laboratory at Houston and first curator of lunar samples, announced he would resign to become head of the geology department at the University of Houston. The astronaut corps lost one of its scientist members when F. Curtis Michel resigned to return to teaching and research at Rice University. Finally, Wilmot Hess, first director of Science and Applications at MSC, left to become director of research at the Environmental Science Services Administration laboratories at Boulder, Colorado.

Each man had his own reasons for leaving NASA, and their near-simultaneous departure seems to have been only coincidental. (King, for example, had made a commitment to the University of Houston more than a year earlier.[51]) Still, all but Hess—who declined to discuss his resignation with the press—expressed some dissatisfaction with the status of science in manned space flight programs. Michel's main reason was to return to research, but he stated his disappointment that NASA had shown "no serious intent to fly scientist-astronauts." King likewise professed a desire to spend more time in research, which he found very difficult to do in an administrative position, and warned that NASA had not yet convinced the scientific community that it "will put together a program that will truly emphasize science." Wise discussed his choice somewhat reluctantly, expressing concern that criticism from the scientific community would intensify science's problems: "With enough screams [from scientists]," he feared "we will fly only five missions instead of ten—this would be the real tragedy." He too pointed out (among other problems) "a lack of understanding of scientific goals at the management level in NASA."[52]

Hess's resignation was perhaps the most serious of all in the eyes of outside scientists, because he had been brought to MSC specifically to give the Houston center some scientific respectability (see Chapter 6). He had, however, found little support from center management and no understanding of the proper role of science. His plans to establish a credible research program at MSC, which

---

*Mitchell held a Ph.D., but it was in engineering (astronautics and aeronautics), not one of the natural sciences.

**However, three of the scientists had spent a year learning to fly, and all five had begun their astronaut training after the next class of pilots, which included all the rest of the Apollo 13 and 14 crews, had been selected.

included a substantial increase in the number of research scientists, had been thwarted by agency-wide cuts in civil-service positions—the result of budget cuts which had not been anticipated when he came to Houston. Apparently seeing little hope of improvement, he had opted for a more promising environment.[53]

Yet one more scientist was to separate himself from Apollo. Eugene Shoemaker, who from the very early days had been a vigorous advocate of science on Apollo (see Chapter 2), had actively participated not only in program planning but also in developing geologic methods for the astronauts to use on the moon. Now he excoriated the program. Speaking to a Pasadena luncheon group in early October, he expressed his strong opposition to the post-Apollo plans recently presented by the President's Space Task Force—specifically the proposal to send humans to Mars. The only justification of that mission, he said, was "to build big, new systems in space." Apollo was that kind of system, built "primarily for the sake of building a big system," and it had turned out to be hopelessly inadequate for scientific purposes. The spacecraft and the mission mode were designed to engineering and operational requirements, and the system was all but useless for any other purpose. As a result, now that it was possible to send humans to the moon, the system could be used for practically nothing else, because the engineers had not considered any purpose for Apollo except to demonstrate that people could go to the moon and return safely. Everything the Apollo 11 astronauts had done, Shoemaker said, including sample collection, could have been done sooner and more cheaply by unmanned spacecraft. It was time to redesign the spacecraft and the space suits, provide surface mobility, and adapt the missions to the tactics of field geology—the only activity that would make the most of humans' inherent superiority to machines.[54]

Shoemaker's criticisms were not entirely without merit, although it was far too late to make the changes he called for, and they were reported in a local paper and then widely circulated by the wire services.[55] Still, it was easy to discount them as the complaints of a discontented participant whose ideas had not been allowed to determine the course of the program. Over the years, Shoemaker's early influence on Apollo science had been gradually preempted by laboratory-oriented scientists (geochemists, geophysicists, petrologists, mineralogists) many of whom would scarcely classify field geology (Shoemaker's own specialty) as a science at all.[56]

Defections by these highly visible scientists in the space agency were—as no doubt some of them had intended—critically noticed by science's advocates in the press. A *New York Times* editorial quoted Elbert King's contention that "there's not enough sympathy with, or understanding of, scientific objectives at the higher levels of NASA." The *Times* editorialist noted that "everything man has learned in this last eventful month about both the moon and Mars makes it plain that scientific objectives must enjoy much higher priorities in NASA's future efforts."[57] *The Washington Post* went further, asserting that Hess's resignation was likely to signal the scientific community that the goal of Apollo was simply "to improve on the techniques of space flight instead of setting the mission of each flight primarily to maximize the yield of basic scientific data. . . . It makes only a little sense to go back to the moon again and again simply to improve our method of getting there. . . . The scientists of space . . . have been forced into the back seat of the manned space program. It is time now to make them the navigators. The choice of missions . . . should be largely in their hands."[58]

Whether these editorials reflected a bias in favor of the intellectual elegance of pure science or merely gross ignorance, they overlooked some basic facts about Apollo's limited capabilities. Maximizing the yield of scientific data *required* improving the techniques of manned space flight. Apollo 11 overshot its aiming point by five miles (eight kilometers), and even the *Washington Post* would likely have agreed that it made no sense to commit scientific missions if they could not reach their targets. Nonetheless, progress toward scientific exploration was too slow to suit some critics, and they evidently felt that their only recourse was to take their case to the public.

The more subjective concerns of scientists were equally important but considerably less easy to understand and more difficult still to implement. King's charge that NASA management did not sufficiently understand or sympathize with scientific objectives (whatever that meant) was seconded by others. Fred L. Whipple, one of the country's foremost astronomers and a member of George Mueller's Science and Technology Advisory Committee, put it to Mueller in early August 1969 in a letter calling attention to the lamentable situation at Houston. "I have yet to talk to a scientist connected with the Apollo project," he wrote, "who feels that he is really welcome there by the engineers. The atmosphere . . . is not hostile but it certainly is negative." In 26 years of working with engineers, Whipple said, he had often had disagreements but had "never encountered the negative type of attitude that persists at the MSC." He had personally heard an astronaut say, in effect, that if only they could put all the scientists in a cage, then the engineers and astronauts could get on with the program. Clearly this was not a milieu in which a scientist could work effectively.[59]

Later in the year, but in response to the same events, Alex J. Dessler, head of the Space Physics and Astronomy department at Rice University, commented on the discontent of the space scientists. In spite of the fact that American space science was recognized as the most productive in the world, and space science was relatively better supported than many other fields of science, the discontent of space scientists was widespread and growing more intense. Dessler noted that their attitude, though in some sense incongruous, was important, because Congress was sure to listen if a number of prominent space scientists began to condemn the Apollo program. Scientists evidently felt that they had no effective advocate at a high level in NASA—in contrast to the engineers, who were quite well represented by George Mueller at the head of a program office. Even when the scientists got what they wanted, Dessler said, they felt frustrated at "the appearance of condescension on the part of the engineers . . . the scientists sometimes feel they are being thrown a bone to shut them up."[60]

Both Whipple and Dessler suggested similar solutions to the problem. Whipple thought that much of the misunderstanding arose from NASA's desire to justify Apollo on scientific grounds, something he believed could not be done. Manned space flight, which he supported, had its own reasons for being, but they were not necessarily scientific. He suggested that the only way MSC could gain the support of scientists was to put a scientist of high repute, acceptable to both outside critics and the engineers and astronauts with whom he worked, in a position "in which [he] has a major role in all of the decisions made with regard to the operation of the Center." Dessler saw a need for a similar scientific heavy-weight at Headquarters; scientists had had no such advocate since Hugh Dryden.* For the manned space flight program to be acceptable to the scientific community, apparently, scientists had to be in undisputed command.

This attitude was not new; even in the unmanned space science program, scientists contended for influence in shaping NASA's programs and felt they did not have enough. Homer Newell, as associate administrator for the Office of Space Science and Applications from 1963 to 1967, had dealt with them time and again. In retrospect he wrote, "Scientists are a contentious lot, . . . and the tremendous opportunities of the space program inspired them to more intense dispute than usual." Much of the tension in the space program, Newell said, "stemmed from the scientists' presumption of special privilege, which at times Congress found irritating."[61] That presumption, sometimes verging on arrogance, was unmistakable in the complaints aired during the summer of 1969, and, like Congress, Apollo's engineers could hardly help finding it irritating. By their own lights, the engineers were doing their best to facilitate scientific exploration of the moon within the limitations of a very complex technology that offered many chances for catastrophic failure. They were reluctant to take extreme risks—as by landing at a site for which they had inadequate data simply because it looked more interesting to the scientists.

The general malaise of the times may also have contributed to this somewhat vague but strongly felt sense of impotence on the part of scientists. For the first time in some years the value of pure research, long extolled by spokesmen for science, was in question. The college generation of young Americans, including many science students, grew concerned for the earth's environment. They often indiscriminately attributed its deterioration to both inhumane science and mindless technology. They also disparaged the academics' devotion to research—most particularly research subsidized by the military—as the major function of the university, demanding more attention to teaching. A small group of radical students forcibly brought these concerns to the attention of scientists late in 1969 at the national meeting of the American Association for the Advancement of Science in Boston. The normally sedate proceedings were disrupted by hecklers, organized demonstrations, and rump sessions, to the considerable distress of the scientific community.[62]

Whatever the origins of the space scientists' unhappiness, their public clamor for more science on Apollo did not go unheeded. George Mueller wrote to MSC Director Robert Gilruth in early September 1969, urging him to give this problem his personal attention. After listing the steps already taken to support the science program within a steadily declining personnel ceiling, Mueller cautioned that

---

*Hugh L. Dryden, director of the National Advisory Committee for Aeronautics (NACA) from 1947 to 1958, achieved his reputation through basic research in aerodynamics conducted at the National Bureau of Standards. His election to the National Academy of Sciences in 1944 attested to his stature among his peers. Although he was not active in research after taking over NACA, Dryden's skills as a research administrator during difficult times earned widespread respect in the agency. He was appointed deputy administrator of NASA in 1958, remaining in that post until his death in 1965. Some pioneer space scientists remember him with something approaching hero worship.

we will certainly detract measurably from the success of Apollo 11, and the missions yet to be flown, unless we meet the challenge [of providing] the support required in the science area.

Noting that "some members of the scientific community are impatient and as you know, are willing to air their views without necessarily relating those views to what is practicable and possible," he stated that

it is our policy to do the maximum science possible in each Apollo mission and to provide adequate science support. . . . we must assure ourselves and the world of science that we are making those adjustments which will provide steadily increasing and effective support for the science area.[63]

It remained to be seen whether the impatient scientists would be mollified by the "adjustments" Mueller promised.

Preparations for the Next Mission

Scientists had been reasonably well satisfied with the way the Apollo 11 samples were processed, although there was considerable waste motion and some confusion attributable to inexperience in the first real exercise of the lunar receiving laboratory. The most serious complaint concerned the long delay in releasing samples to principal investigators. However, many deficiencies in equipment and procedures had shown up during the first mission that required correction before

Jim Lovell (left) and Fred Haise work with the heat flow experiment during training at the Kennedy Space Center.

Fred Haise, lunar module pilot on Apollo 13, practices using the lunar drill.

Apollo 12, scheduled for mid-November. So, while completing their tasks following Apollo 11, scientists in the Lunar Receiving Laboratory were also preparing to incorporate necessary changes.

In early September the Lunar Sample Analysis Planning Team forwarded to Anthony J. Calio, MSC's new Director of Science and Applications, a list of recommended changes concerning matters such as weighing of samples, procedures for transferring samples during quarantine, and photography. They also suggested improving the display of information on sample history, status, and location, since it had proved difficult during Apollo 11 to keep everyone informed of where the individual samples were and what had been done to them.[64]

Also in need of improvement was the vacuum system into which the returned sample containers were first admitted to the laboratory. Problems had been foreseen in handling the samples under high vacuum, and indeed problems had developed during simulations and during the first mission (see Chapter 9). Members of the analysis planning team now felt it was mandatory to open the lunar sample container in an atmosphere of dry nitrogen rather than in vacuum. If two containers were returned, one could be opened in the vacuum system; but rather than conduct the preliminary examination there, the contents should immediately be canned and stored under vacuum for the few investigators who required vacuum-preserved specimens. The vacuum facility posed too many problems, such as rupture of the gloves and contamination by organic materials from the vacuum pumps, to allow continuation of its use for all the lunar samples.[65]

Others who had been involved in handling the Apollo 11 samples offered suggestions for reducing the number of people in the laboratory, separating the functions of the preliminary examination team and the analysis planning team, and eliminating the preparation of two samples for quarantine testing. All these changes would speed up the release of samples to outside investigators.[66]

A meeting in late September settled many of these questions, deciding that a single "biopool" sample would be prepared for the biological tests rather than the two that had been used on Apollo 11 and that the two sample return containers would be processed simultaneously: one in the vacuum system in the physical chemistry laboratory and the other under sterile nitrogen in the biological preparation laboratory. Among other procedural changes adopted at this meeting was a provision that the analysis planning team would begin work only after the preliminary examination team had completed its tasks.[67]

The Interagency Committee on Back Contamination continued to press for the installation of a bacterial filter on the command module postlanding ventilation system. MSC was unyielding although engineers continued to evaluate solutions to the problem.[68] Agreement was reached to discontinue using the biological isolation garments, provided no crew member was ill on return. Recovery teams would provide clean flight suits and respiratory masks for the astronauts to put on before they left the command module.[69]

For the crews, August, September, and October were packed with simulations, briefings, and field trips. In mid-August they got their first briefing on the Surveyor at the Jet Propulsion Laboratory. A month later they spent a day with the geologists in the lunar receiving laboratory, examining the Apollo 11 rocks and discussing plans for collecting samples.[70]

Geologists had high hopes for the Apollo 12 crew. They had been the first to go through a revised geology training program that stressed basic principles of site exploration rather than minutiae such as identification of rocks. By the time they were ready for launch, Pete Conrad and Alan Bean and their backups had well over 200 hours of field work under their belts. Their last field trip, to the volcanic fields of Hawaii, was extremely satisfying to their training officers. They handled every problem put to them, their descriptions and photographs were excellent, and their sampling of the terrain was first-rate. Everyone looked forward to superior results from the first real lunar scientific expedition.[71]

At Kennedy Space Center, preparations for the launch of Apollo 12 went smoothly. The first complete lunar surface experiments package arrived at KSC in late March. Spacecraft and Saturn V were mated on July 1, and the vehicle was moved to launch complex 39A on September 8. A week before launch the recovery quarantine equipment and mobile quarantine facility were ready for shipment to the recovery ship, U.S.S. *Hornet*. From then on, the only hitch in launch preparations occurred two days before launch, when discovery of faulty insulation on a liquid hydrogen tank in the service module required exchanging the tank for one on the Apollo 13 spacecraft. As launch day dawned, the only portent of possible delay was a cold front approaching the Cape from the north.[72]

11

FIRST PHASE OF LUNAR
EXPLORATION COMPLETED:
1969–1970

Lunar exploration began in earnest with the launch of the first "H" mission, Apollo 12. Successful in all respects, Apollo 12 crowned the accomplishments of 1969 by establishing the ability to put the lunar module down on the moon within walking distance of a desired spot. That done, the next mission was planned for a landing on the Fra Mauro Formation, the site highest on the scientists' priority list. Apollo 13 was targeted to land there, but an equipment failure forestalled a lunar landing and came close to costing three lives. Real-time improvisation saved NASA from calamity and demonstrated the versatility of the Manned Spacecraft Center's mission planning and mission control teams.* After a thorough investigation that delayed subsequent missions by almost a year, Apollo 14 was assigned to land at the same site.

Not long after Apollo 12 returned, scientists gathered in Houston to present the results of their investigations of the first lunar samples. In what was surely one of the most exciting scientific conclaves of the century, practically all existing theories of the moon's origin were thrown into doubt in one respect or another, and scientists were more eager than ever for data. Shortly after that conference, management at MSC sat down with lunar investigators and worked out ways of increasing the amount and quality of the data to be returned by future Apollo missions.

Their expectations, however, were soon reduced when the pressures of a tightening federal budget forced cancellation of one of the planned Apollo missions. Faced with drastic reductions in funding, NASA managers chose to allot more of the available money to post-Apollo programs on which continuation of the manned space flight program depended. Apollo 11 had ushered in an era of hard choices, not expanding horizons.

Intrepid Seeks Out Surveyor III

The morning of November 14, 1969, was cold, cloudy, and wet at Kennedy Space Center. Weather radar showed rain showers marking a cold front 80 miles

---

*See Appendix 8.

(130 kilometers) north of the Cape and moving south. With launch only hours away, rain, broken low clouds, and overcast at 3,000 meters (9,800 feet) caused launch officials to consider their options. Reports indicated no thunderstorms or severe turbulence in the area, however, and conditions were better than the minimums specified by launch rules. An hour and 22 minutes before liftoff, a pump replenishing liquid oxygen in the launch vehicle tanks failed. With success depending on a backup pump, launch director Walter Kapryan chose to proceed.[1] The crew had trained intensively for the specific features of the planned landing site, and if the day's launch window closed, that site would not be accessible for another month.

At 11:22 a.m. Eastern Standard Time, President Richard M. Nixon—the only incumbent chief executive ever to witness an Apollo launch—along with 3,000 invited guests and a large crowd of tourists, watched as the Saturn V rose from the pad and accelerated toward the clouds. Just before the vehicle disappeared into the overcast, two streaks of lightning flashed toward the launch complex. In the command module *Yankee Clipper*, mission commander Pete Conrad heard the master alarm. He looked at the caution and warning panel, and "it was a sight to behold." Sixteen seconds later another bolt discharged, and Conrad told Houston, "We just lost the platform [in the inertial guidance system], gang: I don't know what happened here; we had everything in the world drop out."[2] The fuel cells had automatically disconnected and several panel gauges were temporarily disturbed, but otherwise everything seemed normal. After the crew reset the fuel cells, the flight continued while Mission Control tried to figure out what had happened and what the consequences might be. The best guess seemed to be that the launch vehicle, trailing its plume of ionized (and electrically conductive) exhaust gas, had triggered a lightning discharge that otherwise would not have occurred. Except for nine telemetry sensors not essential to the flight, all systems were normal.[3]

Apollo 12 flew smoothly into a normal earth orbit, and after the inertial guidance system was realigned and all systems checked out, Houston gave the signal to fire the S-IVB stage for translunar insertion. Three and a half hours into the flight, command pilot Richard F. Gordon turned command module *Yankee Clipper* around, extracted the lunar module *Intrepid* from its stowage site atop the third stage, and *Apollo 12* continued on its way to the moon.[4]

The only midcourse correction maneuver of the outbound flight was performed the next day, a 9.2-second burn that put the spacecraft on a fuel-saving hybrid trajectory.[5] For the rest of the uneventful three-and-a-half-day trip to lunar orbit, the crew spent their time housekeeping, tending to spacecraft systems, and observing the earth and the moon.[6]

Arriving at the moon 83½ hours after liftoff, Conrad fired the main propulsion engine for almost 6 minutes to go into an elliptical lunar orbit. Five hours later a second burn put the spacecraft into a circular orbit at 60 nautical miles (111 kilometers) altitude, where *Yankee Clipper* would stay until it was time to return to earth. The spacecraft passed over and photographed Apollo 13's landing area in the Fra Mauro formation, and on the tenth revolution Conrad notified Capcom Gerald Carr that "you can tell good Captain Shaky [Jim Lovell, commander of Apollo 13] that he can relax. We've got his pictures."[7]

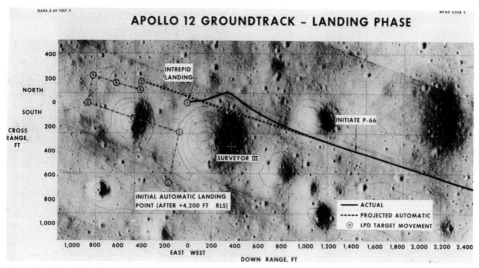

Actual and projected ground tracks for the Apollo 12 landing. Numbered circles indicate successive changes in the landing point designation made by Commander Conrad during the landing.

Six hours into the fourth day, Conrad and Bean prepared to enter and activate the lunar module. Both were having trouble with their biomedical sensors; Conrad's were blistering his skin and Bean's were producing erratic signals. Both men cleaned and reattached their electrodes, then finished donning their space suits and began preparing *Intrepid* for departure.[8]

For the next several hours Conrad and Bean in *Intrepid* and Gordon in *Yankee Clipper* were busy setting up their guidance and navigation computers and exchanging data with the ground. When all was ready, Gordon turned the spacecraft so that the long axis of the command and service module was perpendicular to the flight path with the lunar module outward from the moon, retracted the docking latches, and fired his attitude-control thrusters to move *Yankee Clipper* away from *Intrepid*.[9] The landing craft was 5 miles (8 kilometers) north of its intended ground track—largely as a result of an error in the landing site location and the inability to adequately correct for the moon's irregular gravity field. This and other errors would be removed by the instructions transmitted to the guidance computer after the lunar module headed down for its landing.[10]

On the back side of the moon in the 13th revolution, the computer triggered a 29-second firing of the descent engine, bringing the low point of *Intrepid's* orbit to 8.1 nautical miles (15 kilometers). As the lander passed north of Mare Nectaris, Conrad turned it on its back with the descent engine pointed along the flight path and switched the engine on to begin the final approach. Everything went exactly as expected, and after two minutes Conrad commented that it "feels good to be standing up in a g-field again." Three minutes later the module's attitude-control thrusters began firing busily—more than Conrad thought they should— but Houston assured him that all was well.[11]

After seven minutes *Intrepid* nosed over into a near-upright position and for the first time Conrad could see the lunar surface. The principal landmark identifying his landing point was a pattern of craters the astronauts called "Snowman"; *Surveyor III* lay halfway up the eastern wall of the crater that was the

Snowman's torso, and *Intrepid* was targeted for the center of the crater. As soon as he could see out the window, Conrad cried delightedly, "Hey, there it is [Snowman]! There it is! Son of a gun! Right down the middle of the road!" Then, as Bean called out altitude, velocity, and quantity of fuel remaining, Conrad maneuvered the craft with his hand controller to pick a smooth spot to land on. The engine exhaust began kicking up dust about a hundred feet (30 meters) above the surface and by the time *Intrepid* reached 50 feet (15 meters) the cloud obscured the surface completely. At 1:54:36 a.m. EST on November 20, Pete Conrad made a blind landing—exactly where, he could not tell, but certainly close to the intended spot.[12]

Conrad was naturally anxious to determine where he had set *Intrepid* down, and while he and Bean went through the post-landing check list they occasionally looked out the windows for landmarks that would allow Houston to pinpoint their location, but without success. After changing his mind a time or two, Conrad finally concluded, "I'm not sure that I'm not sitting right smack on the other side of the Surveyor crater, just a little bit past it." Two hours later, Dick Gordon in *Yankee Clipper* confirmed Conrad's guess when he sighted both *Intrepid* and *Surveyor III* through his sextant as he passed over the site. He told Houston that the lunar module was "on the left shoulder of the Snowman . . . , about a third of the way from the Surveyor crater to the head.[13] Postmission calcula-

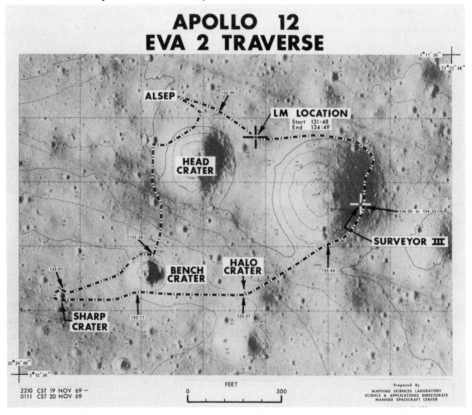

Apollo 12's second traverse, made by Pete Conrad and Alan Bean. "Head" crater was the head of the "Snowman," the pattern of craters used by Conrad to locate his landing site.

tions placed *Intrepid* on the northwest rim of the Surveyor crater, 535 feet (163 meters) from the inert spacecraft.[14] Had there been windows in the back of the lunar module, Conrad could have spotted the Surveyor as soon as the dust settled.

After postlanding checks of systems, Conrad and Bean described what they could see from their spacecraft. *Intrepid* had landed in undulating terrain pocked with craters ranging from a few feet to several hundred feet across, the larger ones rimmed by large blocks of rock. Numerous boulders, up to 20 feet (6 meters) in size, were scattered around the site, most of them angular rather than rounded, many showing fillets of dust around the base. Immediately in front of the landing craft Bean saw an area of "patterned ground"—parallel cracks in the surface soil perhaps an eighth of an inch (3 millimeters) deep. From the lunar module the crew could distinguish no color differences in the rocks or soil; everything seemed the same bright white.[15]

Five and a half hours after landing, Conrad squeezed out the hatch, then clambered down the ladder to the bottom rung. As he stepped off onto the landing

Manned and unmanned lunar exploration meet: lunar module Intrepid, *extreme upper left, rests on the rim of the crater in which* Surveyor III, *right center, landed three years earlier. Apollo 12 astronauts Pete Conrad and Alan Bean took this picture toward the end of their second surface traverse. They photographed the spacecraft and cut off several pieces of its hardware to return to earth for study.*

Apollo 12 Commander Pete Conrad inspects the Surveyor III *spacecraft. Two solar panels at the top of the unmanned spacecraft provided electricity. Projecting in front of Conrad is the Surveyor's surface sampler, extended on its lazy-tongs mechanism. Immediately to Conrad's right is the mirror for Surveyor's television camera, which took hundreds of photographs of the surface. The lunar module* Intrepid, *with its high-gain antenna deployed on the surface, is on the lip of the Surveyor crater, upper right.*

pad Conrad remarked, "Man, that [step] may have been a small one for Neil, but that's a long one for me."* Looking around, he spotted the Surveyor halfway up the opposite wall of the crater. One of the first things Conrad noticed was that he was going to get extremely dirty: the surface dust was finer and deeper than he had expected.[16]

After Conrad had collected the contingency sample Bean joined him on the surface, bringing the television camera with him. A few minutes later Houston reported that the camera was not working. Cursory attempts at trouble-shooting were fruitless, and television coverage for the mission—desirable but not essential—had to be written off.**[17] The explorers pressed on with their other

---

*At 5 feet 6 inches (168 centimeters), Conrad was one of the shorter astronauts; Armstrong was just under 6 feet (183 centimeters). The lowest rung of the ladder was about 2.5 feet (75 centimeters) above the landing pad.

**Later examination determined that the image-tube target in the camera had been damaged by exposure to intense light. Apparently Bean had inadvertently pointed the camera at the sun or a reflection off the lunar module while helping Conrad set up a directional antenna. MSC-01855, "Apollo 12 Mission Report," March 1970, p. 14-50.

chores, apparently enjoying themselves immensely; Conrad chuckled and hummed to himself as he went about examining the lunar module, collecting and photographing samples, and describing the landscape.[18]

The primary objective of their first excursion was to deploy the scientific experiments. Conrad and Bean unloaded the package easily, picked a spot 130 meters (425 feet) northwest of the lander, and laid out the instruments without any serious difficulty.[19] On their way back to *Intrepid* they picked up more documented samples and Bean took a soil sample with the core tube. After nearly four hours on the surface, the astronauts returned to the lunar module, dusted each other off as best they could without brush or vacuum cleaner, and climbed back inside. After a brief evaluation of the day's work and some discussion of the next day's plans, Houston signed off and the two astronauts strung up their hammocks and turned in.[20]

Overnight the geologists in Houston, working from several scenarios prepared before the mission, formulated detailed plans for the geology exercise the astronauts would carry out on their second outing. Next morning, when Bean and Conrad were awake and ready to talk, CapCom Paul Weitz read through the plans with them, discussing the scientists' recommendations for sampling and photography, while the astronauts offered comments based on the previous day's experience. Then they donned and checked over their space suits, recharged with oxygen their portable life-support systems, vented the cabin, and set out to collect samples.[21]

During the next four hours Bean and Conrad covered something more than one kilometer (3,300 feet), following a large-scale photographic map prepared for the traverse and chatting constantly with each other and with Houston.[22] Scientists in Houston followed their progress from their references to the maps they carried. The extensive conversation was intended to substitute for a geologist's field notes.[23] On the nearly featureless lunar surface, sampling proved somewhat difficult; except for size, most of the rocks showed few distinguishing features. Colors and textures were not always easy to determine, and when they were, the astronauts tended to use nonscientific terms in describing them—probably a symptom of their sensitivity to possible misuse of geological terminology. At one point Conrad noted a rock containing a "ginger-ale-bottle green" crystal (which was probably olivine) , and a few minutes later Bean spotted a rock he said "looks almost like a granite," but immediately added, "of course it probably isn't, but it has the same sort of texture." Bean and Conrad documented their samples carefully, photographing many of them and describing the location and bag number for later reference. At "Head" crater, following instructions from Houston, Conrad dislodged a medium-sized rock and allowed it to roll down the slope to determine whether the seismometer some 70 meters (230 feet) away could detect it (it did). Then it was on to "Bench" and "Sharp" craters, where they sampled several large rocks on the surface that might be bedrock thrown out when the craters were formed.[24]

Two hours into the traverse the astronauts were on the edge of the Surveyor crater. The slope of the crater wall was much less than it had appeared the day before, when the low sun cast long, deceptive shadows. Pausing to reload a camera and survey the situation, they decided it would be easier to walk across the slope of the crater wall rather than come down from the rim. Approaching from the side, they could take photographs of the landing-pad imprints and the trenches dug by Surveyor's remotely controlled arm without disturbing the surface. These would be compared with television pictures transmitted from the spacecraft immediately after its landing.[25]

Examination of the Surveyor, the only human artifact ever encountered in lunar exploration, was among the more interesting parts of the mission for Conrad and Bean. What they noticed first was that much of its originally white surface had turned brown—a change they attributed to a deposit of dust when they found it could be wiped off. After photographing the surrounding surface and examining the spacecraft, they removed the Surveyor's television camera and cut off pieces of electrical cable and structural tubing for study by scientists at the Jet Propulsion Laboratory. They decided to remove the trenching scoop as well; then, after collecting soil samples, they headed for another small crater to take a few more samples on their way back to the lunar module.[26]

One of several intriguing mounds in the Apollo 12 landing area in Oceanus Procellarum, reported but not thoroughly investigated by astronauts Alan Bean and Pete Conrad.

Back in their spacecraft, Conrad and Bean had almost six hours before lifting off to rendezvous with *Yankee Clipper*. After throwing out their portable life-support systems they straightened up the cabin, stowing the rock boxes and improvising stowage for the television camera, which Houston wanted to examine. They had carried in considerably more lunar dust than Armstrong and Aldrin had reported; Conrad told Houston they looked like "a couple of bituminous coal miners right at the moment, but we're happy."[27]

Despite the fact that they had trebled the existing record for lunar surface activity, Bean and Conrad were not exhausted and had expended about 10 percent less energy than anticipated. Both returned to the lunar module with almost 40 percent of their oxygen supply remaining on both excursions.[28] Bean suggested to the medical officers that he would have enjoyed an occasional drink of water while working on the surface; not that he felt dehydrated, but he would have been more comfortable if he had been able to relieve the dryness in his mouth.[29]

Intrepid's ascent stage lifted off the moon on time, and an hour and a half later Conrad had *Yankee Clipper* in sight. Back in lunar orbit the dust the lunar explorers had brought in with them began to float, thick enough to be visible in the cabin. After the two spacecraft had docked they attempted to vacuum up the dust, with little success, so they removed and packaged their suits in the lunar module, hoping to minimize contamination of the command module. In spite of their efforts, considerable dust clung to everything they brought back and remained suspended

An eclipse of the sun such as few have ever seen. On earth, a solar eclipse is produced by the moon; as the Apollo 12 spacecraft moved into the earth's shadow on the return trip, the earth eclipsed the sun.

in the atmosphere; the environmental control system seemed not to filter it out as completely as had been expected.[30]

Intrepid, now a useless hulk, still had one more contribution to make to the scientific objectives of the mission. For the benefit of seismologists wanting to calibrate the instrument that Bean and Conrad had just left on the moon, Mission Control now burned the empty spacecraft's remaining fuel to take it out of orbit. At a speed of 1.67 kilometers per second (3,735 miles per hour) the ascent stage plowed into the moon 76 kilometers (47 miles) east-southeast of the instrument package, producing a bizarre response: the seismometer recorded vibrations that persisted almost undiminished for nearly an hour. It was so completely unlike anything ever seen on earth that seismologists had no immediate explanation. One scientist compared the result to striking a church bell and hearing the reverberations for 30 minutes.[31]

Yankee Clipper stayed in lunar orbit for 11 more revolutions, finishing up the "bootstrap" photography and landmark tracking, looking at sites being considered for Apollo 14 and 15. Then the crew boosted their spacecraft out of lunar orbit and settled in for the three-day voyage home. Now and then they chatted with the duty CapCom about something that crossed their minds concerning their exploration. Once in a while scientists wanted to debrief them concerning details of their lunar surface activity. For most of the time, however, they relaxed.[32]

At one point both Conrad and Bean passed along some reservations they had about the field geology exercise. They had found the lunar surface a particularly difficult one for classical field-geological techniques. Bean commented,

> . . . we talked with [the geologists], before we went, about [the fact that] the main objectives of the geology wasn't to go out and grab a few rocks and take some pictures, but to try to understand the morphology and the stratigraphy . . . of the vicinity you were in. Look around and try to use your head along those lines. Well, I'll tell you, there [were] less than ten times I stood in spots . . . and said, "Okay now, Bean, . . . is it possible to look out there and try to determine . . . which is first, which is second, and all that?" . . . That whole area has been acted on by these meteoroids or something else so that all these features that are normally neat clues to you on earth are not available for observation.

Conrad concurred: "I think even a trained geologist would have trouble doing a whole lot of field geology that way on the moon."[33] Long afterward, Bean still felt that astronauts could be most effective on the lunar surface by selecting and documenting as many apparently different kinds of samples as possible rather than attempting on-the-spot geologizing.[34] During a press conference on the last day out, Conrad was asked whether he thought it would be desirable to have a geologist as a member of the crew. He did indeed think so, but pointed out that landing *Intrepid* had required all the piloting skill he had[35]—implicitly echoing Deke Slayton's contention that landing on the moon was not yet a job for a novice.

Early on the morning of November 24 *Apollo 12* splashed down some 600 kilometers (375 miles) east of Pago Pago, 3.5 kilometers (2 miles) from the recovery ship U.S.S. *Hornet*. The landing was rough—apparently *Yankee Clipper* hit a rising wave as it swung on its parachutes—hard enough to dislodge a 16-mm movie camera from its bracket and slam it into Alan Bean's forehead, momentarily stunning him and opening a 1-inch (2.5-centimeter) cut, which Conrad bandaged.[36]

The recovery swimmers soon arrived, tossed respirators and coveralls—replacing the biological isolation garments that the Apollo 11 crew had found so objectionable—into the command module, then assisted the astronauts into the raft. Half an hour later recovery helicopters set the crew down aboard the *Hornet* and they went straight to their mobile quarantine facility. Lunar sample containers and film magazines were removed and flown to Pago Pago and thence to Houston. The astronauts had a longer journey: four days aboard ship to Hawaii, then a nine-hour flight to Houston. On the morning of November 29, Conrad, Bean, and Gordon entered the Lunar Receiving Laboratory for their 11-day stay in quarantine.[37]

That Apollo 12 was a success was apparent even on preliminary evaluation. The procedural changes incorporated to improve landing accuracy had allowed Conrad to put *Intrepid* down within sight of *Surveyor III*, exactly as intended. Lunar exploration had been easy; neither Bean nor Conrad encountered any unexpected difficulties and both had oxygen to spare when they returned. And while they

The last leg of a half-million-mile journey: in the early morning light of November 29, 1969, the Apollo 12 mobile quarantine facility is trucked through the main entrance to the Manned Spacecraft Center on its way to the lunar receiving laboratory and quarantine.

had found it hard to apply their field-geology training on the unrevealing surface of Oceanus Procellarum, they had collected nearly 75 pounds (34 kilograms) of samples, most of them documented. The surface experiments they had set up were returning streams of data, and scientists agreed the astronauts had done a remarkable job. Communication between scientists in Houston and the astronauts on the moon had been well handled by Mission Control.[38] If Apollo 12 was a reliable indicator, the scientific return from the remaining eight missions should be gratifying.

Lunar Receiving Laboratory Operations

The modifications to equipment and procedures (see Chapter 10) necessitated by experience with Apollo 11 were completed and the Lunar Receiving Laboratory was ready for operations before Apollo 12 left the launch pad.[39] When the sample containers arrived in Houston, the one containing the documented samples was transferred to the biological laboratory to be opened under sterile nitrogen, while the bulk sample container was opened in the vacuum system. The first quick look showed that Conrad and Bean had selected larger rocks than Armstrong and Aldrin did, and that there were noticeable differences: the Apollo 12 samples were almost entirely igneous, whereas the earlier specimens had contained about 75 percent breccia (rocks formed by compaction without alteration).[40] Still more samples arrived with the astronauts on November 29. Lacking room in the sample return containers, Conrad had brought back several large rocks as well as the Surveyor parts in the command module.[41]

After samples were taken for biological testing,[42] the Preliminary Examination Team began its work. This time they concentrated on description and photographic documentation of the contents of the rock boxes, saving detailed examination until later, so that the Lunar Sample Analysis Planning Team could begin its work earlier. During this preliminary examination a small cut appeared in one of the chamber gloves. The incident was treated as a "spill" of lunar material that contaminated the laboratory and sent 11 people into quarantine, including several scientists from the Preliminary Examination Team.[43] Providing quarters for that many additional people presented a problem, but by putting extra beds in the available rooms and using the Mobile Quarantine Facility (which could be connected with the Crew Reception Area), all were accommodated. The scientists were disappointed in being cut off from work with the samples, but preliminary examination continued without them.[44]

Other than the fault in the chamber glove, operations in the receiving laboratory went smoothly. A week after the samples were returned, all had been inventoried and given preliminary examination, the sample for biological testing had been prepared, and two samples had been delivered to the radiation counting laboratory for assay of low-level radioactivity.[45]

By mid-December scientists in the Lunar Receiving Laboratory had determined some of the basic characteristics of the Apollo 12 samples. Some members of the site selection board had objected to the Apollo 12 site on the grounds that it was too much like Apollo 11's (see Chapter 10), but these analyses showed otherwise. Chemically, the rocks from both sites were similar, containing the same elements but in somewhat different proportions. Only two rocks from Apollo 12 were breccias, however, which predominated in the Apollo 11 samples. Many of the igneous rocks returned by Apollo 12 were very coarse grained, suggesting that they might have crystallized slowly, probably at considerable depth. While the Apollo 11 samples had been notably rich in titanium, those from Apollo 12 were not. Both types were deficient in volatile elements. Olivine comprised only about 10 percent of the first samples but was abundant in the second set.[46] Petrological examination indicated that the Apollo 12 rocks had crystallized at a high temperature.[47] Preliminary potassium-argon dating suggested that the samples from Oceanus Procellarum had solidified some 2.2 to 2.6 billion years ago, compared to about 4 billion years for those from the Tranquility site.[48] Detailed results from the work of principal investigators would not be available for some months, but these preliminary findings provoked considerable discussion at the first Lunar Science Conference the following month.

Elsewhere in the Lunar Receiving Laboratory, the crew and their companions in confinement had an uneventful three weeks. Debriefings by engineers and scientists took up several hours a day for the first week; preparing the pilot's report required much of the next, and they were grateful to have the uniterrupted time to complete this chore. Visits by their families—through the glass wall in the conference room—plus television and movies and the occasional game of pool or Ping-Pong, provided some diversion. Conrad worked on assembling an FM receiver kit but did not complete it. The scientists managed to get some samples passed in to them in the crew reception area and were able to continue work on the samples to a limited extent. They also got the chance to discuss the lunar traverses with Conrad and Bean, using photographs as they became available.[49]

Periodic medical evaluations showed no change in the condition of either the crews or the quarantined scientists, and on December 10, 36 hours before the quarantine officially ended, they were released, to be kept under medical surveillance until biological tests on the samples showed them to be no threat to life on earth.[50]

Throughout December those tests continued as the Apollo 12 samples were exposed to dozens of plant and animal species. Meanwhile the examination and characterization of the lunar rocks and soil continued, and the Lunar Sample Analysis Planning Team began the task of matching investigators' requirements with the samples. Release of material to outside researchers was scheduled as soon as the biological tests were completed, expected in early January 1970.

The First Lunar Science Conference

The decade of Apollo ended with investigators of the lunar samples preparing to present their first findings to the scientific world. On January 5, 1970, several hundred scientists, including all 142 principal investigators for the Apollo 11 samples, met in Houston for the first of many annual Lunar Science Conferences. For four days they discussed the chemistry, mineralogy, and petrology of the lunar samples, described the search for carbon compounds and evidence of living material, the results of age determinations, and what had been learned from the data returned from the surface instruments.

The flood of information presented at the conference precluded any large generalizations. Some of the rocks brought back by Apollo 11 were basalts, formed by melting, but the samples contained a high percentage of breccias. Most of the rock fragments were similar to the larger rocks; a small number, however, were entirely unlike the rest and may have come from the nearby lunar highlands. The properties of most of the material were consistent with formation at high temperature in the absence of oxygen or water. More than 20 minerals known on earth had been identified, suggesting a similarity of origin for the two planets, but at least three new ones unique to the moon had been discovered. The ages of the lunar rocks and soil seemed to differ: the basalts from Tranquility Base were found to be between 3 and 4 billion years old, while the soil contained particles believed to have been formed 4.6 billion years ago.[51] In light of the recently determined age of the Apollo 12 samples, which were roughly a billion years younger, these numbers indicated that more than one cataclysmic event had shaped the lunar surface.

Other studies showed that some lunar material dug up from below the surface had once been on the surface, indicating that the lunar soil was undergoing constant "gardening." Studies of isotopes produced by cosmic-ray bombardment indicated that some lunar rocks had been on or just below the surface for at least 10 million years.[52]

The search for carbon compounds—possible relics of life forms in the moon rocks—was inconclusive, in part because the methods of detection were extremely sensitive. Carbon and some carbon compounds were detected, but no molecules that could clearly be identified as derived from living organisms. The intensive search for living organisms conducted in the Lunar Receiving Laboratory was negative, as were efforts to identify fossil microorganisms.[53]

On the whole the early work on the Apollo 11 samples was more intriguing than revealing. A few scientists were willing, on the basis of the results presented at the Houston conference, to discard one or more theories of lunar origin, but most were not. One commented, "some still like the moon hot; some like it cold." Many called attention to the limitations of the data and the futility of drawing sweeping conclusions with information from such a small area. One of them put it in perspective with the comment, "what I looked at was an area equal to the size of three postage stamps and a thimbleful of material"; another remarked facetiously, "our results are based on the wanton destruction of two grams [about the weight of a dime] of lunar material."[54]

After four days of discussion, a panel of prominent scientists attempted a summary at a news conference. Robert Jastrow, director of NASA's Goddard Institute for Space Studies, said that "old descriptions in terms of a hot or cold moon are an oversimplification." To Edward Anders of the University of Chicago it was clear that the moon "has been a geologically active body with several episodes of melting and lava flows." For the last two or three billion years, however, it has been almost dead. Jastrow and others doubted that the moon was thrown off from the earth or that it was formed elsewhere and captured later. Gene Simmons, newly appointed chief scientist at MSC, summed up the state of lunar science at the end of the conference. "There is a large amount of undigested data and very little interpretation," he said. "You'll see in the next six months many revisions of statements as to what it all means."[55]

The Lunar Sample Analysis Planning Team best characterized the significance of the Tranquility samples:

> The Apollo 11 samples were collected from a very tiny fraction of the moon's surface. Nevertheless, they have given a vast new insight into the processes that have shaped this surface and have established some significant limits on the rates and mechanisms by which it evolved. The results reported do not resolve the problem of the origin of the moon. However, the number of constraints that must be met by any theory have been greatly increased. For example, if the moon formed from the earth, it can now be stated with some confidence that this separation took place prior to 4.3×10^9 [4.3 billion] years ago. Furthermore, such a hypothesis must now take account of certain definite differences in chemical composition [between the moon and the earth]. . . . There is clear evidence . . . that the surface of the moon is variable in both composition and age. *It is therefore of great scientific importance to obtain materials from a variety of terrains and sites* [emphasis added].[56]

After the conference ended on January 8 the scientists returned to their labs to continue probing the lunar samples, while NASA focused on sampling the variety of terrains and sites they deemed so important. The next mission, Apollo 13, was already targeted for the Fra Mauro Formation,[57] a geologic unit that covered large portions of the lunar surface around Mare Imbrium. The landing site was just north of the crater Fra Mauro, about 180 kilometers (112 miles) due east of the spot where Pete Conrad had landed *Intrepid*.

MSC Increases Emphasis on Science in Apollo

Early in 1970 MSC set about to improve its relations with the lunar science community. As the first lunar science conference was meeting in Houston, Jim McDivitt, manager of MSC's Apollo spacecraft program office, took note of scientists' frequent complaints that MSC was unable or unwilling to accommodate changes in the experiment program. "I would like to take steps to change this impression and to attempt to generate a 'can-do' attitude toward science changes consistent with operational and other constraints," McDivitt said. He intended to establish a schedule for each mission, "to provide information to the science world which will discipline their inputs to our schedule needs," and to establish an Experiments Review Group, which was to consider new or late experiments for the missions to recommend MSC policy on experiment changes. McDivitt himself would chair the group, which included Anthony J. Calio, director of science and applications, Deke Slayton, director of flight crew operations, and Richard S. Johnston, former special assistant to the center director now assigned to McDivitt's office.[58]

The first lunar science conference apparently brought the question of MSC's handling of science to a head, for Director Robert R. Gilruth moved to change his center's antiscientific image at the same time. On February 5, 1970, he and 11 of MSC's highest-ranking managers met with 9 prominent scientists involved in the Apollo program to discuss issues between the two groups.* Several specific problems were discussed and resolved. One major flaw the scientists found in the conduct of Apollo science involved contact between astronauts, principal investigators, and operations planners. Since there was not time to make non-scientists astronauts over into scientists, the investigators insisted on giving experimenters more time to discuss the objectives of the experiment with the astronauts, thus improving the astronauts' understanding of what they could do to enhance the results. Geologists had enjoyed an advantage in this respect, but the experimenters pointed out that other disciplines were involved in lunar investigations and should get more attention in training. Another point involved communication between scientists on the ground and astronauts on the moon; this ought to be made easier, the scientists said, so that surprises could be dealt with when encountered and procedures altered in real time if necessary. When it was pointed out that this would require the principal investigators to learn how to use the existing systems to best advantage, including participation in simulations, the scientists agreed to take on that responsibility.[59]

Finally, participants agreed to clarify relations between outside scientists and MSC offices. In general experimenters should be more forthcoming with explanations of the scientific rationale for experiments and MSC should simplify and expedite reviews of proposed experiments and changes in existing plans. A general improvement in communication between MSC, the scientists, and Headquarters was necessary.[60]

Associate Administrator Homer E. Newell, who attended the meeting, later commented approvingly to Gilruth that "the entire discussion was in a constructive vein," and that "some very complimentary remarks were made . . . about MSC's current approach to handling science." As associate administrator for science and applications, Newell had for years tried to mediate between the scientists and manned space flight officials, and he found the new attitudes gratifying on both sides.[61]

MSC immediately followed up on the February meeting by establishing a Science Working Panel to be the single forum in which science requirements and operational restrictions would be reconciled and adjusted. The working panel would be advised by principal investigators and scientists representing the entire spectrum of science. A science mission manager and a mission scientist (usually

---

*MSC was represented by Gilruth; Deputy Director Christopher C. Kraft, Jr.; George Abbey, technical assistant to Gilruth; Anthony J. Calio, director of Science and Applications; M. Gene Simmons, chief scientist; Donald K. Slayton, director of Flight Crew Operations; Thomas P. Stafford, chief of the Astronaut Office; Harrison H. Schmitt, astronaut; John Zarcaro, chief of the Lunar Missions Office; Ted H. Foss, Geology and Geochemistry Branch; Glynn S. Lunney, Flight Control Division; and Richard S. Johnston, Apollo Spacecraft Program Office. Scientists attending—most of them principal investigators on Apollo—were James R. Arnold, Univ. of California at San Diego; Paul W. Gast, Columbia Univ.; Marcus G. Langseth, Jr., Lamont-Doherty Geophysical Observatory; Frank Press, Mass. Inst. of Technology; Robert Rex, Univ. of California at Riverside; Gordon A. Swann, U.S. Geological Survey; Robert Walker, Washington Univ., St. Louis; and Gerald J. Wasserburg, Calif. Inst. of Technology. Homer E. Newell, NASA Associate Administrator, attended as an observer.

a scientist-astronaut) would be assigned for each flight.[62] When scientists briefed astronauts on the conduct of their experiments, a representative of MSC's Flight Control Division would sit in as well, to ensure complete understanding all around.[63]

The February meeting and its consequences were a considerable relief to scientists involved in Apollo. Newell later commented that "the experimenters' feeling of effectiveness increased steadily with each new Apollo mission until with Apollo 17 . . . the scientists were positively ecstatic."[64]

Personnel and Program Changes

The last months of 1969 brought changes in several key offices in NASA Headquarters and the centers. Thomas O. Paine, who had succeeded James Webb as NASA administrator less than a year earlier, called George M. Low to Headquarters to become deputy administrator, the agency's second-ranking official.* After guiding Apollo through its most difficult years and at long last establishing the first post-Apollo project (Skylab), George Mueller announced his intent to resign as Associate Administrator for Manned Space Flight at the end of the year. The Air Force reassigned Lt. Gen. Sam Phillips, longtime manager of the Apollo Program Office under Mueller, to command its Space and Missile Systems Organization. As Mueller's replacement Paine named Dale D. Myers, Apollo spacecraft manager at North American Rockwell for many years; Phillips's place was filled by Rocco A. Petrone, director of launch operations at Kennedy Space Center. At Houston, Gilruth appointed Col. James A. McDivitt, command pilot on Gemini IV and commander of Apollo 9, to head the Apollo spacecraft project office and Christopher C. Kraft, Jr., MSC's director of flight operations, to be deputy center director.[65] MSC also had a new Director of Science and Applications, Anthony J. Calio, who succeeded Wilmot Hess after Hess's resignation in the fall (see Chapter 10).

New managers faced new problems in the year following Apollo 11's resounding success. Public enthusiasm for lunar missions waned when, with Apollo 12, they began to seem routine. Funding prospects were also bleak. Lyndon Johnson's last budget submittal (in January 1969) requested $3.878 billion for NASA,[66] nearly 25 percent lower than the budget for the peak year, fiscal 1965. When Richard Nixon's first budget was submitted in April, that figure was reduced to $3.833 billion. Paine put the best face he could on the situation; the reductions would require "difficult program adjustments," but in the context of the administration's determination to reduce government spending, "the nation can continue a scientifically effective program of manned lunar exploration" and the capability to produce Saturn V boosters would not lapse beyond recovery.[67] When the space agency's appropriation bill was signed in November, it provided only $3.697 billion, and the adjustments became even more difficult.[68]

---

*At the time Low, who took over as manager of MSC's Apollo spacecraft project office after the AS-204 fire in April 1967, was serving as special assistant to MSC Director Robert R. Gilruth.

Besides being economy minded, the new administration was in no hurry to establish a position on space. Early in 1969 the new President appointed a Space Task Group* to study the space program, calling for a report in six months on alternatives for the post-Apollo period. Predictably, the group's report, submitted on September 15, recommended a balanced program of manned and unmanned space activity. Its most radical suggestion was that NASA should adopt a new long-range goal, comparable to the Apollo goal that had sustained space exploration for eight years, to provide the impetus for new developments. For that goal they suggested manned exploration of the planets, specifically a manned landing on Mars by the end of the 20th century. Three options were proposed: an all-out effort, including a 50-man earth-orbiting space station and a lunar base, culminating with the Mars landing in the mid-1980s; a less ambitious program providing for evaluation of an unmanned Mars landing before setting a date for the manned mission; and a minimum program that would develop a space station and a shuttle vehicle but would defer the Mars landing to some unspecified time before the end of the century. Costs were estimated at between $8 billion and $10 billion per year by 1980 for the most ambitious option and from $4 billion to $5.7 billion annually by 1976 for the least.[69]

Nixon's reaction to the Space Task Group report was not immediately forthcoming. His press secretary declined to predict when the President would make a decision, but said that competing domestic programs and the constraints imposed by inflation would certainly have to be considered in funding any new space ventures.[70]

Public reaction to the proposal of a manned flight to Mars was generally negative. The *Washington Post* suggested that such a trip should be weighed carefully in terms of its potential scientific value: "It is knowledge we seek, not spectaculars. . . ."[71] A point of view shared by many scientists was expressed at the annual meeting of the American Association for the Advancement of Science (AAAS) in December by Gordon J. F. MacDonald, who characterized it as "the utmost folly." Retiring AAAS president Walter Orr Roberts said that the United States should not set a goal of sending men to Mars "now or ever."[72] Even congressional supporters of manned space flight found the proposal unacceptable. The chairman of the House Science and Astronautics Committee told the House that he was unwilling to commit the nation to any specific timetable for sending men to Mars, and the chairman of the Manned Space Flight subcommittee likewise shied away from an endorsement.[73]

Perhaps the thumbs-down reaction to a Mars flight was to be expected, given the projected cost of such a program, which the Space Task Group estimated at $54.1 billion to $78.2 billion during the 1970–1980 decade. Somewhat surprising was that even Apollo, hitherto all but inviolable, was not to escape the effects of NASA's straitening circumstances. On January 4, 1970, following dedication of the Lunar Science Institute at Houston, deputy administrator George Low

---

*Chaired by Vice-President Spiro T. Agnew, the group included Robert C. Seamans, Secretary of the Air Force and former deputy administrator of NASA; Thomas O. Paine, NASA Administrator; and Lee A. Dubridge, Science Adviser to the President. Observers were U. Alexis Johnson, Undersecretary of State for Political Affairs; Glenn T. Seaborg, chairman of the Atomic Energy Commission; and Robert P. Mayo, director of the Bureau of the Budget.

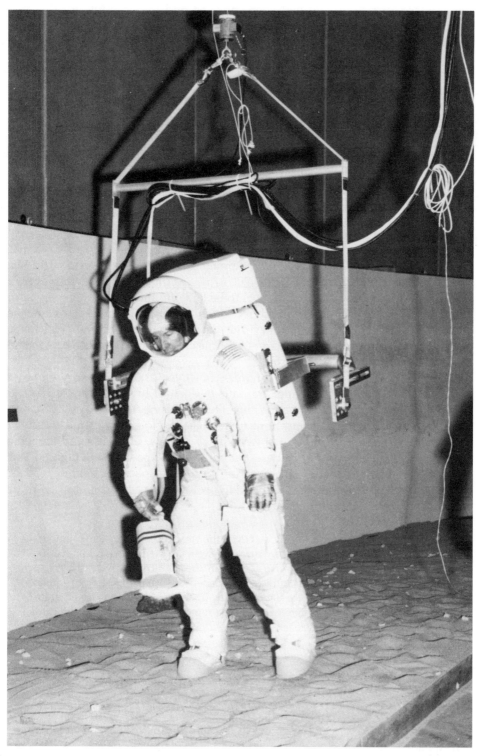

James Lovell, Apollo 13 commander, practices moonwalking in a device that offsets ⅚ of his weight.

announced that Apollo 20 had been canceled and the schedule for the seven remaining flights would be stretched out into 1974. Four would be flown in 1970-1971 at intervals depending on the choice of sites. Lunar exploration would then be interrupted while the three-mission Apollo Applications Program (a rudimentary space station in earth orbit, soon to be renamed "Skylab") was conducted. The last three missions to the moon would be flown in 1973 and 1974. Low denied reports that NASA planned to cancel four Apollo flights, saying that such action would "do away with most of our scientific return and waste the investment we have made." Lunar scientists were reported to have been pleased by Low's announcement;[74] they had evidently feared that even more missions would be deleted.

Ten days later, after preliminary discussions on the fiscal 1971 budget, administrator Thomas O. Paine revealed more changes in space exploration. Saturn V launch vehicle production was to be suspended indefinitely after the fifteenth booster was completed, leaving NASA with no means of putting really large payloads into earth orbit or continuing lunar exploration. The last Saturn V was reassigned from Apollo 20 to Skylab. Unmanned explorations of Mercury and Mars were reduced or deferred. Some 50,000 of the estimated 190,000 employees of NASA and its contractors would have to be laid off, and many university scientists would find their projects without funds. Though the new plans imposed real austerity, Paine noted that they did provide for a start on the next project, development of a reusable spacecraft to shuttle crews and payloads between earth and a space station in earth orbit.[75]

Apollo still had supporters in Congress, however, and they tried their best to add $130.5 million to the administration's budget for lunar exploration in fiscal 1971. But the Senate would not go along, and after a vigorous debate, the conference committee reported an authorization bill containing an increase of only $38 million for Apollo.[76] After the first two lunar landings, Congress and the nation were ready to get on to other things—but nothing so expensive as a flight to Mars. A congressional historian who served on the House space committee described the dawn of the 1970s as "the worst of times for the space program," and then summarized the budget debates:

> By hindsight, it seems unlikely that even the strongest and most adept mobilization of the supporters of more manned flights to the Moon could have successfully overcome the adverse feeling in the country in the early 1970's. Congress and the Nation could be persuaded to support Skylab, the Space Shuttle, and a modest level of activity by NASA in many other areas. But . . . Von Braun's dream of a manned flight to Mars was not in the cards for the 20th century, at least.[77]

Mission to Fra Mauro

Whether any Americans were going to Mars in this century or not, the crew of Apollo 13—Jim Lovell, Ken Mattingly, and Fred Haise—were going to the moon, and they were too busy preparing for that mission to worry about anything more

Apollo 13 astronauts James Lovell (left) and Fred Haise in training at Kennedy Space Center. This photo gives an excellent view of the lunar module hatch (square opening above the ladder), through which the suited astronauts had to crawl backwards, and the height of the ladder above the lunar surface—which prompted Pete Conrad to remark on Apollo 12, "that may have been a small [step] for Neil, but it was a big one for me."

spectacular. Their Saturn V was trundled out to launch complex 39A on December 15, 1969, anticipating launch on March 12, 1970.[78] On January 8, however, Headquarters announced that Apollo 13 had been rescheduled for April 11 to allow more detailed analysis of specific plans.[79]

The target for Apollo 13 was a spot some 180 kilometers (112 miles) east of Apollo 12's landing site, just north of the crater Fra Mauro. The mission's primary task was to sample the geologic unit called the Fra Mauro Formation, which covers a large area around the Imbrium basin. The Fra Mauro Formation was generally believed to consist of material ejected when Imbrium, one of the largest impact basins on the moon, was formed. Scientists hoped that samples would allow them to date the "Imbrian event," which would establish the time relationships of related and adjacent features. Finally, they expected that the Fra Mauro site would provide samples that came from deep within the moon, excavated by the Imbrian event.[80] The mission would also emplace a second set of lunar surface instruments, differing from those flown on Apollo 12; a heat-flow experiment and a charged-particle environment detector were substituted for the solar wind spectrometer, the magnetometer, and the suprathermal ion detector. A second calibration of the lunar seismometers was to be accomplished by crashing the Saturn V's third stage at a preselected location on the moon.[81]

Preparations for the launch proceeded without major problems through February and March. Three days before launch, Jack Swigert took Ken Mattingly's place as command module pilot, after the prime crew had been exposed to German

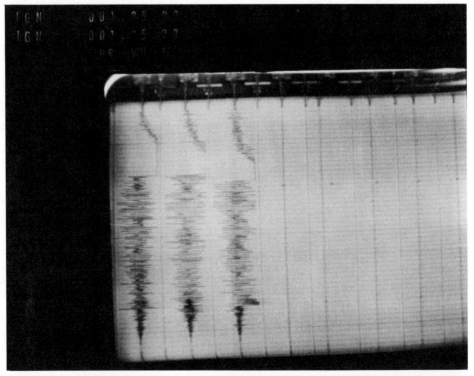

Apollo 13's only contribution to lunar science: a seismometer in Mission Control recorded these tracings after the Saturn S-IVB stage struck the moon.

measles and Mattingly was found to have no immunity. Apollo 13 lifted off on schedule at 2:13 p.m. Eastern Standard Time on April 11, and for two days operations were routine; Capcom remarked at one point that flight controllers were bored to tears. A few hours later, however, an oxygen tank in the service module ruptured, depriving the spacecraft of most of its electrical power and oxygen, and the mission had to be aborted. Only a heroic effort of real-time improvisation by mission operations teams saved the crew.[82]

Almost lost in the drama of the mission was the one piece of scientific information that Apollo 13 was able to provide. Shortly after command module pilot Jack Swigert had extracted the lunar module from atop the S-IVB stage, ground controllers fired the auxiliary propulsion system on the big rocket, putting it on a course to crash into the moon. Three days later the 30,700-pound (13,925 kilogram) hulk struck the lunar surface at 5,600 miles per hour (2.5 kilometers per second) some 74 miles (119 kilometers) west-northwest of the Apollo 12 landing site, releasing energy estimated as equivalent to the explosion of 7.7 tons (7,000 kilograms) of TNT. Half a minute later the passive seismometer left by Apollo 12 recorded the onset of vibrations that persisted for more than four hours. Another instrument, the lunar ionosphere detector, sensed a gas cloud that arrived a few seconds before the seismic signal and lasted for more than a minute. Seismologists were baffled by the moon's response to shock, but welcomed the new means of generating data.[83]

In the year that it took to discover and correct the cause of the Apollo 13 failure, the scope of the remaining missions was altered. Apollo 14 would visit the site intended for exploration by Apollo 13, but it would go there as the last of the intermediate exploration missions. The flights that remained would stretch the capabilities of the Apollo systems virtually to their limits, providing longer visits to the lunar surface and increased mobility for the astronauts. Unfortunately for lunar science, two missions were to be cut from an already minimal program.

12

APOLLO ASSUMES ITS FINAL FORM:
1970–1971

As budgets tightened and public support for lunar missions faltered in 1970, NASA managers faced tough choices. On the one hand, the successful Apollo exploration missions had given lunar scientists an appetite for more and more samples—more, in fact, than even the six missions remaining on the schedule were likely to provide. Against that stood the need to get moving on post-Apollo programs, the first of which (Skylab) had reached a stage where it required substantial funds to stay on schedule. Given the unpalatable alternatives, NASA chose to cut back on lunar exploration and apply the savings to its future.[1]

Cutbacks and Program Changes

In February 1970 NASA took to Congress a budget proposal for fiscal 1971 for $3.333 billion, down more than $500 million from the previous fiscal year. Administrator Thomas Paine listed for the House Committee on Science and Astronautics the "hard choices" he had been forced to make to stay within that budget while still making progress: suspending production of Saturn V launch vehicles and spacecraft assigned to those vehicles, "mothballing" the Mississippi Test Facility and reducing the work force at the Michoud Assembly Facility where Saturn Vs were assembled, closing the new Electronics Research Center at Cambridge, Massachusetts, and canceling Apollo 20.[2] During the following summer it appeared that Congress was in a mood to cut space funding even more. By July rumors were circulating that more Apollo missions, perhaps as many as four, were candidates for cancellation.[3] The first bill containing NASA's appropriation (vetoed in August by President Nixon, who objected to the funds provided for another agency) gave the space agency $3.269 billion—a figure that came out of weeks of debate in House and Senate and conference committee and was clearly all the space agency could expect.[4]

On July 28, while Congress debated the funding issue, Tom Paine submitted his resignation as NASA administrator, effective September 15, to President Nixon. He planned to return to General Electric, the company he had left in early 1968 to become deputy administrator. Paine denied that NASA's immediate funding problems had influenced his decision.[5] Probably more decisive was the fact that he saw no prospect of immediate change. From the start of his tenure as administrator, Paine's ideas for the future of the space program were far more ambitious than those of the President and the Bureau of the Budget.

On September 2 Paine called a press conference to announce NASA's interim operating plan for fiscal 1971. Cutbacks and stretchouts in most major programs were the order of the day, among them the cancellation of two more Apollo missions, 15 and 19. The remaining missions, redesignated Apollo 14–17,* would be flown at six-month intervals in 1971 and 1972. After that would come the three manned Skylab missions (which under earlier schedules had interrupted the Apollo program for two years). Paine noted that he and his managers had made this decision reluctantly and in spite of the conclusions of special reviews by the Space Science Board of the National Academy of Sciences and the Lunar and Planetary Missions Board, both of which had strongly recommended that all six remaining missions be flown. Paine's main consideration had been "how best to carry out Apollo to realize maximum benefits while preserving adequate resources for the future post-Apollo space program."[6] For whatever consolation scientists might get from it, one of the canceled missions (Apollo 15) was the last of the limited exploration missions the three remaining after the cut would all have considerably extended capability, including a surface roving vehicle. Left unanswered was the question of what was to be done with the three Saturn V boosters that no longer had missions to fly.[7]

Criticism of the decision to shorten Apollo was immediately forthcoming. A *New York Times* editorialist called the decision "penny-wise, pound-foolish," noting that the savings would amount to less than 2 percent of NASA's fiscal 1971 budget and less than 0.25 percent of the total investment in Apollo. "The budgetary myopia which forced this . . . decision can only vindicate the critics who have insisted that Apollo was motivated by purely prestige considerations, not scientific goals," said the *Times* writer. While the *Times* had for years opposed Apollo's high priority, "now that these huge sums have been spent the need is to obtain the maximum yield, scientifically and otherwise, from that investment."[8]

This argument was typical of the reaction voiced by many after the cutback. Particularly upset were lunar scientists. Astronomer Thomas Gold of Cornell University likened the decision to "buying a Rolls Royce and then not using it because you claim you can't afford the gas." The grand old man of lunar science, Harold Urey, said the $40 million saved on operational costs was "chicken feed" compared to the $25 billion already spent.[9] To the *Washington Post* he wrote,

> It cost us . . . one half of one per cent of our gross production. . . . Now we wish to finish a job which has been beautifully begun. And we get stingy. Because of an additional cost of about 25 cents per year for each of us we drop two flights to the moon recommended by scientific committees composed of men who personally profit from the expenditure little or not at all. How foolish and short-sighted from the view of history can we be?[10]

Gerald Wasserburg, a highly respected Caltech geochemist and member of the lunar sample preliminary examination team, protested in the *Los Angeles Times*,

---

*At the time, Apollo 14 had not yet flown. New mission numbers were 15, 16, and 17, with science plans corresponding approximately to those of the old 16, 17, and 18.

> . . . this decision threatens our existing investment in planetary exploration during a period when we are obtaining maximum returns and makes doubtful the scientific justification of the manned space program. The total effect appears to indicate a return to the dark ages of planetary science. . . .[11]

None of the protesters, however, could offer a reasonable alternative to the cutbacks. The *New York Times* editorial writer suggested that NASA seek participation, including financial contribution, by the Soviet Union or our European allies in long-term lunar exploration—surely an unrealistic suggestion at best.

These complaints, well founded though they were, were far too late. By mid-1970 NASA was operating on its lowest budget since fiscal 1965, under an administration determined to reduce the federal deficit, while trying to get started on a new generation of spacecraft and programs toward which administration budget officers were less than enthusiastically committed. Complicating the problem of living within a restricted budget, NASA's record in holding program costs to preprogram estimates was not good, a fact often brought up in congressional budget hearings. (That had very nearly cost the agency the Mars landing project, Viking, at the end of 1969.[12]) Public and congressional support for purely scientific missions, no matter how spectacular or important, had begun to erode before Apollo 11. Whatever the fact may say about the nation's commitment to the space program, NASA now lived or died by the perceptions of Congress and the administration of the short-term value of its projects. Political pressure on the space program was shifting in the direction of using space to solve earth's problems.[13]

In mid-September 1970, 39 scientists who had long been associated with Apollo's science program formally protested the cutbacks in a letter to the chairman of the House Committee on Science and Astronautics, Representative George P. Miller. They pointed out that the reduction in the number of missions might cause the lunar science program to "fail in its chief purpose of reaching a new level of understanding" of the origin of the earth and moon. As best they could tell, the money saved in the "approved and scientifically most fruitful lunar program" was intended to support a post-Skylab manned earth-orbital project—"an as yet unapproved program for whose scientific value there is no consensus, and whose purpose is unclear" (i.e., the space shuttle). They concluded by expressing the hope that the decision to cancel two Apollo missions was not yet final.[14]

In reply, Miller called the scientists' attention to his committee's long record of support for Apollo, pointing out that the committee had tried to get $220 million more for lunar exploration into the authorization bills for fiscal 1970 and 1971. "However, the [Nixon] Administration, in realigning national priorities, has relegated the space program to a lesser role." Undoubtedly the two canceled flights would have provided additional scientific information, Miller said, but that had to be balanced against assuring the rational development of future space programs. In conclusion he mildly reproved the scientists: "Had your views on the Apollo program been as forcefully expressed to NASA and the Congress a year or more ago, this situation might have been prevented." But it was too late to offer any serious hope of reinstating the two missions.[15]

As the launch date for the next flight approached, lunar scientists faced hard choices. Plans for Apollo 14 had reached the stage where no major changes were possible. With three missions remaining instead of five, landing sites had to be chosen to give the best hope of answering the most important scientific questions. Experimenters and flight planners had to make the best possible use of the extra payloads that the last three missions would be able to land on the moon.

To Fra Mauro, At Last

The loss of the scientific information Apollo 13 would have returned from Fra Mauro made it necessary to reevaluate objectives for later missions. At its meeting in early March the Apollo site selection board had recommended Littrow (Figure 4, site 21) as the target for Apollo 14.[16] After Apollo 13 the board's scientific advisers almost unanimously agreed that the Fra Mauro site still rated the high priority it had been given; they recommended sending Apollo 14 there instead of Littrow, and the board agreed.[17] Since a number of changes in the service module were expected as a result of the Apollo 13 investigation, Apollo 14 was rescheduled for the third time in six months. Launch readiness was targeted for no earlier that January 31, 1971.[18]

In the nine months between the third and fourth lunar exploration missions, the last of the year's budget shocks hit the Apollo program. With the loss of Apollo

Apollo 14 commander Alan Shepard practices towing the mobile equipment transporter while the KC-135 airplane flies a parabolic path that reduces gravity to the level experienced on the moon. Lunar module pilot Ed Mitchell follows the "rickshaw."

15 and 19, Apollo 14 became the last of the intermediate (''H''-type) missions. Its basic objectives were the same as those of Apollo 13, but several changes were made in details.

Apollo 13 had carried a surface science package consisting of a passive seismometer, an atmospheric detector and a charged-particle detector, and a heat flow instrument. Apollo 14's package left off the heat-flow experiment and added three more: an active seismic experiment, a laser reflector like the one carried on Apollo 11, and an ionosphere detector like the one carried on Apollo 12. The active seismic experiment consisted of a set of detectors (geophones) to be laid out on the surface and two devices for producing calibrated shocks. It would aid

The central processing station for the lunar surface experiments package on Apollo 14. The object behind and to the left of the station is one component of the active seismic experiment, a mortar which was intended to lob four explosive charges whose shocks would be recorded on the seismometer. However, it was placed so close to the central station that experimenters feared that firing it would cover the central station with dust, so the experiment was not performed.

in the interpretation of seismic data by giving seismologists a measure of the velocity of seismic waves in the near-surface layer of the moon.[19]

After earlier crews reported problems in carrying all the gear required for sampling on the moon, MSC's Technical Services Division designed and built a two-wheeled modular equipment transporter for the Apollo 14 mission. The rubber-tired "rickshaw," as it was nicknamed—actually it more resembled a golf cart—weighed only 18 pounds (8.2 kilograms) but could carry a 360-pound (163-kilogram) load; either astronaut could pull it with one hand. Providing storage space for core tubes, sample bags, cameras, maps, and hand tools, it folded up for storage on the lunar module at launch, alongside the experiment packages.[20]

Significant changes were made in pre- and postflight medical operations on Apollo 14. A prelaunch flight crew health stabilization program was established by medical officials to prevent the crew's exposure to communicable disease, which had caused problems on Apollo 7, 8, 9, and 13. For three weeks before launch, prime and backup crewmen were isolated from close contact with everyone except their families and ground-support personnel whose duties required them to work closely with the astronauts. All these "primary contacts" were examined, tested for immunity to common diseases, and kept under medical surveillance. Crewmen were restricted to specified areas at the Cape and the air-conditioning system in their quarters was fitted with biological filters to minimize exposure to airborne disease organisms.[21] Postmission quarantine arrangements were also modified to shorten the return to the lunar receiving laboratory. Since the recovery zone was some 600 miles (965 kilometers) from the U.S. air base in Samoa, plans called for the crew to be transferred by helicopter from the mobile quarantine facility aboard the recovery ship to Samoa. There they would enter a second isolation trailer, which would be flown to Houston.[22]

After the Apollo 13 accident, preparations for the next flight were delayed while the service module oxygen tanks were modified and tested. Meanwhile, several other minor changes to spacecraft and launch vehicle were made at the Cape.[23] The new tanks—including a third, isolated from the other two, which would assure that the command module had enough oxygen to supply the crew—were installed and tested by January 18, and no further problems showed up as launch date approached.[24]

Launch day, January 31, was cloudy and rainy; eight minutes before the scheduled liftoff, the launch director stopped the countdown to wait for the heaviest clouds to move across the Cape. Forty minutes later Apollo 14 was on its way.[25] The trip to the moon was uneventful until the time came to remove the lunar module from the S-IVB stage. Five attempts to dock the command module with the lunar module failed for no apparent reason—a worrisome anomaly, to say the least—but the sixth was successful.[26] The spent S-IVB stage was then put on a course to crash on the moon some 100 miles southwest of the Apollo 12 landing site.[27] Command module *Kitty Hawk* and lunar module *Antares* braked into lunar orbit 82 hours after liftoff. Two hours later *Kitty Hawk's* main engine lowered both spacecraft to the altitude from which *Antares* would begin its descent.[28] This maneuver was one result of the refinement of mission techniques that planners had been working on since Apollo 12, designed to conserve fuel in the lunar module and give the crew more time to hover before landing if they needed to look for a suitable site.

Commander Alan Shepard walks toward the mobile equipment transporter (on which the camera was mounted) while Ed Mitchell lays out the Apollo 14 lunar surface experiments. Note the checklist attached to Shepard's left cuff.

After mission commander Alan Shepard and lunar module pilot Edgar Mitchell had checked out *Antares*, command module pilot Stuart Roosa pulled *Kitty Hawk* away and *Antares* began its descent to the surface. Last-minute course corrections sent up from Houston were entered in the guidance computer and Shepard piloted the spacecraft to a routine landing about 350 miles (563 kilometers) west-southwest of the center of the moon's visible side. *Antares* was only 175 feet (53 meters) from its targeted landing site.[29] Meanwhile Roosa had boosted *Kitty Hawk* back up into a higher, circular orbit, where he had a number of tasks to perform while his colleagues explored the Fra Mauro Formation.

The terrain on which *Antares* sat was gently undulating, with numerous craters but comparatively few boulders. Mitchell commented that there was "more relief [i.e., variations in elevation] than we anticipated from looking at the maps,"[30] a characteristic that would cause them some difficulty later on. Having given Houston a description of what they could see, Shepard and Mitchell put on their space suits and prepared for their first excursion.

Shepard's first words as he stepped on to the moon were inspired by his 9 years, 10 months, and 10 days of waiting from Mercury-Redstone 3, when he had been the first American in space, to the day he stepped on the moon.* "It's been a

---

*Shepard was the only one of the "Original Seven" astronauts to make that journey, only the second to fly in the Apollo program.

The lunar module Antares *on the surface of the moon. The crewman to the right of the LM cannot be identified. Note the rolling terrain, which often obscured landmarks and caused the Apollo 14 astronauts considerable difficulty in finding their way.*

long way," he said, "but we're here."[31] Mitchell joined Shepard on the lunar surface and they unloaded the rickshaw and experiments and picked a spot some 500 feet (150 meters) west of *Antares* for the instruments. After laying out the geophones for the active seismic experiment, Mitchell fired the explosive charges in his hand-held "thumper" as they walked back to the lunar module. On the way Shepard stopped to collect a comprehensive sample of rocks and fine surface material from a representative area, found two "football-sized" rocks, and collected some other surface samples. After more than four and a half hours they were back in the lunar module. Houston then had half an hour's worth of questions from the scientists in the back room, and then it was time to turn in.[32]

Shepard and Mitchell did most of the mission's geological field work on their second traverse. Their biggest problem was in determining their location from the landmarks shown on their map. More than once they changed their minds about where they were. At the time and later, they attributed this to the rolling terrain and the relation of their line of sight to the sun: craters might be visible in one direction but not in another. Without familiar objects for reference, they found it difficult to estimate distances. A prime objective was to sample the rim of "Cone" crater, about a thousand meters (3,300 feet) from the spacecraft. By the time they got there, however, they had spent considerable time and were not positive that they were in the right place. As it turned out, they stopped just a few meters short of the rim, but at the time they were not certain they were on the slope of Cone, and Shepard was concerned with the tasks they had yet to accomplish and the time available.[33] They turned back, completed the planned traverse, and returned to *Antares* after another 4½-hour excursion. They had collected nearly a hundred pounds (45 kilograms) of samples and taken hundreds of photographs documenting many of the rocks, boulders, and sampling sites, including several panoramic views of the landing site.[34] Before climbing back into the lunar module, Shepard took out of his suit pocket "a little white pellet that's familiar to millions of Americans"—a golf ball—and dropped it on the surface. Then, using the handle for the contingency sample return container, to which was attached "a genuine six-iron," he took a couple of one-handed swings. He missed with the first, but connected with the second. The ball, he reported, sailed for "miles and miles."[35]

During the 33½ hours Shepard and Mitchell were on the moon, Stuart Roosa had several important tasks to perform in *Kitty Hawk*. Continuing what had begun on Apollo 12, he photographed one of the remaining candidate landing areas (Descartes) and made numerous observations of prominent lunar landmarks to provide data that would improve landing accuracy on subsequent missions.[36]

Back in *Antares*, Shepard and Mitchell stowed their samples and discarded their expendable equipment. Houston then passed up questions for half an hour concerning details of their visual observations, which brought out some of the difficulties they had experienced on the traverse. Geological features had been subtle, occasionally they had had too little time to observe and comment on details, and the rolling terrain had sometimes blocked their view of features only a few meters away.[37]

Liftoff from the moon came at 1:48 p.m. EST on February 6. Mission planners had worked out a "direct" rendezvous scheme—that is, the ascent trajectory was programmed to meet the command module at its highest point, with necessary corrections being made during ascent—which they used for the first time. (On previous missions several maneuvers had been necessary to adjust the LM's orbit before bringing the spacecraft together.) Two and a half hours after liftoff, *Antares* and *Kitty Hawk* docked; three hours later, having sent the lunar module crashing to the lunar surface, *Kitty Hawk* headed home.[38]

Along the way the crew performed some "inflight demonstrations"—experiments exploring some zero-g techniques that might offer useful application of technology in space: electrophoresis (the migration of charged molecules in solution under the influence of an applied voltage), transfer of liquids between two containers, heat transfer, and casting of various materials from the molten state.[39] Results were promising enough to warrant further investigation on Skylab and, later, on space shuttle missions. After finishing those demonstrations, Shepard, Mitchell, and Roosa had little to do for the rest of the mission. *Kitty Hawk* made a normal reentry and landed 0.6 miles (965 meters) from its targeted point in the South Pacific near the aircraft carrier U.S.S. *New Orleans* in the early morning light of February 9. Three days later the astronauts in their quarantine trailer arrived at the lunar receiving laboratory at MSC, where they spent 15 days in quarantine.[40]

Apollo 14 successfully concluded the intermediate stage of lunar exploration, closing a period in which the progress made in mission planning and operations exceeded expectations. Armstrong had overshot his target by five miles (eight

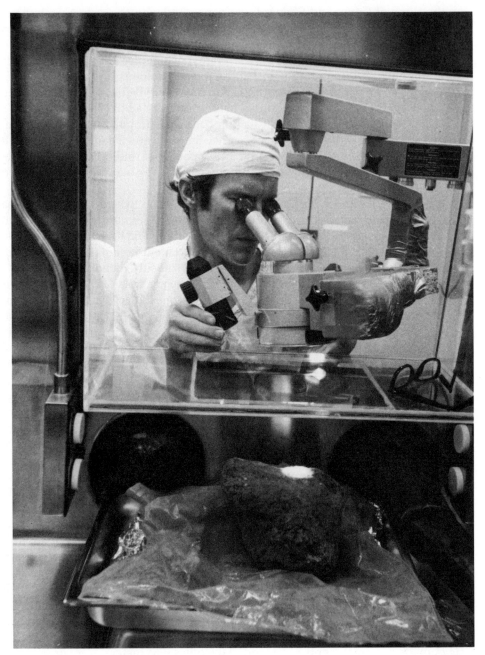

A scientist views an Apollo 14 lunar sample through a binocular microscope in the Lunar Receiving Laboratory at the Manned Spacecraft Center.

kilometers). Conrad and Shepard, aided by improved techniques, had landed within a quarter of a mile of theirs (400 meters) —an accuracy that MSC's mission planners had expected to achieve after three or four tries, but scarcely hoped for on the second. *Eagle* had stayed on the moon for 21½ hours, *Intrepid* for 31½,

Technicians examine Apollo 14 samples in the sterile nitrogen atmosphere of a processing cabinet in MSC's lunar receiving laboratory. The cabinet is at atmospheric pressure. Earlier processing in vacuum had to be abandoned, in part because the gloves through which technicians handled the samples could not stand up to the large pressure differential between the cabinet and the outside.

Antares for 33½. Armstrong and Aldrin's 2½ hours on the surface was more than doubled by Conrad and Bean and extended nearly ninefold by Shepard and Mitchell. Apollo 12 brought back 50 percent more lunar material than Apollo 11, and Apollo 14 returned 25 percent more than that. About the only remaining improvements to lunar exploration would come from the addition of extra supplies and a powered vehicle to save time in exploring the lunar surface.

Changes for Extended Lunar Missions

In the first half of 1969 the Office of Manned Space Flight (OMSF) had anticipated the requirements of the extended ("J") missions, Apollo 16–20, by ordering studies on modification of the lunar module and development of a powered

lunar-surface vehicle (see Chapter 10). After studies by the lunar module contractor, Grumman Aircraft Engineering Company, were reviewed by OMSF's
Management Council, MSC authorized Grumman on June 9 to proceed with
modifications on LM-10 and subsequent spacecraft. Changes to the landing module would allow it to carry additional payload (the lunar roving vehicle) to the
lunar surface, return more lunar samples to earth, and provide consumables (oxy-

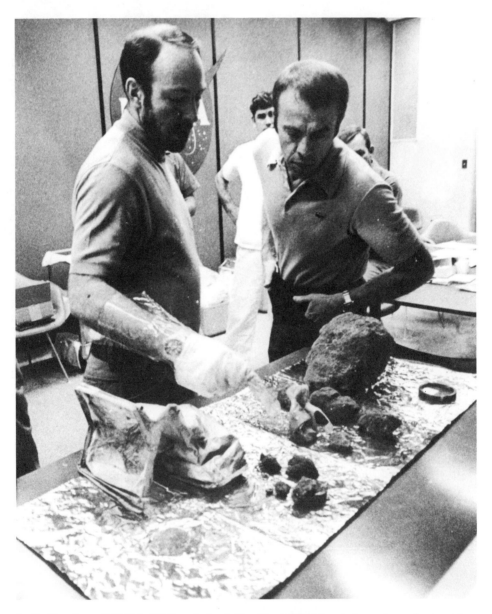

*Apollo 14 explorers Ed Mitchell (left center) and Alan Shepard (right center) examine and comment on
some of the samples they collected. The specimen in front of Shepard was, at the time, the largest rock
brought back from the moon.*

gen, water, and electrical power) to support the crew on the moon for periods up to 78 hours. The first "extended" lunar module was to be delivered in time for launch in April 1971.[41] Major changes included enlargement of the fuel and oxidizer tanks on both ascent and descent stages, extension of the descent engine nozzle to improve its efficiency, addition of batteries and solar cells to recharge them, and rearrangement of stowage space.[42] At a critical design review in mid-September, MSC cut out some of the changes to minimize cost increases, discarding the solar cells and reducing the maximum duration of stays on the moon to 54 hours. The extended LM would weigh 36,000 pounds (16,330 kilograms), compared to 32,000 (14,515 kilograms) for earlier versions.[43]

A large part of this extra payload was the lunar roving vehicle, which would extend the astronauts' range on the moon. In May 1969 a surface-traversing vehicle was chosen rather than a flying unit as the most suitable means of carrying the astronauts and their tools to sites of scientific interest, and Marshall Space Flight Center was assigned responsibility for its development.[44] That fall, the Boeing Company won the contract to develop the lunar roving vehicle (LRV or "rover"). The rover was to be a battery-powered vehicle weighing 400 pounds (181 kilograms), capable of carrying a 970-pound (440-kilogram) load as far as 120 kilometers (75 miles).[45] Boeing's facility in Huntsville, Alabama, was to build the chassis and assemble the complete vehicle from components manufactured by Boeing in its Seattle plant and by AC-Delco Electronics. Development time was short: the schedule called for delivery of the first of four vehicles by April 1, 1971, for flight on Apollo 17, then scheduled for the following September.[46]

The Apollo 16 crew training on the lunar rover.

The weight increases necessary to extend the lunar missions forced a number of changes in the launch vehicle as well. Rocket engineers, whose experience had made them familiar with the tendency of payloads to grow, usually built their engines to a somewhat higher capacity than the design requirement, so that they could increase the thrust by making relatively minor changes. So it was with the Saturn V.[47] In August 1969 MSC flight planners and Marshall engineers began discussing means of enabling the Saturn V to launch 106,500 pounds (48,310 kg).[48] The F-1 engines on its first stage could be readjusted to a higher thrust level after acceptance firing. Timers could be reset to reduce the amount of unburned propellant left in the tanks at engine cutoff. These changes, plus modifications to the S-II second stage, would raise the payload by more than 700 kilograms (1,540 pounds). An equally significant change was reduction of the fuel reserves provided to make up for unpredictable variations in launch conditions and engine performance. Seven manned launches had given engineers enough performance data to indicate that these reserves could be reduced without affecting crew safety, adding another 700 kilograms of payload capacity.[49]

Changes in some launch constraints could increase the launch capability of the Saturn V as well. For example, lowering the earth orbit into which the spacecraft and S-IVB were initially injected (the parking orbit) by 10 nautical miles (11.5 statute miles, 18.5 kilometers) could allow another 300 kilograms (660 pounds) to be added to the spacecraft, at the cost of some slight additional heating of the spacecraft and a loss of some radar-tracking data. Finally, some 200 kilograms (440 pounds) of payload could be added by optimizing the azimuth on which the Saturn V was launched.[50] After some months of analysis and discussion, several of these changes were made in preparation for Apollo 15, enabling the Saturn V to launch 107,000 pounds (48,535 kilograms).[51]

Landing Sites for the Last Missions

In July 1969 the Apollo Site Selection Board had compiled a list of 10 landing sites for Apollo 11-20, subject to further evaluation and revision in light of the results of early landings (see Chapter 10 and Table 2). In August and September, responding to a request from Headquarters and MSC program managers for detailed science mission plans, members of the Group for Lunar Exploration Planning and its Site Selection Subgroup, working with MSC operations planners and U.S. Geological Survey scientists, suggested some minor changes, based on a reevaluation of geologic features at one of the sites and the expected availability of good photography at another. These were endorsed by the Group for Lunar Exploration Planning on August 23.[52]

At the end of September, however, while many other complaints were being registered concerning Apollo (see Chapter 10), some members of the Science and Technology Advisory Panel and the lunar panel of the Lunar and Planetary Missions Board objected that their opinions were not receiving sufficient consideration in site selection. Harold Urey complained that he did not know who was making decisions regarding landing sites, nor what the reasons for choosing the

Table 2. **Lunar landing site assignments.***

Mission	Site	Mission	Site
G-1	Site 2 (Tranquility)	J-1	Copernicus peaks
H-1	Site 5 or 4 (Oceanus Procellarum)	J-2	Marius Hills
H-2	Fra Mauro	J-3	Tycho
H-3	Rima Bode II	J-4	Rima Prinz I
H-4	Censorinus	J-5	Descartes

*Recommended by the Group for Lunar Exploration Planning; minutes of Apollo Site Selection Board Meeting, July 10, 1969.

various sites were: the correspondence he received was unintelligible, full of unexplained acronyms and couched in unreadable bureaucratic jargon. He could not take the time to attend meetings of the site selection subgroup to argue his case, however, because of academic commitments.[53] Another meeting was held in mid-October in the hope of considering the advice of the objectors, but they were unable to attend. Gene Simmons, MSC's chief scientist, was annoyed and disappointed. He wrote to the chairman of the Science and Technology Advisory Committee,

> To those of us who have been intimately involved with the scientific recommendations on Apollo sites for some time, it was particularly disappointing that only one of these "senior scientists" showed up in time for the meeting. . . . The entire decisionmaking process of NASA on the landing sites for the H and J series missions was postponed in order for these individuals to meet with us and provide the kind of counsel that only they, presumably, could give. . . . [Nonetheless,] we as a group felt that we should proceed with the reexamination of the proposed landing sites.[54]

The meeting went on without the objecting members, thoroughly reviewing the sites and the scientific rationale and operational restrictions for each. After two days of intensive review, the group endorsed the proposed list of sites without alteration (see Table 3).[55]

The Site Selection Board, meeting at the end of October, discussed this list in light of the improvements (lunar roving vehicle and added payload) expected for the J missions. Marius Hills, Descartes, and Hadley appeared to present no problems of accessibility, but Copernicus and Tycho were only marginally acceptable. Marius Hills was accessible only during two summer months. Descartes, Hadley-Apennine, Davy Rille, and Censorinus were not adequately covered by available site photographs. MSC was prepared to accept the available photographic coverage for Fra Mauro, Littrow, Marius Hills, Copernicus, and Hyginus, but planners would have to sacrifice some fidelity in terrain models for training. Tycho was discussed at some length. It was the most difficult site on the list to reach, but in many ways was scientifically attractive. Finally the board relegated Tycho

Table 3. **Landing Sites and Alternates.***

Mission		Prime Site	Alternate 1	Alternate 2	Alternate 3
H–2	(13)	Fra Mauro	Alphonsus**	Alphonsus**	Fra Mauro**
H–3	(14)	Littrow	Littrow	Littrow	Littrow
H–4	(15)	Censorinus	Fra Mauro	Fra Mauro	Censorinus
J–1	(16)	Descartes	Censorinus	Censorinus	Descartes
J–2	(17)	Marius Hills	Marius Hills	Marius Hills	Marius Hills
J–3	(18)	Copernicus	Copernicus	Davy Rille	Davy Rille
J–4	(19)	Hadley	Hadley	Hadley Rille	Hadley Rille
J–5	(20)	Tycho	Tycho	Copernicus	Copernicus

*Recommended by the Group for Lunar Exploration Planning, October 16–17, 1969.
**Hyginus Rille is the alternate site for the H-2 mission if operational constraints prevent landing at Fra Mauro or Alphonsus.

Table 4. **Sites recommended MSC.***

Mission		Prime Site	First launch opportunity	Alternate site
H–2	(13)	Fra Mauro	Mar. 12, 1970	Hyginus
H–3	(14)	Littrow	July 8, 1970	Littrow
H–4	(15)	Censorinus	Oct. 30, 1970	Fra Mauro
J–1	(16)	Descartes	Mar. 29, 1971	Censorinus
J–2	(17)	Marius Hills	July 30, 1971	Marius Hills
J–3	(18)	Copernicus	Feb. 19, 1972	Davy Rille
J–4	(19)	Hadley	July 14, 1972	Hadley
J–5	(20)	Tycho	Feb. 7, 1973	Copernicus

*Minutes, Apollo Site Selection Board Meeting, October 30, 1969.

to the last J mission, which would leave ample time to look for ways to overcome the operational limitations. Houston's representative proposed yet another sequence of missions (Table 4), which the board approved as a basis for continuing evaluation.[56]

Assignment of sites to specific missions was tentative at this stage of the project because the necessary information was still sketchy. By the end of 1969, however, much of this information was becoming available. Apollo 12 demonstrated the ability to make precision landings, a requirement for the later H missions. As the initial scientific study of samples from Apollo 11 and 12 yielded information on chemical composition and age, the basis for choice among the remaining sites became somewhat more clear-cut. On the other hand, in January 1970 Apollo 20 was canceled, and the prospect that others might be dropped clouded the picture.

In February 1970 the Group for Lunar Exploration Planning met at Houston to reassess the list of sites in light of recent developments and to make new recommendations to the Site Selection Board. Tycho was reluctantly dropped from consideration because of its operational problems, although the group indicated that location was still interesting. The remaining sites on the list approved in October were endorsed, with minor changes. Fra Mauro remained the highest priority on the list because it offered the chance to sample material that originated deep within the moon. The group wanted to move Marius Hills from the second to the first J mission, interchanging it with Descartes. It specified that the target on the fourth "H" mission should be as near the highland terrain around Davy Rille as possible, to be reasonably sure of sampling highland material; otherwise Censorinus was preferred. Finally, the group recommended that the site for the landing at Hadley (J-4) be moved from the west side of Hadley Rille to the east side, to allow the astronauts to reach the Apennine Front and sample more than one type of terrain.[57]

MSC's Science and Applications Directorate evaluated this list with three guiding principles in mind: information gained from previous missions should weigh heavily in selection of later sites; since only a few missions remained, sites should be chosen to answer as many scientific questions as possible and missions should have multiple objectives; and the sites of undisputed scientific interest should be scheduled as early as possible. On this basis, MSC recommended shifting Copernicus to Apollo 16 and the Marius Hills to Apollo 18. The objectives at the Marius Hills site might be satisfied at either Davy or Copernicus, and some of the instrumentation desired for a Marius Hills mission might not be developed in time to fly on Apollo 16. Should that be the case, a delay would imperil the sequence, because the useful payload for a Marius Hills mission fell off by as much as 6,000 pounds (2,720 kilograms) after midyear.*[58]

After the Apollo 13 mission had to be aborted, the Group for Lunar Exploration Planning reconsidered its recommendations and endorsed Fra Mauro (the planned Apollo 13 site) for Apollo 14. Its importance in dating the Imbrian event remained, and it offered advantages for placement of another passive seismometer, an active seismometer, and a third laser retroreflector. For Apollo 15, the Group recommended a site near the Davy crater chain, assuming adequate photography could be obtained on 14.[59]

As it turned out, the photos from Apollo 14 would come too late to allow certification of Davy as a site for Apollo 15, and in June the site selection subgroup convened once more to evaluate candidate sites for Apollo 15 and 16. By that time it was already clear that more missions might be canceled, which further

---

*This was the result of the mission rule requiring daylight launches and of the annual variation in the geometric relationships of the earth, sun, and moon.

complicated the subgroup's deliberations. After a long discussion that reexamined 14 sites* for which reasonably adequate photographs were available, the subgroup agreed on a recommendation that the Marius Hills be the candidate for Apollo 15. Littrow was the chosen alternate. For Apollo 16, Descartes was the landing site of choice. In the event that Apollo 15 was canceled, Marius Hills should be shifted to Apollo 16 and Descartes to 17. The remaining sites were ruled out, either because MSC had determined that they were unsuitable for landing or because they appeared to offer insufficient new scientific information to justify consideration.[60] Following this meeting only Littrow, Descartes, Hadley-Apennines, and the Marius Hills remained as strong candidates for the last missions.

In the next three months, MSC began to try to overcome the difficulties in getting a consensus on landing sites among the various scientific groups.** Anthony Calio decided to ask three groups of scientists to examine the question separately and make independent recommendations. After receiving their opinions, Calio summarized the criteria for selection of an Apollo 15 site. It should offer a major advance in the study of the moon and a high probability of answering essential scientific questions and satisfying the objectives of more than one scientific discipline; it should be certifiable on the basis of existing photography and not dependent on photographs to be obtained by Apollo 14; it should be operationally feasible without further analysis; and it should be appropriate for either a walking mission or a rover mission, so that it could be flown on Apollo 15 or (if 15 should be canceled) 16. Calio then described the geologic features of the Hadley-Apennine site, located on the eastern rim of Palus Putredinus, nearly 30 degrees north of the moon's equator. Among the sites still under consideration, Hadley was unique in offering direct access, with or without a roving vehicle, to a mountainous highland, a mare surface, and a sinuous rille. Furthermore, it appeared on most of the priority lists produced by the disciplinary groups. It would offer the advantage of establishing the high-latitude arm of a well dispersed array of geophysical instruments, essential to investigating the moon's interior (by seismology) and its orbital librations (by measurements from the laser retroreflectors).[61]

After the cancellation of Apollo 15 and 19 in early September 1970, MSC presented the case for Hadley at the Apollo Site Selection Board's meeting later that month. Arguments for Hadley and Marius Hills were fairly evenly matched, both from scientific and operational standpoints, and the debate between the two was virtually deadlocked until astronaut David Scott, recently picked to command Apollo 15 (see below), said that he preferred Hadley although he thought he could land at either site. Scott's opinion tipped the balance. The Board recommended Hadley for a launch date between July and September 1971, Descartes for Apollo

---

*Censorinus, Littrow, Alphonsus, Hyginus Rille, Rima Bode II, Hipparchus, Mösting C, Dionysius, Sinus Medii, Flamsteed P, Gassendi, Copernicus, Descartes, and Marius Hills (see Table 1). At this meeting, MSC ruled out Copernicus and Censorinus as unsuitable for landing because of rough terrain. The subgroup dropped Gassendi, which appeared to present the same disadvantage.

**According to an MSC participant, consensus among the scientists at a site selection meeting was nearly impossible to obtain. Each individual repeatedly voted for the site of his choice, and in the end, NASA (i.e., MSC) had to have made the decision. Brooks, Grimwood, and Swenson, *Chariots for Apollo*, p. 365.

16 between January and March 1972. The choice of a site for Apollo 17 was left open; Marius Hills and Copernicus were the leading candidates, but others (e.g., Littrow) were still in the running, and a new site might be found in future orbital photography.[62]

The September meeting wrapped up the Apollo Site Selection Board's unfinished business for the time being. After Apollo 14, scheduled for the following April, it would reconvene to examine the list of sites in light of the results of that mission.

Crews for the Last Missions

In the weeks following Apollo 11, when scientists and others found fault with many aspects of the program, NASA was frequently castigated for its failure to assign a scientist to a moon mission at the earliest possible date. When crews were named for Apollo 13 and 14 in August 1969, the absence of a scientist-astronaut on either crew was pointedly noticed by the press.[63] Scientists, eager to have a scientist assigned to a crew, argued that one pilot in the lunar module was enough: the second person was largely a bystander during the landing anyway. But Deke Slayton, director of flight crew operations at MSC and the person responsible for selecting crews, was adamant. At this point only one landing had been made, and it had required almost instinctive action by the pilot in the last seconds before touchdown. Slayton was not yet willing to entrust that responsibility to a less experienced astronaut. Besides, he had to be prepared for the worst case. "Both guys have to be able to fly it. . . . It would sure be a crime to have one guy as a passenger and the commander break a leg and then you lose two guys instead of one, plus the vehicle."[64] Slayton had nothing but good things to say about the scientist-astronauts and admitted that one of them, geologist Harrison H. ("Jack") Schmitt, might be about ready for a flight, but he was not going to be rushed.[65] By the end of the year, however, "reliable sources" were saying that Schmitt would be named to the backup crew for Apollo 15.[66]

On March 26, 1970, two weeks before Apollo 14 was launched, Slayton announced the names of the crewmen selected for Apollo 15. As he had done twice before, Slayton picked a backup crew from a prior mission*: commander David R. Scott, command module pilot Alfred M. Worden, and lunar module pilot James B. Irwin, all Air Force officers. As backup crew he named Richard F. Gordon, who had piloted the command module on Apollo 12, commander; Vance D. Brand, command module pilot; and Jack Schmitt, lunar module pilot.[67] Three scientist-astronauts made up the support crew: astronomers Karl G. Henize and Robert A. Parker and physicist Joseph P. Allen IV. Schmitt's assignment signaled MSC's intention to send a scientist to the moon before Apollo was finished:

---

*While it could not always be followed, Slayton's policy was to keep crews together. He normally assigned backup crewmen to prime crews for the third mission following their backup assignment, as was the case for Apollo 7 and Apollo 9. Neil Armstrong and Buzz Aldrin of Apollo 11 had both been on the backup crew for Apollo 8; Mike Collins, initially on the prime crew for Apollo 8, had to drop out when he developed a medical problem that required surgery. He returned to active status and six months later was assigned to Apollo 11.

in the normal course of events he could expect to draw the assignment as lunar module pilot on Apollo 18, which at the time was the next-to-last mission on the schedule.

As the only geologist among the scientist-astronauts, Schmitt was the logical choice to be the first scientist assigned to a lunar landing mission. Logic aside, he had developed his flying skills assiduously and taken care to make himself useful around the astronaut office. Having participated in the geology training since joining the program, he had been instrumental in reorienting the course to suit the rather specialized purposes of lunar exploration, bringing in outside experts who emphasized the larger picture rather than the minute details of field geology that could easily get in the way of astronauts interpreting what they could see. He had spent many hours with the crews of Apollo 11, 12, 13, and 14, helping them sharpen their observational and sampling skills. He had also made an effort to get acquainted with the people in flight operations, learning as much as he could about their problems and capabilities and helping to coordinate science with flight operations.[68] As the one scientist-astronaut whose professional specialty was particularly suited to the objectives of Apollo, Schmitt was able to satisfy his need to stay current in his profession while carrying out the duties demanded of an astronaut. That, plus his general determination to do whatever seemed likely to get him a crew assignment, kept him from getting on the wrong side of Slayton and the test pilots.

Yet Schmitt's reservation on a moon-bound lunar module was by no means confirmed. One mission already had been dropped, and additional cuts were a strong possibility when Schmitt was named to a backup crew; when Apollo 15 and 19 were canceled in September 1970, it appeared that the backups for Apollo 14—Gene Cernan, Ron Evans, and Joe Engle, named in August 1969—would be the prime crew of Apollo 17 and thus the last people to explore the moon.

Scientist-Astronauts' Dissatisfaction Surfaces

The scientist-astronauts had kept out of the public discussions of whether pilots or scientists should draw assignments to lunar missions. But in the year and a half after the first lunar landing they grew increasingly restive and their discontent became known within NASA. In January 1971, Associate Administrator Homer Newell spent a day in Houston, privately hearing two or three scientists at a time air their grievances. Practically all of them said, in effect, that they could not keep up their scientific competence under the existing organization and leadership of the Astronaut Office. Most of them expressed no animosity toward Deke Slayton; they simply thought his criteria for crew selection were wrong. He gave no consideration to the scientific activity that most of them had tried to work into their schedules. Even among the pilots, proficiency in science seemed unimportant: one nonscientist who was regarded as the best geology student on his crew was designated command module pilot and thus would never set foot on the moon. They were almost all pessimistic about prospects for science on Skylab and Shuttle unless something was done to change the way crews were selected.

Yet none of them felt that resigning in protest was a good way to change the system, and none wanted to leave the astronaut corps.[69]

Newell jotted down a few points after the meetings that he would later incorporate into a list of recommendations to Administrator James C. Fletcher: that a change in the method of and responsibility for selecting crews should be considered; a geologist should be assigned to a lunar mission as soon as possible; and if feasible, two scientists should be assigned to each Skylab mission. In addition the training program should be restructured to allow the scientist-astronauts to give more attention to their scientific careers.[70]

After reporting his findings to Fletcher—calling them tentative because he had heard only one side of the question—Newell discussed the problem at some length in the following weeks with Dale D. Myers, associate administrator for manned space flight, and Robert R. Gilruth, director of MSC. Gilruth's (and Slayton's) view that lunar missions required the skills of experienced test pilots prevailed. Myers decided he could not "commit casually to the flight of scientist astronauts on an Apollo mission": a failure on a lunar flight could mean the end of manned space flight for a long time. On March 1 Myers advised the Chairman of the Space Science Board, Charles H. Townes, of the conclusions he and Gilruth had reached:

1. NASA should, if possible, fly a geologist on Apollo;
2. Dr. Harrison H. Schmitt is our geologist candidate;
3. Flight crews are normally announced after they have worked together on backup crews, as Jack is now doing on Apollo 15;
4. If his training continues to progress satisfactorily, . . . and if all other aspects of crew selection are satisfied, he will be chosen for Apollo 17;
5. We will make that decision in the same time frame as for previous crew selections, which for Apollo 17 will be no earlier than August 1971 (after Apollo 15). I will review our decision with you at that time.[71]

This was the best Newell could get for the scientist-astronauts. He continued to work to get two scientists on each Skylab mission, but without success.[72]

So Schmitt's place on a lunar landing mission was assured, barring some highly unlikely occurrence, but the announcement would be another six months in coming. Meanwhile, a month after Alan Shepard and his crew returned from Apollo 14, Slayton named the crew for Apollo 16: commander John W. Young, command module pilot Ken Mattingly, and lunar module pilot Charles M. Duke, Jr., who had trained together for Apollo 13. Young had logged more time in space than any other eligible astronaut,* having flown on Gemini III, Gemini X, and Apollo 10. Mattingly had come within three days of flying on Apollo 13 but had been taken off for medical reasons (see Chapter 11). Duke, a member of the third class of astronauts chosen in 1966, had not yet had a flight assignment. Veterans Fred Haise, Stuart Roosa, and Edgar Mitchell made up the backup crew.[73]

---

*Jim Lovell, the record holder for time in space, was not eligible, having announced before the launch of Apollo 13 that it would be his last flight.

Second Apollo Lunar Science Conference

On January 11, 1971, the second symposium on Apollo results convened in Houston. Some 400 scientists gathered to hear further findings from the Tranquility samples and the first from Oceanus Procellarum. An added feature of the conference was the presence of Academician Aleksandr Pavlovich Vinogradov, vice-president of the Soviet Academy of Sciences and director of the Vernadsky Institute for Analytical Chemistry in Moscow, who presented the results of examination of lunar samples returned by the Soviet Union's *Luna 16* from a site almost 1,000 kilometers (625 miles) east of Tranquility Base.* Like its predecessor of 1970, the lunar science conference of 1971 was rich in detail but poor in large-scale conclusions. This was only to be expected in view of the small number of sites sampled, the complexity of the material returned, and the short time scientists had been studying the samples.

Vinogradov's report indicated nothing especially novel about the site where *Luna 16* landed. It was covered with a fine soil essentially similar to the samples Apollo 11 and 12 had brought back.[74] Other papers, however, presented fundamentally new information. One of the puzzling aspects of the Apollo 11 investigations had been that lunar rocks seemed to be younger than the soil ("regolith" is the technical term) around them, by as much as a billion years, and differed in chemical composition as well. Material in some of the Apollo 12 samples appeared to offer an explanation of the differences. MSC scientists found some fragments rich in potassium (chemical symbol K), rare-earth elements, and phosphorus (leading to the acronym "KREEP"), best explained as debris from an old bedrock.[75] Other investigators found a rock fragment of similar composition, which, they speculated, represented material which originally was located near the surface of an ancient lunar crust.[76] These discoveries supported the idea that very early in its history, the moon had had a molten surface layer that slowly cooled to produce a primitive crust, which was then subjected to later episodes of melting. The scientists conjectured that the KREEP material, which showed evidence of having been subjected to considerable shock, was thrown onto the plain of Oceanus Procellarum by the impact of a large meteorite in the highlands.[77] This view was consistent with the "chemical anomaly" of the KREEP component compared to the neighboring rocks. It was relatively low in iron and relatively high in aluminum, and the differences decreased the farther the regolith samples were from the highlands, as would be expected if the material responsible had been thrown out by impact.[78]

The second lunar science conference provided no ultimate key to the origin and history of the moon, any more than the first had done. Besides adding some

---

*The unmanned lunar probe *Luna 16* landed in Mare Fecunditatis (lunar latitude 0°41' south, longitude 56°18' east, approximately 100 kilometers [63 miles] west of crater Webb) on Sept. 20, 1970. An electrically driven core drill penetrated some 35 centimeters (14 inches) into the lunar soil, collecting about 100 grams (3.5 ounces) of sample, which was dumped into an upper stage of the spacecraft and returned to earth four days later. NASA, *Astronautics and Aeronautics, 1970: Chronology on Science, Technology, and Policy* (NASA SP-4015) , pp. 299, 316, 325, 349. Vinogradov's presentation at the conference was one aspect of the cooperation in space exploration that was developing at the time, which culminated in a joint American-Soviet manned mission in 1975. See Edward C. Ezell and Linda N. Ezell, *The Partnership: A History of the Apollo-Soyuz Test Project*, NASA SP-4209 (Washington, 1978).

200-odd papers to the scientific literature on lunar science, it showed that future missions might be expected to provide more scientific surprises. Future conferences, however, would present new information in a growing context of facts about the moon, established in the first two conferences and rapidly expanding as investigators published their results elsewhere.*

Although the results of the second lunar science conference came too late to influence the choice of a target for Apollo 14 (which was launched two weeks later), they indicated that the Fra Mauro site selected for that mission was likely to be a good choice. Its intrinsic interest as the first nonmare landing site was heightened by the discovery of KREEP, for if that material were found in substantial quantity at Fra Mauro, the tenable hypotheses about the origin and early history of the moon would be considerably restricted.

The End of Quarantine

Astronauts and lunar scientists alike had chafed under the requirement to quarantine spacecraft crews and lunar samples until it could be proved that no infectious organisms had been brought back from the moon. Early in 1970 the Interagency Committee on Back Contamination (ICBC) reviewed quarantine requirements in light of the results from the first two Apollo missions. Considering the conditions that had been found to exist on the moon—and probably had existed for billions of years—the requirement for quarantine now appeared superfluous. After lengthy discussion, the committee agreed to recommend to the NASA administrator that crew quarantine be discontinued. They believed, however, that biological examination of the lunar samples should be continued and that biological containment practices in the lunar receiving laboratory should continue, since among other things they assured the integrity of the samples. It was agreed that the Space Science Board would be advised of this recommendation.[79]

In February 1970 the Space Science Board discussed the matter and found no reason to discontinue quarantine for Apollo 13, which was to fly in two months, since it was targeted for a highland site that was different from the mare sites sampled by Apollo 11 and 12. Nonetheless it agreed with the ICBC that crews on future missions need not be quarantined unless anticipated differences in landing sites seemed to require it.[80]

After the aborted flight of Apollo 13, the quarantine remained in force for 14, since it was scheduled to take a deep lunar core sample, which might be different from surface samples.[81] So Alan Shepard, Ed Mitchell, and Stuart Roosa spent their time in the quarantine area of the Lunar Receiving Laboratory, the last lunar explorers to do so. When no further evidence of exotic diseases or organic material appeared, the requirement was finally revoked, although the ICBC would continue to function through Apollo 17.[82]

---

*The moratorium on publication before the conference, stipulated by NASA and agreed to by the principal investigators of the Apollo 11 samples (see Chapter 9), was lifted after the first lunar science conference. It had been imposed to avoid "a wild scramble for publication immediately after samples became available." Philip H. Abelson, "The Moon Issue," *Science* 167 (1970): 447. There was evidently a strong feeling that scientists were not immune to the temptation of the publicity that would accrue to the first publication of the results of investigating a lunar sample.

13

LUNAR EXPLORATION CONCLUDED

When the spacecraft *Kitty Hawk* returned from Fra Mauro on February 9, 1971, the preliminary stage of lunar exploration ended. Project Apollo, severely truncated in the budget cuts of 1970, entered its last phase as NASA prepared to exploit the potential of earth-orbital manned flight. Skylab, an earth-orbiting laboratory scheduled for three long-duration missions between 1973 and 1974, would lay the groundwork. Shuttle, the new reusable spacecraft that carried NASA's hopes for a continuing manned program, had yet to receive the blessing of the Nixon administration's budget officials and appeared to be facing stiff opposition in Congress. After that, it would face a long period of development that seemed likely to push the next set of manned missions into the late 1970s or early 1980s.

Apollo's mission planners, looking ahead to only three more lunar missions, intended to make them as scientifically productive as the limitations of the system allowed. Modifications to the lunar module were under way and the first flight model of the lunar rover was undergoing tests. Landing sites for the next two missions had been selected and a group of sensors for remote sensing of the moon from lunar orbit was under development (see Chapter 12).

Apollo 15: Great Expectations

Apollo 15, the first of the "J" missions, was the most complex mission yet attempted, one of which much was expected. Its landing site, 26 degrees north of the lunar equator, was the first outside the original "Apollo zone." Commander David R. Scott would bring the lunar module to the surface along a steeper approach path (25 degrees) than had been used previously, coming in over the Apennine front north of the 3,500-meter (11,500-foot) ridge of Hadley Delta. His landing site was about 5 kilometers (3 miles) northwest of Hadley Delta, 11 kilometers (7 miles) southwest of the foot of 5,500-meter (18,000-foot) Mount Hadley, and about a kilometer (half a mile) east of the edge of Hadley Rille.[1]

Apollo 15 was the first mission to fly the extended lunar module with the increased payload it could carry: expendable supplies sufficient to support the astronauts for 67 hours, a complete lunar surface experiments package (see Appendix 5), and the first lunar roving vehicle. Excursions were planned to the foot of the Apennine Front and along Hadley Rille. In addition to the usual geological investigations, the crew would take core samples to a depth of 2.5 meters (8 feet), using an electrically powered drill. Temperature sensors inserted into the drill

This photo of the Apollo 15 command and service module, taken from lunar module Falcon *while still in lunar orbit, clearly shows the arrangement of the instruments in the scientific instrument module (SIM) bay of the service module.*

holes would enable scientists to record the rate at which heat flowed from the moon's interior to the surface.[2]

Starting with Apollo 15, the orbiting command and service module had a much larger role to play in the lunar science program. While Scott and LM pilot James Irwin were on the moon, command module pilot Alfred Worden's flight plan called for him to operate a group of instruments carried in the scientific instrument module (SIM) in the service module. Two cameras would collect high-quality photographs of the lunar surface. A laser altimeter was provided to give a precise profile of the lunar surface and to allow photogrammetrists to correct the data from the cameras, from which accurate maps of the moon could be produced. Besides cameras, the SIM bay housed instruments for remote sensing of the lunar surface and a scientific subsatellite to be released in lunar orbit for long-term measurement of particles and magnetic fields. On the trip back to earth Worden would make the first extravehicular sortie in cislunar space to bring film from the cameras back into the command module.[3]

Apollo 16 crewmen Charles Duke (right) and John Young (behind the vehicle) observe the deployment of their lunar roving vehicle during a test at Kennedy Space Center.

The Lunar Rover and New Experiments

The lunar roving vehicle was among the more important new developments for the extended Apollo missions, since it would allow the astronauts to sample a much larger area around the landing site. When Boeing won the rover contract in November 1969, it had little time to produce the novel vehicle: at the time the schedule called for flight of the first rover on Apollo 16 in April 1971.[4] Many details of design were still to be settled, and much of 1970 was spent in ironing them out.

What evolved from the program was a unique four-wheeled vehicle remarkable for its versatility and compactness. Smaller than a subcompact car, the rover had a 2.3-meter (7.5-foot) wheelbase and measured 3.1 meters long, 1.8 meters wide, and 1.14 meters high (10 x 6 x 3.75 feet). For its trip to the moon, all four wheels folded inward and the rover folded double for stowage in the descent stage of the lunar module. It was powered by two 36-volt silver-zinc batteries with sufficient power for a range of 65 kilometers (40 miles) at speeds up to 17 kilometers per hour (11 miles per hour). If either battery failed the other could

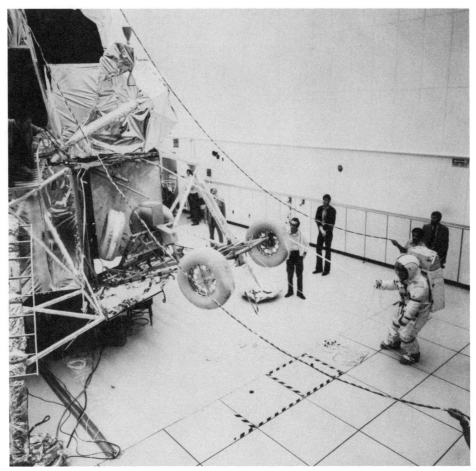

Apollo 16 commander John Young practices unstowing the lunar roving vehicle from its storage space in the lunar module.

carry the entire load. Four separate motors, one at each wheel, drove the vehicle; any wheel could be cut out of the circuit and allowed to ''free-wheel'' if its drive mechanism developed problems. Tires of spring-steel wire mesh carried treads of titanium-alloy chevrons for traction; ''bump stops'' inside the wire mesh prevented collapse of a tire in case of severe shock. Two collapsible seats were mounted on the chassis, along with racks for equipment, tools, and sample bags.

All the rover's driving functions—forward and reverse speed, braking, and steering—were controlled by a T-handle mounted between the seats. Besides four-wheel drive, the rover featured redundant four-wheel steering. Either front or rear wheels could be turned while the other pair was locked straight, or both could be used simultaneously, allowing the rover to turn completely around within its own length. It could clamber over obstacles 30 centimeters (1 foot) high, climb and descend slopes of 25 degrees, and park on slopes as steep as 35 degrees.

The rover navigated by a dead-reckoning system. At the start of a traverse the astronauts oriented the system's navigational gyroscope with reference to the sun. During the trip, odometers recorded the distance traveled and the system con-

Driver's view of the control panel for the lunar roving vehicle. Power monitors and controls are in the center; the steering control is the handle at the extreme right. The navigational system readout (upper left) shows the current heading of the rover and the distance and bearing to the lunar module. The speedometer is at the upper right.

tinuously computed and displayed the direction and distance to the lunar module.

Communications from the rover were relayed to earth through the lunar module, but could be sent directly to earth when the lunar terrain blocked the line of sight between the two vehicles. Mounted on the front of the rover were a television camera and a high-gain antenna. At each stop the astronauts would activate the camera and orient its antenna toward earth. For the first time, scientists at Mission Control would be able to look over the astronauts' shoulders during surface exploration. Unlike the rover itself, the camera could be controlled from earth, allowing scientists to direct the lunar activity, if necessary, on the basis of what they could see. Alternatively, they could use the camera to survey the area for interesting features while the astronauts carried out their preplanned activities.

Fully equipped and ready for flight, the lunar rover weighed just over 200 kilograms (440 pounds) and could carry about 490 kilograms (1,080 pounds), including the astronauts, their life-support systems, their tools and equipment, and 27 kilograms (60 pounds) of lunar samples that could be brought back to the lunar module on each trip. Models to be flown on missions were designed to support this weight on the moon; in all prelaunch tests they had to be supported on special racks, since if anyone inadvertently sat down on one it could collapse.[5]

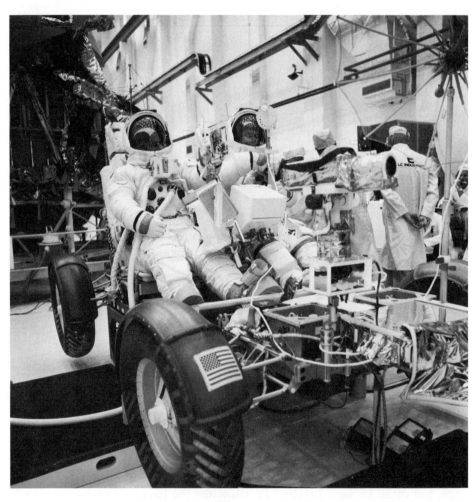

Apollo 17 astronauts Jack Schmitt (left) and Gene Cernan during the checkout of their lunar roving vehicle. The roving vehicle used on the moon had to be supported on earth because it would not support the astronauts' weight in normal gravity; note the wheels off the surface.

Development of the rover was not without its problems. Although the project got some schedule relief with the postponement of the first flight from April to July 1971, at one point it was estimated to be two months behind schedule.[6] The deficit was soon made up, however. The one-g training vehicle, a structurally strengthened version equipped with conventional tires for use on earth, was delivered on November 18, 1970, and crews immediately began training with it.[7] The first vehicle to be flown to the moon was delivered to Kennedy Space Center the following March for prelaunch testing. Eight weeks later the rover was stowed in the lunar module and both were loaded into the Saturn V stack.[8]

In January 1971 Apollo 15's scientific instruments began arriving at Kennedy Space Center, where they were subjected to two months of extensive preflight testing. KSC engineers had more than a few problems with the instruments, as they usually did when a new component was added to the spacecraft. Some arrived late; principal investigators or contractor representatives often wanted to make small changes or conduct last-minute tests. Mechanical problems complicated the process as well. Two of the instruments had to be extended from the service module on long booms to avoid interference from spacecraft systems. The booms, designed for zero-g operation, would not support the instruments in earth gravity, so on-the-spot fixes had to be improvised, with less than completely satisfactory results. But calibrations were finally completed and the SIM was returned to North American's California plant for rework. When retesting was completed in late April, the package was installed in the service module. The lunar subsatellite arrived in May and after checkout was loaded into the space craft on June 9.[9]

To the Mountains of the Moon

Thunderstorms visited launch complex 39A during the month preceding Apollo 15's launch. Four lightning strikes damaged some ground support equipment but the spacecraft and launch vehicle were undamaged.[10] Otherwise the countdown for Apollo 15 was, as launch director Walter Kapryan described it, "the most nominal countdown that we have ever had."[11] At 9:34 a.m. on July 26, just 187 milliseconds behind schedule, the Saturn V carrying command module *Endeavour*** and its crew blasted off. Twelve minutes later Scott, Irwin, and Worden were in orbit some 92 nautical miles above the earth, and after two revolutions their S-IVB stage reignited to send them on their way to the plain at Hadley. After extracting the lunar module *Falcon* (named for the Air Force Academy's mascot by the all-Air-Force crew) from its stowage atop the S-IVB, they settled down for an uneventful 3-day trip to the moon. Only two small midcourse corrections were required along the way.[12]

---

*Conscious of the importance of science on their mission, the crew named their spacecraft for the ship commanded by Lieut. James Cook, who had sailed from England to Tahiti in 1768 to observe the transit (passage between the earth and the sun) of the planet Venus in August 1769. Apollo Prime Crew Press Conference, June 18, 1971, transcript.

Four hours before settling into lunar orbit, the crew jettisoned the cover of the scientific instrument module in preparation for the lunar-orbital science to be conducted later. *Endeavour* went smoothly into lunar orbit. Scott and Irwin entered *Falcon* about 40 minutes early, checked out its systems, and had ample time to eat lunch before beginning powered descent. After they undocked, Worden put *Endeavour* into a circular orbit suitable for gathering scientific data.[13]

Ten hours into their fourth day, Scott powered up *Falcon* for the approach to Hadley. As had been the case throughout the mission so far, everything went well during the landing approach. At about 9,000 feet (2,750 meters) above the surface Scott noted the peak of Hadley Delta to his left; until he reached 5,000 feet (1,500 meters) the only other landmark he could spot was Hadley Rille. The terrain was less sharply defined than he had expected from simulations. After entering several redesignations of the landing site into his control computer, Scott

Astronaut Jim Irwin is aboard the lunar rover waiting for Apollo 15 commander Dave Scott to finish photographing the landing site before departing on their first lunar traverse. The view is toward the south. St. George crater, about 5 kilometers distant, appears immediately behind Irwin. Hadley Delta rises behind the lunar module, which is visibly tilted because Scott landed on the rim of a shallow crater.

brought *Falcon* down through blinding dust and touched down at 6:16 p.m. Eastern Daylight Time on July 29. Not sure of his exact location, Scott was sure he was well within the boundaries of his designated landing zone. The lunar module came to rest tilted back and to its left; two of its landing pads were just over the edge of a small crater that Scott had not been able to see as he approached the surface.[14]

During the next 67 hours Scott and Irwin were to ride and walk over more of the lunar surface than any of their predecessors. The first chore was a "standup EVA," to survey the landing area for the benefit of the scientists in Houston. Scott removed the upper hatch of the lunar module and stood on the engine cover, describing what he could see and taking photographs that would allow construction of a panorama of the landing area. One remarkable photograph showed clear evidence of stratigraphy in a prominent landmark ("Silver Spur") to the east of Hadley Delta—a feature the explorers were unable to see later because of unfavorable lighting. While Scott and Irwin carried out their assignments on the surface, Al Worden, orbiting above in *Endeavour*, had activated the scientific instruments and cameras and was busy gathering data. On his 15th revolution he spotted *Falcon* on the surface and relayed its position to Houston.[15]

The next morning Scott and Irwin were up early and out on the surface. They unloaded their roving vehicle without difficulty, climbed on and buckled their seat belts, and set out on the first traverse. Scott soon discovered that the rover's front wheels could not be steered. He could not correct the anomaly, but went ahead, relying on the rear-wheel steering alone.[16]

Their first goal was "Station 1," a spot on the rim of a medium-sized crater ("Elbow") located on the edge of Hadley Rille at the point where it makes a sharp bend. On the way Scott had to get the feel of driving the rover, which he found to be an excellent vehicle. Driving it on the moon, however, was not quite like driving the 1-g trainer, and among other things he had to keep his eyes on the surface ahead, especially when heading toward the sun, because surface features were not always readily visible. The rover's maneuverability was good, but at 8 to 10 kilometers per hour (5 to 6 miles per hour) the ride was a bit bouncy. Driving took all of Scott's attention, so Irwin provided most of the descriptions for Houston's scientists.[17]

Proceeding along the edge of Hadley Rille, the explorers had an excellent view of the entire width of the canyon (about 1.5 kilometers or almost 1 mile) and down its length to the sharp bend by "Elbow" crater. Landmarks along the route were not always easy to identify, but their course was laid out by range and bearing and they were never as uncertain about their position as Shepard and Mitchell had been on Apollo 14. Less than half an hour after leaving the LM, and having traveled something over 3 kilometers (about 2 miles) they reached their first stop, dismounted, and turned on the television camera to give Houston a view of the area. Using a telephoto lens, Scott photographed the far side of Hadley Rille where layering was obvious in outcrops not far below the rim. Panoramic photography and sample collection completed their work at Site 1.[18]

Remounting their vehicle, Irwin and Scott drove around "Elbow" crater to Site 2, about 500 meters (1,600 feet) farther down at a point where ejecta from St. George crater encroached on the edge of the rille. Here they began to encounter large blocks lying on the surface. One that excited Irwin particularly was a meter or so (3 to 4 feet) across, its downslope edge buried in the loose soil and its upslope edge free of the surface. He photographed and sampled it; then once again they took panoramic photographs and soil samples and drove a double core tube into the edge of a crater.[19]

With no more stops scheduled, Scott now turned the rover on to the heading prescribed by his navigation system to return to the LM. On the way he cautiously tried some maneuvers with the rover. Attempting a turn on a downslope, he discovered that "you can't go fast downhill in this thing, because if you try and turn with the front wheels locked up like that, they dig in and the rear end breaks away, and around you go; and we just did a 180 [degree turn]."[20] Near Elbow crater Scott noted that there was a "neat place to go down into the rille," but Capcom reminded him that "we'd rather that you don't take that option."[21]

Hadley Delta slopes upward and to the right in this photograph taken from the Apollo 15 lunar module Falcon's upper hatch shortly after landing. The feature in the background, 20 kilometers southeast of the landing site, is "Silver Spur."

Riding back toward *Falcon*, Scott and Irwin frequently saw features they would have liked to stop and examine, but they had no time, so they briefly described them to Houston and continued.[22] When they first caught sight of the LM they estimated its bearing from the rover at 15 degrees, but the navigation system showed 34, indicating some drift in the system. They reached the lunar module 2 hours and 15 minutes after they had left, parked the rover, and unloaded the surface experiments package. Scott had picked a spot about 110 meters (360 feet) west-northwest of the *Falcon* to deploy the instruments. Apollo 15's experiments package included a new heat flow experiment, which required drilling two holes in which temperature sensors would be placed. The experiments were deployed without difficulty, but the second heat-flow drill hole caused problems when the drill stuck and proved difficult to remove. By this time the astronauts were approaching the limit of their life-support systems, so Mission Control directed Scott to leave the drill in the hole and get on to other tasks. After deploying the laser reflector and the solar-wind collector he and Irwin returned to *Falcon* and closed the hatch. They had been out for six and a half hours and covered 10.3 kilometers (5.6 miles) on the lunar surface—more than twice as far as the Apollo 14 astronauts had traversed (on foot) during their entire mission.[23] A 40-minute debriefing with Houston completed their first day of lunar exploration.[24]

In the press briefing immediately following the first day's surface activity, flight director Gerald Griffin expressed satisfaction with the astronauts' performance. Apart from the failure of the rover's front-wheel steering, which had not significantly impaired the planned activity, everything had gone well. Scott had shown a somewhat higher metabolic rate than anticipated, resulting in faster consumption of oxygen than had been allowed for. This was not a serious concern, but it would have to be taken into account in planning the remaining excursions. About the only changes anticipated for the next day's activity were to attempt to solve the rover's steering problem and to try to extract the drill core for the heat-flow experiment.[25] However, the science planning team still had to review the day's results; plans for the second traverse could well be changed in light of what had been accomplished.

Next morning, after Scott and Irwin had finished breakfast and attended to some chores in the LM, Capcom Joe Allen briefed them on the upcoming day's work. Science planners had decided to shorten the traverse somewhat to provide more exploration with less travel time. A low-priority sampling stop had been tentatively deleted and the route at the foot of Hadley Delta curtailed. The scientists were giving the astronauts considerable freedom on this traverse, as Allen's instructions showed:

> We're going to depend very much on the observations from the two of you, and it's going to be . . . your choice on exactly where you'd like to range and where you'd like to carry out your major sampling tasks. . . . We're looking now, primarily, for a wide variety of rock samples from the [Apennine] Front. You've seen the breccias already. We think there may very well be some large crystal[line] igneous [rocks], and we'd like samples of those and whatever variety of rocks which you're able to find for us—but primarily, a large number of documented samples and fragment samples. . . . I'll [stop] now and ask for any more questions.[26]

Scott replied, "No, no questions, Joe. You're really talking our language today."[27] During their extensive geological training, Scott had come to enjoy field geology and considered himself a serious amateur,[28] and this was the kind of freedom he wanted. It was the kind of exploration the geologists had wanted as well, and although they would have preferred to have one of their own on the moon, apparently they were confident enough of Scott's and Irwin's training to give them a much freer hand than any previous team of lunar explorers.

Suited up and out on the surface, Scott and Irwin got aboard the rover and belted in, pausing before they started to flip the circuit breaker for the front steering open and then closed again. When Scott moved the hand controller the front wheels moved. Nobody knew why, but he now had full four-wheel steering.[29] Then they set out, on a southerly heading, for the sloping terrain at the foot of Hadley Delta. Rolling along at 8 to 9 kilometers per hour (5 to 6 miles per hour), they described the craters and rocks along the route. In about 40 minutes they were in the vicinity of "Spur" crater, one of their sampling stops, and they parked the rover to reconnoiter and collect specimens.[30] With continuous commentary for Houston, the astronauts roamed the area for the next hour, taking pictures and bagging samples. Then, as they examined a boulder that had attracted their attention, Scott remarked, "Guess what we just found." Then Irwin came over for a look:

"Silver Spur," photographed from a distance of some 10 miles. The stratigraphy in the peak is evident. Hadley Delta rises to the right of Silver Spur, its sloping flank visible in the foreground.

Irwin: I think we just found what we came for.

Scott: Look at the plage [plagioclase] in there. . . . I think we might [have found] ourselves something close to anorthosite, because it's crystalline, and . . . it's just almost all plage. What a beaut!

Irwin: That really is a beauty. And . . . there's another one down there.[31]

"What we came for" was a specimen of the primitive lunar crust—anorthosite, the rock that some scientists believed was the first material that solidified from the molten outer layer of the moon. The possibility of finding this material was one reason the Hadley-Apennines site had been chosen; scientists thought that if anorthosite was to be found on the moon, a highland site (and one adjacent to a large collisional mare) would be the best place to look for it. Scott put the sample into a bag by itself; back on earth it was dubbed the "Genesis rock."[32]

The "Genesis rock" as Apollo 15 commander Dave Scott spotted it in its "come-and-sample-me" position atop a small mound of soil (upper center). Scott quickly identified it as anorthosite, a rock many scientists believed made up the primitive lunar crust. The vertical-locating gnomon, with attached color chart, is just to the right of the specimen.

237

After a few more minutes of sampling and photographing the site, they climbed back on the rover and started back to the lunar module. Since there was time available, Houston directed them to stop at site 4 (the one that had been deleted before the traverse started) where they took more samples. Then it was back to *Falcon* to unload their cargo and return to the surface experiments site. Scott made another attempt to remove the core drill he had left in the ground the day before, but succeeded only when Irwin helped him pull it out. The second hole for the heat-flow experiment was drilled and the temperature sensors were emplaced in both holes. After performing a few more experiments in soil mechanics, Scott and Irwin drove back to the lunar module, where they had one more task to perform: planting the American flag at the landing site. That done, they transferred the samples to the LM, and closed the hatch, having completed the most productive lunar science traverse of the program so far.[33]

Flight director Gerald Griffin expressed that evaluation more strongly at the change-of-shift press briefing: in his opinion, "we probably have just witnessed the greatest day of scientific exploration . . . in the space program that we've ever seen."[34] Allowing for his understandable enthusiasm, Griffin's statement had solid foundation in fact. Lunar missions had made advances in two years' time that few would have expected. The equipment had functioned virtually without flaw, with margins of expendables to spare. Scientists and network audiences on earth had received live color television pictures far surpassing the crude black-and-white images transmitted by Apollo 11. The astronauts had wasted no time, they had provided excellent descriptions of what they could see, they had exercised their judgment in collaboration with the ground-based scientists, and had documented and collected more than 100 pounds (45 kilograms) of rocks, soil, and core samples. Observers in Mission Control thought they could tell that Scott and Irwin adapted to lunar gravity on the first traverse and as a result were more sure-footed and confident on the second.[35] One more traverse remained, a trip to the edge of Hadley Rille, but 15 would have been rated a success without it. Capcom Joe Allen passed the word from scientists to the crew just after they returned to the LM:

> I'm told that we checked off the 100 percent science completion square some time during EVA-1 or maybe even shortly into EVA-2. From here on out, it's gravy all the way, and we're just going to play it cool, take it easy, and see some interesting geology. It should be a most enjoyable day.

"Okay, Joe," Scott replied, "Thank you. We're looking forward to it."[36]

Their work was not the mission's only contribution to lunar science. While they had been dashing around the surface, Al Worden in *Endeavour* had been circling the moon operating the cameras and instruments. In *Endeavour's* highly inclined orbit he was able to see features that no previous observers had laid eyes on, and he provided detailed descriptions each time he crossed the earth-facing side of the moon.[37]

Early the next morning Irwin and Scott left the lunar module on a west-northwesterly heading to get a good look at Hadley Rille. After describing, photographing, and televising the features of the broad canyon, including evident layering in the walls, they completed their third traverse in just under 5 hours.[38]

Back at their base, they collected the solar-wind experiment, after which Irwin produced a postal cover carrying a new stamp commemorating ''a decade of achievement'' in space, and applied a first-day cancellation provided by the Postal Service.[39] A few minutes later Scott stood before the rover's TV camera to conduct a scientific demonstration:

> . . . In my left hand, I have a feather. In my right hand, a hammer. . . . One of the reasons we got here today was because of a gentleman named Galileo a long time ago who made a rather significant discovery about falling objects in gravity fields. . . . The feather happens to be, appropriately, a falcon feather, for our *Falcon,* and I'll drop the two of them here and hopefully they'll hit the ground at the same time. [They did.] . . . This proves that Mr. Galileo was correct in his findings.[40]

Scott then drove the rover about 300 feet (90 meters) away, where he parked it with the television camera pointed at *Falcon,* aligned the antenna with earth, and walked back to help Irwin load sample bags into the lunar module.

From St. George crater at the foot of Hadley Delta, this view looking north up Hadley Rille shows the floor of the gorge littered with rock fragments that have broken off the layers higher up the sides.

For the next three hours the crew of the *Falcon* were occupied with stowing equipment and sample containers and configuring the LM for takeoff. Then, while the television camera watched, *Falcon's* ascent stage shot up from the surface in a shower of fragments of insulation, visible for only a second or two. Flight controllers had intended to follow *Falcon* with the camera, but decided against it when problems developed in the camera's control system.[41] The result contrasted sharply with the majestic rise of a Saturn V; with a "quick pop" and "a shower of sparks [that] looked more like something left over from the Fourth of July," as one columnist put it, *Falcon* quickly disappeared from the TV screen.[42]

Falcon and *Endeavour* linked up an hour and 34 minutes later, and after jettisoning the lunar module the three astronauts settled in for two days of additional lunar-orbital data-gathering.[43] On their last orbit they released the scientific subsatellite, then headed for home. While the spacecraft was still nearly 200,000 miles (320,000 kilometers) from earth, Al Worden carried out the last extravehicular excursion of the mission, a 38-minute "space walk" to remove film cassettes from the cameras in the scientific instrument module.[44] The long voyage home ended on August 7, when *Endeavour* dropped into the Pacific Ocean about 320 miles (515 kilometers) north of Hawaii. The landing was a little harder than normal because one of the three parachutes failed to open fully, but no damage or injury resulted.[45]

Apollo 15's Science Dividend

It was clear even before Scott and his crew returned that Apollo 15 had surpassed all prior missions in both the quantity and variety of information it returned. More than 350 individual soil and rock samples aggregating 77 kilograms (170 pounds), ranging in size from 1 or 2 grams (0.03 to 0.06 ounce) to more than 9.5 kilograms (21 pounds) and taken from two distinct selenological regions (the mare plain and the foot of Hadley Delta), were the richest geological haul of the program.[46] The seismometer and the heat flow probes emplaced with the surface experiments package were important in obtaining data on the moon's interior, while the instruments Al Worden had operated in lunar orbit gathered information on the chemical composition of large areas on the front and back sides of the moon.

If the scientists had been unhappy with MSC's cooperation in the early missions, they were overjoyed with the way Apollo 15 was conducted, and several graciously congratulated MSC on it. Gerald Wasserburg of Caltech wrote to Robert R. Gilruth to offer his congratulations "on one of the most brilliant missions in space science ever flown." He particularly praised the fairness and wisdom of the decisions made during the astronauts' explorations, even though he had not always got what he wanted.[47] Larry A. Haskin, vice chairman of the lunar sample analysis planning team, told Gilruth that

> Scott and Irwin had not been on the surface very long before . . . we felt comfortable and confident that the scientific aspects of the mission were in competent hands. . . . We always felt that the decisions made in real time . . . were

fair with respect to priorities given to various activities and that the mission was well balanced.

Haskin stressed that the inclusion of scientists in planning before the mission and decision-making during the mission (through the science working group) was a breakthrough in MSC's relations with the scientific community.[48]

Preliminary examination of the samples began as soon as the containers were returned. Documentation of these samples was unusually complete, except for some taken on the last traverse when time ran short. Taking advantage of the fact that the astronauts were not quarantined after this mission, the preliminary examination team asked Scott and Irwin to be present while the containers were opened, to provide whatever information they could remember about some of the less well-documented specimens while the circumstances were still fresh in their minds.[49]

The preliminary examination team concluded that three distinct types of structures had been sampled at Hadley: the Apennine Front, which had been produced by collision of an asteroid-sized object with the moon; a series of horizontal lava flows that filled part of the impact basin; and a ray of ejecta from a large young crater produced after the basin was filled. Samples from the mountain front were very diverse, including both breccias and igneous rocks. Those from the mare plain, where *Falcon* landed, were of two basic types: basaltic rocks and glass-coated breccias. The "Genesis Rock," so designated by the press, did turn out to be

Dave Scott and scientist-astronaut Joe Allen (rear) examine the "genesis rock" that Scott and Irwin found on Apollo 15. The rock was almost pure anorthosite, possibly a remnant of the original lunar crust. The gadget suspended above the rock displays identifying numbers when the sample is photographed.

an extremely pure anorthosite substantially altered by shock, which suggested that it was involved in a severe impact after it crystallized. Its age was determined to be 4.15 billion years, with a possible error of 0.2 billion either way, making it the oldest whole rock so far returned from the moon.[50]

The core tubes, one containing a 2.4-meter (almost 8-foot) section of the soil, were not opened during the preliminary examination, but x-ray examination indicated 58 individual layers, from 0.5 to 21 centimeters (0.2 to 8.25 inches) thick, showing that the soil contained a substantial stratigraphic record.[51]

Instruments in the service module provided information about the large-scale chemical variations across the lunar surface. The x-ray fluorescence experiment measured the x-rays emitted by light elements (magnesium, aluminum, and silicon particularly) when struck by x-rays from the sun. From these data scientists could determine the ratio of aluminum to silicon over large areas under the orbit of the command and service module. Initial examination of the results found distinct chemical differences between the maria and the highlands. The highlands showed an aluminum/silicon ratio characteristic of anorthosite, which, according to some theories, was formed in the first major event of the moon's geologic evolution. The maria, on the other hand, were chemically different; sample analysis showed them to be basalts formed by lava flows, and the x-ray fluorescence results were consistent with this interpretation.[52]

The preliminary examination team completed the first stage of its work with descriptions of more than 50 rocks larger than 50 grams (1.7 ounces) correlated with the astronauts' descriptions and photographs at the end of August. MSC recommended release of samples to 20 research teams as soon as possible, since their results would be useful in planning future lunar surface activity.[53] The rest of the samples were less time-critical and were released to principal investigators two months later.[54]

A Geologist for Apollo 17

As was his custom, Slayton waited until Apollo 15 had returned from its flight to announce his choice of crewmen for the last lunar mission. On August 13, 1971, he named Eugene Cernan, Ronald Evans, and Jack Schmitt as the prime crew for Apollo 17, backed up by Dave Scott, Al Worden, and Jim Irwin, just returned from Apollo 15.[55] The decision to put Schmitt on the crew instead of Joe Engle—who had trained as lunar module pilot with Cernan and Evans on the backup crew for Apollo 14—was, like all those before it, Slayton's, and no one in the NASA organization put pressure on him to make it.* He noted later, however, that "we might have gotten some [pressure], I suppose, if we hadn't made the right decision down here. [But] there wasn't any doubt . . . that we

---

*But see Chapter 12. Earlier in the year, Dale Myers, chief of manned space flight, and Robert Gilruth, MSC director, had agreed to Schmitt's appointment to the Apollo 17 crew, subject to his satisfactory completion of training, and Slayton was undoubtedly a party to that decision. The decision was evidently a well kept secret, however, for as late as August 3, Schmitt was quoted as admitting that it seemed unlikely he would get the chance to explore the moon. Stuart Auerbach, "Apollo 15 Leaves Moon Orbit At End of Historic Mission," *Washington Post*, Aug. 5, 1971.

Apollo 16 commander John Young salutes the colors while leaping clear of the lunar surface.

were going to have to do that when they canceled [Apollo 15 and 19] , and we did it."[56]

Engle, who had been out of town during Apollo 15 on personal business, learned that his plans had been changed on August 10, when he called in to the Astronaut Office to see if he had any messages. It was a tremendous disappointment; he had made the hard decision to leave the X-15 program in 1966 because he thought that going to the moon was the only way he could surpass his past accomplishments. But, he said, "when something like this happens, you can do one of two things. You can lay on the bed and cry about it . . ., or you can get behind the mission and make it the best in the world." Engle's choice was to support the mission. He would try to help Schmitt fit into the crew that he himself had trained with for so long. After that, he hoped to test-fly the space shuttle.[57]

At the customary news conference the week after the announcement, the first question challenged Schmitt's assignment to the crew instead of Engle. Schmitt said, "There's no question that Joe Engle is one of the most outstandingly qualified test pilots in the business," but he was confident of his own abilities: "as far as my qualifications to fly the spacecraft are concerned, I will attempt to compete with anybody in the program." Cernan agreed: "Jack isn't sitting here as part of this crew for any other reason than that he has rowed hard, he's earned it, and he deserved it." He noted that Engle and Dick Gordon (command module pilot on Apollo 12) had been assigned to the new space shuttle project and commented (perhaps with a trace of envy), "Those guys moving into shuttle right now are probably going to contribute . . . a lot more than maybe even we can contribute by a lunar mission.''[58]

On its final flight to explore the moon, then, Apollo would send a scientist—the one astronaut indisputably qualified to make the observations scientists had long wanted to make. It remained to be seen whether, under the constraints that limited lunar exploration, he could apply his experience to enhance the quality of the results.

To the Lunar Highlands: Descartes

Two landing sites had been seriously considered for Apollo 16, the crater Alphonsus, 300 miles (480 kilometers) south of the moon's center, and the region some 340 miles (550 kilometers) east-southeast of the moon's center, north of the ancient crater Descartes. In both places geologists thought they would find highland material differing in composition from the Fra Mauro samples and the basalts filling the maria. The wall of Alphonsus was, some argued, pre-Imbrian highlands material, while dark craters on its floor were thought to consist of relatively young volcanic material that might have originated at great depths. North of Descartes, two formations (Cayley and Descartes) were of major interest. Evidence indicated that both were volcanic but of different types and ages. The area is the highest topographic region in the highlands on the visible face of the moon, indicating that the Descartes volcanics represent remobilized highlands. Analysis of these materials was expected to clarify the basic processes that formed the highlands.[59]

Preliminary discussions among interested scientists considered both sites but did not entirely agree on either.[60] Alphonsus remained a strong candidate, but in view of the fact that it would be preferable to have more results from Apollo 14 and 15 before landing there, Descartes was preferred for Apollo 16. With one more mission left, Alphonsus could still be visited. The Apollo Site Selection Board approved Descartes as the landing site for Apollo 16 at its meeting on June 3, 1971.[61]

Science plans for the Descartes mission were much the same as they had been for Hadley Rille: to inspect, survey, and sample materials and surface features in the landing area; emplace and activate the lunar surface experiments; and conduct photography and remote sensing of the moon from lunar orbit. The surface package included two new instruments: a magnetometer and, in place of Apollo 15's suprathermal ion detector, an active seismometer that would be excited by several explosive charges after the astronauts left. Another new experiment was an automatic far-ultraviolet camera and spectrograph to photograph several galaxies from the moon's surface. Its objective was to study the distribution of interplanetary and intergalactic hydrogen.[62]

Apollo 16 blasted off from Kennedy Space Center at 12:54 p.m. Eastern Standard Time on April 16, 1972. Command module *Casper* and lunar module *Orion* arrived in lunar orbit three days later. All systems functioned well until *Orion* separated from the command module; a malfunctioning component in the main propulsion system caused Houston to delay the lunar module's descent for nearly

Apollo 16 commander John Young, identified by the stripes on his legs, arms, and helmet, returns from laying out the lunar surface experiments at the Descartes landing site. In the distance behind Young is the lunar triaxis magnetometer; to his left is the radioisotope thermal generator (black), which powered the experiments. Other apparatus on the surface is not identifiable.

six hours while it was checked out. When Mission Control was satisifed, *Orion* fired its descent engine and landed easily on the plain at Descartes at 9:33 p.m. EST on April 20.[63]

In the next 71 hours mission commander John Young and lunar module pilot Charles Duke laid out the surface instruments and conducted three traverses in their lunar rover, covering in all some 27 kilometers (nearly 17 miles). While they were busy on the surface, Ken Mattingly in *Casper* was occupied with operating the instruments in the service module. The only serious mishap on the surface occurred when Young tripped over the cable to the heat-flow sensors, pulling it loose from the central station and incapacitating the experiment.[64]

Young and Duke finished their exploration, loaded the 96 kilograms (210 pounds) of samples they had collected into *Orion*, and rejoined Mattingly in lunar orbit on April 23. They released the moon-orbiting subsatellite, but because of recurring problems with the service propulsion system, the spacecraft was not in the optimum orbit for the satellite. As a result, the satellite crashed into the moon after only five weeks.[65] During the four-day return flight they conducted additional experiments with electrophoresis, a technique that offered advantages for separating certain biological preparations that could not be efficiently done in a gravity field. A normal landing in the Pacific, north of Christmas Island, completed the mission on April 27.[66]

Even before the Apollo 16 rock samples had been returned to Houston it was apparent that premission interpretations of the Descartes site had not been accurate. The service module instruments showed abnormally low radioactivity and high aluminum-silicon ratios in the area—both characteristic of typical lunar highlands. Furthermore, the laser altimeter found that the extensive plateau on which the site lay was almost four miles higher than the surrounding terrain, an elevation that seemed impossible to achieve by buildup of volcanic material.[67] Preliminary examination of the samples confirmed what scientists had suspected as soon as Young and Duke began to explore: the site was not volcanic. Most of them appeared to be impact breccias rather than basalts.[68] This result, confirmed by later analyses, "forced a reevaluation of the process of photogeology and site selection," because the majority opinion before the mission had affirmed that the site was volcanic in origin.[69]

The more samples returned from the moon the less clear the picture of its origin and evolution became to scientists. At the third lunar science conference, which included presentations of information from Apollo 15 and the Soviets' Luna 16, a more comprehensive picture of the moon was evident, but fewer investigators would claim to understand its evolution. The best summary at the time seemed to be that the moon is now relatively cold and inactive but that it has

Apollo 16 commander John Young collects a rake sample, screening out fragments smaller than about 1 centimeter.

gone through a complex sequence of melting and resolidifying by internal or external heating.[70] Toward the end of 1972 one baffled scientist, Gerald Wasserburg, summarized the frustrations of the lunar scientists: "we've got answers but not the questions," he said. "I'm not sure we're asking the questions in exactly the right way."[71]

Apollo 17: The Close of an Era

Following completion of Apollo 15, with 16 in the planning stages and only one mission remaining after that, one scientist expected that the process of choosing a site for the last flight would be "a dilly of an affair."[72] Perhaps with that thought in mind, Rocco Petrone, Apollo program director at NASA Headquarters, started the procedure early, in October of 1971. After a site evaluation document had been circulated among selected interested scientists for their critique and response, an ad hoc site evaluation committee would meet no later than December 30 to consider site proposals and make recommendations to the Apollo Site Selection Board in early 1972.[73]

Actually, site selection for Apollo 17 was not as difficult as might have been expected, considering that it was the last chance for many years to obtain material and data from the moon. Preliminary review of possible sites produced a clear consensus of objectives for Apollo 17: to investigate the pre-Imbrian highlands

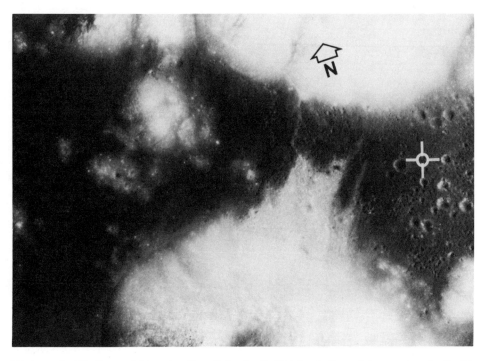

Photograph of the Taurus-Littrow landing site for Apollo 17 taken by the metric camera on Apollo 15. The landing site, marked with a white cross, lies between two massifs (north and south).

as far as possible from the Imbrium basin; to sample "young" volcanics; to obtain photographic and remote-sensor coverage of areas not previously investigated; and to provide the best coverage for the new traverse geophysics experiments that were to go along. Application of all these considerations reduced the interesting sites to three: Taurus-Littrow, a region on the southeastern rim of Mare Serenitatis; Gassendi, a large crater on the northern rim of Mare Humorum; and (ranking a distant third) Alphonsus, previously considered as an alternative to the Descartes site. Of these three, the ad hoc site evaluation committee concluded that Taurus-Littrow was the best. It was a two-objective site (highlands and young volcanic material); it offered a reasonable contingency for a walking mission in case the lunar rover should fail; and it would provide the greatest amount of new information from the orbital sensors in the service module.[74]

Operational considerations, taken up at the February meeting of the site selection board, gave no site a clear advantage. Problems existed at all three, although Alphonsus, already well studied, was acceptable. At Gassendi there was a good chance that the astronauts would not be able to get to their prime objective if they had to land beyond the nominal 3-sigma ellipse. For Taurus-Littrow, detailed studies at first indicated that under "worst-case" conditions the site was unsuitable; the 3-sigma ellipse could not be fitted into it. When trajectory designers were urged to reconsider their calculations in light of the results of Apollo 15, however, they produced some less conservative numbers and the site was accepta-

Geologist Jack Schmitt moves around a huge broken boulder at Taurus-Littrow (Apollo 17). He has sampled the surface material where scoop marks are visible on the lower left front flank and is preparing to take more samples. The boulder rolled down from the North Massif.

ble. With operational considerations out of the way, the board picked Taurus-Littrow because of its greater scientific potential.[75]

The Taurus-Littrow site was a flat-floored valley some 7 kilometers (4.3 miles) wide, a dark, bay-like indentation in the broken mountain chain on the eastern edge of Mare Serenitatis, at 20 degrees 10 minutes north latitude, 30 degrees 46 minutes east longitude. Bounded on three sides by high mountain massifs, the valley contained numerous craters of possible volcanic origin, as Al Worden had noted from orbit on Apollo 15.[76]

Premission data from the landing site indicated it to be geologically complex. Rock types appeared to vary in age, composition, and probable origin, and the mountains displayed structural features that might be correlated with several major landforming events. The valley floor on which Apollo 17 was to land appeared to be volcanic and thus might have originated deep within the moon. A landslide intruded on one side of the valley floor, and boulder tracks down the mountainsides added to scientists' interest in the site.[77]

Until early 1971 Apollo 17 had been scheduled for June 1972; in February the launch schedule was revised and both Apollo 16 and 17 were moved forward, to March and December 1972, respectively.[78] The new launch date for Apollo 17 required one more launch constraint to be lifted, because Taurus-Littrow was not accessible by translunar injection (departure of the spacecraft and S-IVB from the earth parking orbit) over the Pacific Ocean. Early in the program Pacific injection was adopted as the rule, since the alternative—injection over the Atlantic—required a night launch at some times of the year. Although Atlantic injection offered an economy in fuel and thus an increase in allowable payload, the potential problems incident to a night launch (see Chapter 6) tipped the scales.[79] By the time the site selection board met in February 1972, however, "the bullet had been bitten on night launches"[80]; experience had made those problems much less worrisome.

So it was that Apollo 17 was scheduled to leave the Cape at 9:38 p.m. EST on December 6, 1972.[81] From August 8, when the launch vehicle was moved to the pad, preparations went well. Some minor hardware problems cropped up but were handled without delaying the schedule. Launch day was bright and warm; the temperature was in the mid-80s at midafternoon. All looked well for the last Apollo mission until 30 seconds before launch, when the automatic sequencer—the oldest and most reliable piece of automation at the launch complex—shut the system down. The first launch delay in the Apollo program caused by failure of equipment amounted to 2 hours and 40 minutes while Cape and Marshall Space Flight Center engineers worked around the malfunction. At 12:33 a.m. Eastern Standard Time on December 7, command module *America* and lunar module *Challenger* were off for Taurus-Littrow.[82] Flight controllers made up the lost time during the translunar coast.[83]

America and *Challenger* entered lunar orbit on schedule and all preparations for landing went smoothly. After the two spacecraft separated, Cernan and Schmitt took a spectacular photograph of their landing site. Three hours later Cernan powered up *Challenger* for its descent. It touched down within 200 meters (650 feet) of its targeted landing point amid a field of craters at 3:15 p.m. EST on December 11.[84]

For the next 75 hours Cernan and Schmitt conducted the longest, and in many ways the most productive, lunar exploration of the Apollo program. During three trips from their base they laid out the surface experiments, drove the lunar rover about 36 kilometers (22 miles) in all, ranging as far as 7.37 kilometers (4.5 miles) from *Challenger*, and collected roughly 243 pounds (110 kilograms, of soil and rock samples along with more than 2,000 documentation photographs.[85]

Like their predecessors, Cernan and Schmitt were somewhat constrained by the preplanned sequence of activity. Still, before they left the lunar module and while unloading the Rover and the surface experiments, Schmitt found time to give Houston's back-room scientists both large- and small-scale descriptions of the landing area and the surface under his feet.[86] Schmitt found the environment in the landing area "superb" for observation: the lighting was excellent and the rock surfaces generally clean, and he found little difficulty in distinguishing mineralogical and textural differences. For the most part he based his decisions on taking samples on visually detectable differences or similarities.[87]

Besides collecting and documenting samples, on their traverses the explorers laid out explosive charges for a seismic profiling experiment (the charges would be set off after they left), took readings on a portable gravity meter at various points along the route, and set up an instrument to measure electrical properties of the lunar surface. At the site for the surface experiments package they drilled two 2.54-meter (8.3-foot) holes for heat-flow sensors, took a deep core sample,

A view of the lunar orbital experiments in the scientific instrument module of the service module, taken from lunar module Challenger *before it docked with* America *on return from the lunar surface.*

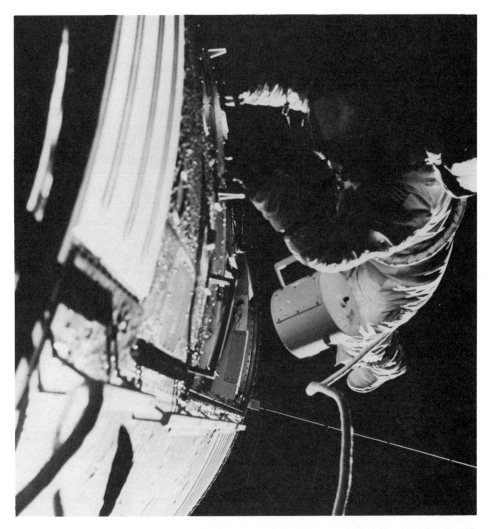

Command module pilot Ron Evans outside the spacecraft America, *recovering film from the scientific instrument module (SIM) bay of the service module. The cylindrical object with the handle, near his knees, is a film cassette.*

and set up the geophones (detectors) for the seismic profiling experiment.[88] The first excursion was largely taken up by these chores.

After an overnight rest and a discussion with Houston concerning plans for sampling, Cernan and Schmitt set out on their second trip to collect specimens from boulders along the lower slopes of the South Massif and to sample the lighter-colored soil that overlay the western part of the valley. It was a long trip—an hour by rover to the first major sampling stop—and would stretch their life-support systems almost to the limit. The last couple of kilometers up the slope taxed the rover, too, but it brought them to their objective in fine style. Schmitt took samples from three boulders which, as best he could tell, had come from layers visible farther up the South Massif.[89]

The obviously interesting features of their first site prompted Houston to lengthen their stay there and cut some time from later stops. On their way back they stopped to take an unscheduled reading on the traverse gravimeter and sampled soil at a couple of crater rims, one of which drew considerable attention. During routine examination of the surface around the crater called "Shorty," Schmitt suddenly called out, "Oh, hey—wait a minute— . . . There is orange soil!" Cernan confirmed it. "He's not going out of his mind. It really is." While looking for the limits of the orange deposit, Schmitt remarked, "if there ever was something that looked like fumarole alteration, this is it."[90] He was excited because orange soil (characteristic of oxidized iron, at least on earth) indicated volcanic activity, probably recent, a feature not previously discovered on the moon. While Houston kept reminding them they were almost at the limits of their walk-back capability—time was running out—they dug a trench, took a core sample and several scoop samples, and took photographs. Then they mounted the rover to head back to *Challenger*. Schmitt talked about his discovery all the way back.[91]

The last sampling trip of the Apollo program was a traverse to the foot of the North Massif, where they found two large boulders that had obviously rolled down from outcrops higher on the mountain; their tracks were visible in the soft soil. After covering 12 kilometers (7.5 miles) and picking up 63 samples (137 pounds, 62 kilograms), Cernan and Schmitt returned to the lunar module.[92] Schmitt picked up a symbolic rock sample in honor of a group of foreign students touring the United States; it would be divided up to provide samples for each country represented. Then Cernan unveiled a plaque on *Challenger*'s landing strut, which commemorated the completion of the first exploration of the moon by humans. Then he made some final dedicatory remarks:

> This is our commemoration that will be here until someone like us, until some of you who are out there, who are the promise of the future, come back to read it again and to further the exploration and the meaning of Apollo.

He then parked the rover at a spot where its television camera could watch their takeoff, and Apollo's last two explorers finished their last tasks on the moon. Cernan closed out the surface activity with the comment that "I believe history will record that America's challenge of today has forged man's destiny of tomorrow."[93] Then they packed up their samples, discarded the tools they would no longer need, and climbed back into *Challenger*.[94]

Next day, December 14, they blasted off to join *America* in lunar orbit. As television audiences on earth watched, the rover TV camera, directed from Houston, followed their ascent stage until it was out of sight, then slowly scanned the now-deserted lunar surface.[95] The awareness that no living person was around made the scene all the more impressive. It was almost possible to hear the silence.

After hooking up with the moon-circling command module, Schmitt and Cernan transferred their samples and data. *America* still had a day's work to do, completing the photographic and remote-sensing work that Ron Evans had been doing while his crewmates were on the surface. Their work completed, the crew of Apollo 17 left the moon with a blast from their service propulsion engine at 8:42 p.m. EST on December 16. A routine transearth coast brought them back to a landing about 300 kilometers (200 miles) east of Pago Pago at 2:25 p.m. EST on

December 19, 1972.[96] Apollo's exploration of the moon, ''one of the most ambitious and successful endeavors of man,''[97] was over.

While scientists awaited the samples from Taurus-Littrow, editorial writers offered varying evaluations of the lunar exploration project. Perhaps the bitterest farewell to Apollo was expressed by William Hines, syndicated *Chicago Sun-Times* columnist, who had opposed the project from the start: ''And now, thank God, the whole crazy business is over.''[98] The *Christian Science Monitor* cautioned that ''Such technological feats as going to the moon do not absolve people of responsibilities on earth.''[99] The *New York Times* decided that ''Man's entire perspective on the universe and on his place in it has been radically changed. Man evolved on the earth, but he is no longer chained to it.''[100] *Time* magazine noted that after the magnificent effort to develop the machines and the techniques to go to the moon, Americans lost the will and the vision to press on. Apollo's detractors, said *Time*, were

> prisoners of limited vision who cannot comprehend, or do not care, that Neil Armstrong's step in the lunar dust will be well remembered when most of today's burning issues have become mere footnotes to history.[101]

Scientist-astronaut Jack Schmitt steps out of the command module America *floating in the Pacific Ocean. The large spherical bags are the uprighting system, which rights the command module if it lands nose-down. A flotation collar has been attached by the frogmen of the recovery team.*

14

PROJECT APOLLO: THE CONCLUSION

At the start of 1973 scientists eagerly awaited the last samples from the moon. Apollo was all but forgotten, however, at NASA's manned space flight centers, which had already shifted gears and were preparing for the next manned projects. In August 1972, when Apollo 17's Saturn V rolled out of the Vehicle Assembly Building at Kennedy Space Center, two launch vehicles in adjacent bays were being prepared for Skylab's first flight, nine months away. Barely a month after Apollo 17 returned, Skylab program officials completed the design certification review for modifications required at the launch complex. At Houston, Skylab crews—officially named in January 1972—had been in training since late 1970.[1] Crews for Apollo-Soyuz, the joint mission with the Soviet Union, scheduled for mid-1975, were named in early February 1973,[2] the same week the first lunar samples from Apollo 17 were released.[3]

The Manned Spacecraft Center now had nearly 380 kilograms (836 pounds) of samples from the moon—all the lunar material the scientists could expect to get for many years. Besides the samples, there was a mass of data from the lunar-orbital experiments and a continuing flow of information from the lunar surface experiments, most of which were still functioning (see Appendix 5). Scientists now began to concern themselves with the preservation, description, and cataloging of the samples and the provision of adequate support for a continuing program of research.[4]

Apollo 17 Samples

The Lunar Sample Analysis Planning Team met at MSC on January 29 to plan for allocation of Apollo 17 samples. In keeping with the custom established on the previous two missions, five key samples were released very early for age-dating, including the orange soil that had so excited Jack Schmitt. During February samples were sent to 116 of 185 principal investigators, even before the preliminary examination was completed, so that some results could be reported to the fourth Lunar Science Conference scheduled for March.[5]

Schmitt's selection and documentation of samples taken from the large boulders examined at Taurus-Littrow presented a new problem for the team. Considering that issuing samples to investigators working in isolation would waste a unique opportunity to study the relationships between different rock types, the team recommended that each boulder (or set of possibly related boulders) be examined by a small interdisciplinary group working in close collaboration. After these groups had examined the samples and extracted as much information as possible, the specimens would be cut up and distributed for study.[6]

Fourth Lunar Science Conference

The fourth annual lunar science conference convened on March 5, 1973, with more than 700 scientists in attendance.[7] NASA's Deputy Administrator, George Low, opened the four-day meeting by reviewing the Apollo lunar exploration program and the progress made in understanding of the moon in three and a half years. He expressed NASA's gratification with the willingness of the scientific community to participate effectively in analyzing the data and planning the succeeding missions, citing the Group for Lunar Exploration Planning and the Science Working Panels, the support provided during the missions by the scientists in the back rooms, and the help others had given in training the crews and selecting sites. Low told the group that NASA was "firmly committed" to continue support of lunar studies based on the Apollo samples.[8]

Few results from Apollo 17 were presented at the conference, since so little time had elapsed since the samples were released. One of the first studies had determined that the orange soil, which consisted mostly of tiny beads of brown to reddish glass of a composition similar to lunar basalts, was probably of volcanic origin. While it could also be explained as the product of a meteorite impact, the soil was age-dated at 3.5 to 3.7 billion years, an age that could not be correlated with any major basin-forming event and therefore with any major impact.[9] Other papers at the conference reported basic data on the nature, chemistry, and ages of the Apollo 17 samples that had been released. [10]

While analysis of lunar data continued unabated (the proceedings of the fourth lunar science conference, like those of the three before it, filled three large volumes) by early 1973 the major constraints to a model of lunar evolution were beginning to be clear. Geological, geochemical, and geophysical evidence made it clear that the moon is quite different from the earth in structure and composition and therefore probably in the process of its formation.[11] A coherent picture of the moon's history gradually emerged over the next few years (see below).

This conference marked the end of Apollo's voyages, and perhaps that was responsible for the "notable spirit of amity between lunar scientists and NASA's engineer-managers" that some observers perceived at the meeting. In response to George Low's acknowledgement of their contributions to Apollo and his pledge of continuing support for lunar studies, the conferees adopted a motion that acknowledged some "awkward moments" over the years but praised the space agency for its overall execution of a project that had "already revolutionized ideas of the solar system's evolution." In a less formal but probably more meaningful gesture, a small group of the scientists who had been most intimately involved in planning the Apollo missions dug into their own pockets to honor selected NASA managers and engineers at a private banquet. One anonymous scientist said it was their way of saying thanks to "some of the higher people who played ball and to some of the lower echelon people—beautiful guys who would come along at the right time, stand up to their bosses, and help us get an experiment on board."[12] It was a considerable change from the prevailing attitude only three and a half years before (see Chapter 10), but much else had changed in those years as well.

The New Moon

In 1974, following the fifth lunar science conference, Robert Jastrow* attempted a brief synthesis of what science had learned from the Apollo studies for the *New York Times*. After outlining the gross features of the moon's evolution, he

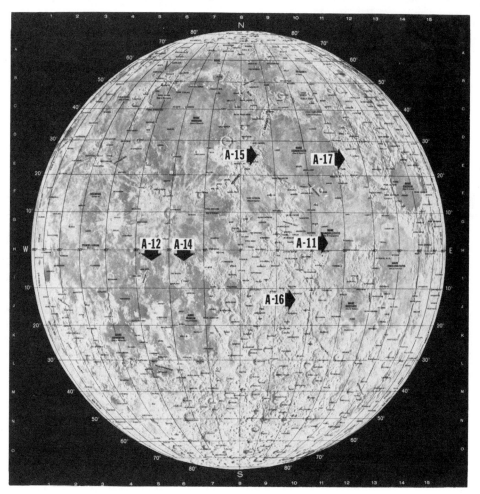

The Earth-facing side of the moon, showing the locations of the six Apollo landing sites.

*Jastrow, the first director of the Theoretical Division in the Office of Space Sciences (see Chapter 2), later headed NASA's Goddard Institute of Space Studies in New York City.

acknowledged that "Apollo has left at least one great cosmological question unanswered: Where did the moon come from?" For years before Apollo 11 landed, Jastrow said,

> scientists argued the merits of various theories with great intensity, but the battle ended in a stalemate. Each theory suffered from at least one major defect. Everyone expected that the lunar landings would promptly settle the debate. It seemed obvious that as soon as we found out what the moon was made of, we would be able to tell where it came from: it would have either the same or different chemistry from that of the earth.
>
> These hopes were not realized. Analysis of the moon rocks has shown that . . . the moon did not come from the earth. But it didn't suggest any alternatives.
>
> The origin of the moon remains as much a mystery as it was before Apollo. . . .[13]

But if the moon's origin remained mysterious, the early stages in its evolution became clearer as the results of Apollo science accumulated.

Arriving at a model of lunar evolution was not easy, since so many different and often unrelated scientific disciplines contributed essential information. Interpretation of the data from geological studies concerning volcanic activity, for example, had to be consistent with the seismic scientists' findings about the moon's internal structure. Heat flow measurements had to be interpreted in light of the geochemical abundances of radioactive elements. Gerald J. Wasserburg, professor of geology and geophysics at Caltech, who participated in the lunar studies from the beginning, commented that this required most investigators to change their approach to science. For the most part, Wasserburg said, they had come to the first lunar science conference as specialists, "without any conception of the relationship of their particular observations to the global problems of a planet." As the work proceeded, however, "specialists began to recognize the interrelationships between their own work and studies done by other individuals in completely different fields. So a broader, truer planetary science [began] to emerge."[14]

Before Apollo 11, considerable information about the moon had been available—principally its physical characteristics, but including some basic chemical facts as well.[15] For many years it had been known that the moon was only 60 percent as dense as the earth (3.3 grams per cubic centimeter as against 5.5 for the earth) and that its rotational properties were almost identical with those of a homogeneous sphere. In the 1960s, space probes yielded more information. *Surveyors V, VI,* and *VII* carried instruments that measured the proportions of key chemical elements in the lunar surface. They showed that the maria were different from the highlands and that the moon was chemically different from both the earth and the (presumed) primordial material of the solar system. The terrestrial material most closely resembling the maria was basalt, an igneous rock, indicating the maria had been flooded with molten rock that had later solidified. Geochemists deduced from its density that the lunar basalt could not represent the average composition of the entire moon, which allowed the inference that the moon was not homogeneous.

Later, the *Lunar Orbiters* returned photographs that showed the moon's hidden side to be topographically quite different from the visible side. Besides providing information essential to the selection of Apollo landing sites, *Orbiter* photographs contained a wealth of geological clues. They convinced geologists that at least some of the moon's surface features had been produced by volcanic activity. Analysis of the spacecrafts' orbits provided evidence for anomalous concentrations of mass ("mascons") under some of the maria, suggesting that the lunar crust was thicker and more rigid than was previously supposed. Had the crust been more plastic, the mascons would have settled deeper into it over geologic time, eliminating the gravitational anomalies that affected *Lunar Orbiter V*. Thus the outer layer of the moon had been cold and rigid for a very long period, yet lava flows large enough to fill the basins had occurred as well. This too implied a nonuniform structure for the moon and probably a complex evolutionary history.

To this sketchy body of knowledge about the moon Apollo contributed an overwhelming flood of new information, out of which scientists began to contruct new models, some to be discarded quickly, others to be modified and retained. The geophysical data may have been the easiest to integrate into a coherent picture. Apollo's four seismometers detected infrequent, weak tremors originating within the moon, at depths of 800 to 1,100 kilometers (500 to 700 miles), much deeper than those on earth. Forty-three moonquake zones were identified, each showing periodic activity correlated with lunar tides. Artificial seismic events (impacts of LM ascent stages and spent S-IVB stages)—plus the fortuitous collisions of a large (1,100 kilograms, 2,400 pounds) meteorite with the moon[16]—produced signals that suggested a crust about 60 kilometers (37 miles) thick, overlying a different homogeneous layer extending down to about 1,000 kilometers (625 miles). Deeper still a partially molten core may exist, which, assuming it to be a silicate rock, would be at a temperature of about 1,500° (2,700°F).

Results from the laser altimeters carried on Apollo 15 and 16 confirmed that the moon's center of mass does not coincide with its geometrical center. Some scientists suggested this was due to the presence of the low-lying maria on the near side. Others suggested that the crust on the far side was thicker than that on the near side, which would make the absence of maria on the far side easier to explain: a thicker crust could have prevented the extrusion of molten material except in the very deepest craters, such as Tsiolkovsky.[17]

Geological and geochemical data were harder to interpret; only the broadest general conclusions could be stated after Apollo 17. The returned samples showed that the moon was chemically different from the earth, containing a smaller proportion of volatile elements (those that are driven off at low to moderate temperatures) and more radioactive elements than the cosmic average. Geochemical evidence indicated that three types of rocks predominate on the lunar surface: basalts rich in iron covering the maria; plagioclase or aluminum-rich anorthosites characteristic of the highlands; and uranium- and thorium-rich basalts that also contain high proportions of potassium, rare-earth elements, and phosphorus ("KREEP" basalts). Some scientists felt, however, that these could not represent the structure and texture of the primordial lunar rocks; it seemed likely that those characteristics had been virtually obliterated in the massive bombardment that the moon has obviously undergone.[18]

259

In spite of the difficulties, scientists—as is their habit—began to formulate models for lunar evolution as soon as the first results became available. Each successive mission brought a surprise, however, and it was only after Apollo 15 that the broad outlines of the moon's history emerged fairly clearly.[19]

The oldest rocks found on the moon appear to have been chemically assembled around 4.5 billion years ago, in the late stages of formation of the solar system. How the materials of the moon came together is still an unanswered question; the evidence indicates that the moon aggregated out of the debris left over from the formation of the sun. According to a widely held view, it was never completely molten; only its outer layer, perhaps to a depth of 320 kilometers (200 miles, roughly one-fifth of its radius) was melted. This sea of molten rock, agitated by a continuing rain of fragments, lost its heat to space and began to solidify. As it did so, different minerals crystallized at different temperatures. Convection currents set up by cooling at the surface brought deeper, hotter material to the surface and at least partially remelted the surface crust. As cooling continued, crystals of different composition separated, giving rise to the chemical segregation observed in the lunar crust. Eventually, probably after some 200 million years, a rigid crust of considerable thickness formed, composed mostly of light-colored minerals rich in calcium and aluminum (plagioclase). Beneath the crust a mantle of iron- and magnesium-rich material settled, consisting predominantly of the minerals pyroxene and olivine. At the center of the moon a core of dense, partially melted material may have formed, rich in iron and sulfur. The presence of a semiliquid core is indicated by the behavior of seismic waves passing through the moon and by the presence of residual magnetism in lunar rocks: at some time in the past, it appears, the moon had a magnetic field (thought to be produced by a liquid metallic core) which has since almost entirely vanished.

For perhaps 300 million years after the crust formed, fragments of primordial material, some of them 50 to 100 kilometers (30 to 60 miles) in diameter, pelted the moon. The impacts caused local melting of the crust and chemical alteration of the original material, scattered debris over thousands of square miles of the moon's surface, and fractured the crust to a considerable depth.

Below the solidified crust the moon's mantle remained partially molten, owing to the insulating properties of the crust and also probably to considerable heating by disintegration of radioactive elements (uranium and thorium). Toward the end of the massive bombardment period, from around 4.1 to 3.9 billion years ago, the large basins (Imbrium, Serenitatis, Crisium) were gouged out, probably by objects almost as large as planets. This was followed by episodic flows of basaltic material into the basins, filling them to approximately their present levels. The youngest igneous (heat-formed) rock so far found on the moon is 3.16 billion years old, but major surface manifestations of internal lunar heat may not have stopped for another two billion years.

This generalized picture of the moon's history has changed little in the years since the end of Apollo. Subsequent study has extended the onset of mare volcanism backward in time somewhat, and its duration is now thought to have been considerably longer. It has been recognized that many of the craters saturating the highlands are secondary, having been produced by large fragments thrown out by impacting objects. The need to understand lunar craters (and, in subsequent years, those on other celestial bodies, e.g., Mercury, Mars, and the satellites of Jupiter and Saturn) has stimulated interest in the mechanics of crater formation, leading to laboratory studies using high-velocity particles. Finally, the impact theory of the moon's origin has recently been revived, with modifications. Collision of an object the size of Mars with earth has been computer modeled and found to be not unattractive. Newer thinking about the origin of the moon considers that more than one process may have contributed significantly, that is, collision followed by accretion of additional material by both bodies. As yet, the available evidence does not permit a clear-cut choice among the various postulates.[20]

So far as the Apollo results have shown, the moon's surface has undergone no large-scale changes for something like 3 billion years. Continuous slow erosion of large features by the impacts of small meteoroids has added to the dust layer on its surface, now and then a substantial meteoroid has produced another crater, and occasionally a boulder, dislodged by an internal tremor or an external shock, has rolled down a slope, leaving its track in the dust. Today the moon looks very much as it looked to the first humans, and it must have changed very little since the age of the dinosaurs.

The Scientific Accomplishments of Apollo

The importance of Apollo in the advancement of planetology is self-evident. If only a single manned landing had been accomplished, the return of samples from another body in the solar system would have established Apollo as a milestone in the history of science. That the project was only a beginning goes without saying; it could hardly have been anything else. As Gerald Wasserburg wrote after Apollo 15, because the Apollo missions had all been targeted for areas that are not representative of the most widespread lunar surface features, "even on a coarse scale, *we must consider the Moon as unexplored* [italics in the original.]"[21] The last two missions filled important gaps without altering the basic truth of this assessment. But even if all ten missions could have been flown as originally planned, they could not have adequately sampled an area as large as the moon and as heterogeneous as it turned out to be.

The materials returned by Apollo have not fulfilled the early hopes of the most optimistic scientists: that they would yield an understanding of the origin and evolution of the moon, the earth, and the solar system. This is no reflection on Apollo; those hopes—unrealistic, as it turned out—were based on excessive optimism and inadequate firm knowledge about the moon. "As is natural," Wasserburg commented, "the paucity of data did not inhibit the scientific community from either serious thought or rampant speculation on the origin of the Moon. The Apollo missions proved most of the speculation wrong."[22] It was apparent after Apollo 12 that the moon's surface was far more complex than most cosmologists had suspected, and hopes for an explanation of the moon's origin rapidly waned. Whether the origin of the moon can ever be deduced from the Apollo samples is still an open question.

But whatever its shortcomings with respect to some hypothetical ideal, Apollo produced a store of scientific treasure whose value, although recognized, has only begun to be exploited by the scientific community. The lunar samples were, in Wasserburg's words, "the crown jewels of the scientific return of the Apollo missions." Before 1969 the only samples of extraterrestrial material known on earth were meteorites, which, as Wasserburg wrote, were less than satisfactory: they "come from unknown sources [and] were often terribly abused. [Yet] A large fraction of all human knowledge about the origin and age of the solar system . . . has been derived from the study of these objects."[23] The Apollo samples changed all that. It seems probable that nothing like them will be available for many years to come.

No other effort to explore the moon compares with Apollo, although all added to the scientific knowledge of the moon. Ranger, which produced photographs showing craters and rocks as small as 1 meter (3 feet) in size, clarified some of the uncertainty about the nature of the lunar surface.[24] The five Surveyor spacecraft, which landed in different types of terrain, carried instruments that demonstrated differences in the chemical makeup of lunar soil and identified the soil as "basaltic." Results from Surveyor allowed scientists to infer that the moon's interior was different from its surface.[25]

That unmanned samplers could be useful in lunar exploration was proved by Soviet space scientists, who sent two unmanned explorers (*Luna 16*, landed September 20, 1970, and returned to earth September 24, and *Luna 20*, landed February 22, 1972, returned February 25). Each returned approximately 100 grams (3 ounces) of lunar soil from the eastern edge of Mare Fecunditatis.*[26] In bringing back selected and documented samples from six locations on the moon, however, Apollo far surpassed all other efforts in whatever terms are used for comparison. Furthermore, the Apollo samples now stored on earth allow lunar research to continue, based on new concepts and using improved techniques and instruments. New concepts of the moon will be restricted only by the limited number of sites from which samples were taken.

Major Issues in Apollo

Science in the Project

Most of the issues over which NASA and the external science community wrangled grew out of the necessity to define, virtually from scratch, the scientific content of lunar exploration. One lunar scientist pinpointed the difficulty in midprogram:

> . . . Apollo 11 marked the beginning of a new generation of lunar science. . . . born in the hot arc of one of the greatest technical achievements in the history of society. *Full recognition of, and attention to, the scientific aspects*

---

*In 1976 one more *Luna (Luna 24)* returned a much larger sample from the southern edge of Mare Crisium in the moon's northeastern quadrant. *Astronautics and Aeronautics, 1976, A Chronology,* NASA SP-4021 (Washington, 1984) pp. 182–83.

of the mission w[ere] for some time lost in the management and excitement of the larger
enterprise both by NASA and an insufficiently involved and unprepared scientific com-
munity [emphasis added].[27]

To appreciate why attention to the scientific potential of Apollo was neglected
for so long, it is necessary only to recall where the nation's manned space flight
program stood at the time President Kennedy issued his challenge:

> When lunar landing became the Apollo objective in May 1961, the United
> States had only 15 minutes of manned flight experience in space and a tenta-
> tive plan for a spacecraft that might be able to circumnavigate the moon. No
> rocket launch vehicle was available for a lunar voyage and no route (mode)
> agreed on for placing any kind of spacecraft safely on the lunar surface and
> getting it back to earth. Nor was there agreement within NASA itself on how
> it should be done.[28]

Given those circumstances, it is easy to understand the reaction of one on whom
much of the responsibility would fall:

> Acutely aware that NASA's total manned space flight experience was limited
> to one ballistic flight and that he was being asked to commit men to a 14-day
> trip to the moon and back, [MSC Director] Robert Gilruth said he was simply
> aghast.[29]

It can hardly be doubted that Gilruth's reaction was shared by his line engineers
at the Manned Spacecraft Center—and by managers in the Office of Manned Space
Flight as well. The 8- or 9-year time limit given them by Kennedy* would have
daunted even the most optimistic engineer who was familiar with the state of
the art in aerospace engineering and the complexities of a moon landing. Charged
with that responsibility, they had to put first things first, and their single-minded
concentration on essentials left science with a low priority. No evidence has been
found that Bob Gilruth and his engineers actively opposed the incorporation of
science into Apollo. They knew that it would have been absurd to land men on
the moon if they did not at least leave their spacecraft to explore the surface—
and, if possible, collect samples and emplace scientific instruments.

Still, the feeling developed among some scientists that MSC was obstructive
toward science, and as much as anything else, the Houston center's narrow focus
on nonscientific aspects of the program was responsible. Homer Newell, associate administrator for space science, knew well that MSC could be difficult to deal
with, but he could at least perceive a reason for it:

> [MSC's] need to be meticulously careful in the development and operation
> of hardware for manned spaceflight, plus [the center's] general disinterest in
> the objectives of space science as the scientists saw them, led to extreme difficul-
> ties in working with the scientific community.[30]

Not that other branches of NASA found the scientific community easy to work
with. In its early days the space agency skirmished more than once with the

---

*The phrase, "before this decade is out," was deliberately chosen by the President to allow for
some flexibility of interpretation. It could without serious equivocation be construed as meaning either
1969 or 1970. See Theodore C. Sorenson, *Kennedy* (New York: Harper & Row, 1965), p. 525.

National Academy of Sciences and its Space Science Board over the question of who should decide the content of the space programs. The executive director of the Space Science Board went so far as to urge that NASA rely exclusively on the outside scientific community for its science program, which would effectively have reduced NASA's role to providing launch vehicles and operational support for science (besides, of course, financing the experimenters). Newell later recalled that gradually a working arrangement evolved putting NASA "in the driver's seat, [with] the scientific community serving as navigator, so to speak, . . . with a mixture of tension and cooperation that is best described as a love-hate relationship."[31]

Science would eventually become the navigator for lunar exploration as well, but only after the primary objective had been attained. During the early years the Manned Spacecraft Center at least made some significant gestures toward eventual accommodation of science, establishing an office to coordinate experiments, providing space in the lunar module to carry scientific equipment,[32] and starting to train astronauts in the principles of field geology. It was less concerned with developing a full program of lunar exploration. MSC, in fact, could do little in that regard, because strategic planning for lunar exploration was the responsibility of the Office of Space Sciences and Applications (OSSA) at Headquarters.

For a considerable time following the establishment of Apollo, no organized lunar and planetary science lobby existed. So in formulating a program of lunar science, OSSA first sought statements of broad objectives from ad hoc groups, such as the Sonett committee, which established the basic criteria for science on the Apollo missions in 1962 and the training required for astronauts to execute the science program (see Chapter 2). Considered and endorsed in principle at the 1962 Summer Study at the State University of Iowa, these criteria were then used by discipline-oriented planning teams to define specific scientific investigations. By mid-1964 these teams had listed the experiments expected to be the most productive—some to be conducted on the moon, some to be done on returned samples, and some to be carried out by instruments left on the lunar surface (see Chapter 3).

In 1965, the Summer Study at Woods Hole, Massachusetts, sharpened the focus by formulating 15 key questions about the moon that lunar science studies should be designed to answer. Following the Woods Hole sessions, a study group met at nearby Falmouth to make the first attempt to explicitly define an evolutionary program of lunar science for the next 10 to 15 years (see Chapter 3 and Appendix 3). The Falmouth report became the basis for much of the planning that followed.[33]

Thus while the Manned Spacecraft Center was working toward its primary aim of landing people on the moon, OSSA was preparing the ground for the scientific work to be done when they got there. In 1966, after years of encouragement from Headquarters, MSC established a Science and Applications Directorate, which assumed much of the responsibility for Apollo science and, more important to the science community, put a research scientist into the management structure at MSC (see Chapter 6). The new office had no significant effect on Apollo plans immediately, but its influence was to become stronger as the project went on.

One of the first actions taken by Wilmot N. Hess, MSC's first Director of Science and Applications, was to convoke a summer study to translate the plans devised at Falmouth into scientific requirements for Apollo exploration: mission duration, lunar surface mobility, and scientific use of the command module in lunar orbit. Participants did their job well: in the preface to their report Homer Newell noted that "the plans are optimistic and exceed the capability of the agency to execute."[34] Even so, the most important recommendations that came out of the 1967 study at Santa Cruz (see Appendix 3) were carried out on the Apollo missions as soon as the required engineering and operational changes could be incorporated into the system.

Few questions were as important in the ultimate success of Apollo as making certain that the external science community had a voice in the planning of the missions. In the summer of 1967 Wilmot Hess created three science teams that would operate throughout the duration of the project: the Group for Lunar Exploration Planning (superseded in 1970 by the Science Working Panel), the Lunar Sample Analysis Planning Team and the Lunar Sample Preliminary Examination Team. The specific responsibilities of these groups were probably not more significant than their role in promoting cooperation between MSC and the science community. Establishment of the two teams concerned with lunar samples placed responsibility for sample distribution in the hands of scientists; the Group for Lunar Exploration Planning was intended to reassure scientists that their concerns were at least being considered in developing plans for later missions.

Yet when the first lunar landing mission succeeded in bringing back nearly 50 pounds (21 kilograms) of lunar samples for study, some scientists once more went public with the complaints that NASA was "not responsive to the needs of science" (see Chapter 10). It is not easy to determine exactly what was meant by this charge, expressed as it was in such general terms. Nor is it easy to grant it much validity, considering that NASA began, immediately after Apollo 11, to incorporate some of the high-priority improvements in the Apollo system suggested by the Santa Cruz conference—the lunar roving vehicle, the extended lunar module, and modifications of mission plans to accommodate larger payloads— all of which would improve the scientific return. Trajectory designers at MSC set out to improve the landing accuracy of the second mission, explicitly if not solely for the benefit of science. What more NASA could have done for science in Apollo, short of turning the program over to scientists, is difficult to imagine.

The complaining scientists overlooked or ignored the fact that manned space flight officials could not be *certain* that the first attempt at a lunar landing would succeed. (Michael Collins, command module pilot on Apollo 11, confessed that on launch day he would not have given better than even odds that the whole complex mission would be carried out without a failure.[35]) That being true, it was not prudent to begin modifying the spacecraft before the first landing was accomplished.* If Neil Armstrong had been forced to abort his landing, George Mueller,

---

*MSC's refusal to anticipate scientific requirements and modify the spacecraft to accommodate them was one of the scientists' most serious complaints. The head of the Experimental Planetology Branch, Solar System Exploration Division, at Johnson Space Center pointed out in criticizing the draft of this book that "It require[d] at least a year of lead time for even the simplest of new ideas to be introduced into a program as complicated as Apollo. Therefore, waiting until after the first one or two successful landings to begin accommodating the scientists' wishes was guaranteed to delay any implementation for a long time." He also cited Gilruth's waiting until early 1970 to work out MSC's differences with the scientific community as a similar source of irritation for scientists. Wm. C. Phin-

chief of manned space flight, was prepared to send missions to the moon at the shortest intervals launch teams could manage until someone landed on the moon and returned safely to earth before the decade of the 1960s was out. As soon as that was done, Mueller urged MSC Director Bob Gilruth to do his best to accommodate science. Gilruth's response, following the first lunar science conference in January 1970, was to bring his principal lieutenants and a group of leading scientists to the table to work out their major differences, with the result that scientists' input to the later missions was effective and satisfying to the scientists (see Chapter 11).

Scientists as Astronauts; Astronauts as Scientists

An issue that sharply divided the scientists and manned space flight officials from first to last was the question of whether a scientist should go to the moon. Even before they knew what the lunar surface was like, scientists took it for granted that only one of their fraternity could properly conduct a scientific survey while on the moon. Analogously, Deke Slayton, Director of Flight Crew Operations at MSC and the man most responsible for assigning crews to missions, was confident that only an experienced test pilot thoroughly familiar with spacecraft systems could be depended upon to react correctly in whatever emergency might develop during a landing. The scientists seem to have felt that piloting a spacecraft was a skill much like driving a car and could be learned by any reasonably intelligent person in a comparatively short time.[36] Slayton considered it easier to teach an astronaut a few basic scientific tasks than to make a pilot out of a scientist[37]—hence if either skill took precedence, it should be piloting.

Both were right and both were wrong. If lunar explorers had only a little time on the moon, science would be best served if one of the crew was a scientist of considerable experience who could make the most of what little he could see and do. On the other hand, an emergency in a lunar mission might leave no time for conscious thought. The ability to take the proper action instinctively would be expected of an experienced test pilot but might be difficult to instill into a person having little experience of acting decisively in emergencies.

As long as spacecraft were regarded as experimental rather than operational, Slayton's view—shared by Bob Gilruth—prevailed, modified by the addition of some basic instruction in geology to the astronaut training program. Eventually MSC yielded on the point of accepting scientists for astronaut training, but did so with the stipulation that they must also qualify as jet pilots, if necessary by taking flying training before starting in the astronaut program.

---

ney to William Waldrip, "Review of Apollo History," Mar. 31, 1987. The point is certainly valid, but it presumes, as scientists typically did, that science should have taken precedence as early as Apollo 12. The present author does not agree. The last 2½ years before Apollo 11 were spent recovering from the AS-204 fire, which totally occupied spacecraft engineers and program managers. Problems with the lunar module lingered through 1968. To have begun modifying the untried lunar module to suit the purposes of science a year or more before the first lunar flight would have been to invite trouble.

Fortunately, an astronaut's ability to react appropriately in a time-critical emergency was never really tested in Apollo.* No lunar module pilot ever had to take over from a disabled commander during a lunar landing or had to bring back a lunar module from the moon alone. Whether a scientist, properly trained, could have performed as well as a test pilot was never determined. For what it is worth, Jack Schmitt was sure that he could have done it, and his commander, Gene Cernan, was satisfied to have Schmitt along.

On the other hand, 11 of the 12 men who walked the moon were pilots who did have to play the role of scientist, at least to a limited degree, in selecting lunar samples. As might be expected, their performance varied, but at least publicly none of the scientists ever seriously faulted the results. Jack Schmitt, who took an active interest in geology training before he was assigned to Apollo 17, later rated the performance of the crews as generally good:

> We got excellent sampling, we got excellent photography, . . . but until Apollo 17 we did not get very much good, solid descriptive work, with one exception— that exception being Neil Armstrong. . . . He was probably the best observer we sent to the moon, in spite of very limited training; he just had a knack for it.[38]

The best-prepared crew, in terms of time spent in preparation for exploration, was that of Apollo 15. Scott, Irwin, and Worden put in more training sessions on lunar surface operations than any crew (see Appendix 7), and, with Jack Schmitt on their backup crew, they had more tutoring than the others as well. Dave Scott took a special interest in the lunar science,[39] and it paid off; Apollo 15 was acclaimed at the time as the most productive mission yet flown (see Chapter 13) and, except for Schmitt's descriptive work on Apollo 17, was not surpassed by the last two missions. The scientists' delight with the results of Apollo 15 may have been influenced by the fact that it was the first of the extended missions, with all the extra work that allowed, but that does not detract from the performance of Scott and Irwin.

Whether all the missions would have benefited from the presence of a geologist was not settled in Apollo. By the time Jack Schmitt flew on Apollo 17 it had become clear that geological clues were not as obvious on the lunar surface as they are on earth. Possibly an experienced field geologist could have found more than the astronauts did. Still, the conditions of working on the moon (i.e., limited time and the restrictions on mobility and dexterity imposed by the space suit) suggest that the advantage would not have been as great as some believed. A longer program might have provided more time for that kind of work, but by the time the missions began to fly, scientific emphasis had shifted away from the field and into the laboratory.

Apollo experience provided no final answer to the question of the relative capabilities of scientists and astronauts in space exploration. Manned space flight programs later evolved to the point where specialists of both kinds could go into space, so the debate has lost some of the relevance it had before Apollo.

---

*Apollo 13 was indeed an emergency, but the responsibility for saving the mission was in the hands of experts on the ground. The success of the rescue did not depend on the crew's ability to act swiftly in a critical situation.

Lunar Sample Management and Quarantine

Early in Apollo geoscientists at MSC recognized that the samples collected on the moon required special handling to avoid contamination. The extreme sensitivity of some methods of chemical analysis* made it necessary to exclude all earth-borne traces of the substance being sought, if the analysis was to be valid. Ideally, lunar sample containers would have been sealed on the moon and their contents kept in the highest possible vacuum until they reached the laboratory where they were to be investigated. Practically, preserving the vacuum and avoiding contamination while examining and dividing the samples was extremely difficult. So it proved in the Lunar Receiving Laboratory (LRL); although systems and procedures for manipulating sensitive materials in a protective enclosure were well known, they proved less than completely satisfactory for handling the lunar samples. The heavy gloves used to manipulate specimens inside the vacuum system leaked; contamination proved unavoidable; and scientists soon abandoned the use of the vacuum system except for those samples that absolutely required it. Instead, an atmosphere of chemically inert and noncontaminating gas (nitrogen) was used (see Chapter 9).

The most troublesome requirement imposed on the management of lunar samples, however, was biological containment. From the construction of the LRL to the conduct of mission operations, the requirement to prevent contact between the earth's biosphere and the objects and persons who had been to the moon made Apollo more costly and more complex.

The added cost and complexity were unavoidable after a conference sponsored by the Space Science Board issued a formal statement in 1965 that the earth must be protected from back-contamination by any organisms that might be brought back from the moon (see Chapter 4). The cult of extraterrestrial life, although lacking the smallest shred of positive scientific evidence to support it, had many followers among scientists and the public.[40] NASA could not refute such a warning, nor could it afford to ignore it. What if the believers were right? Granted the infinitesimal probability that living organisms existed on the moon, and the real question of whether they might survive on earth and be dangerous, the consequences of releasing them on earth were so potentially enormous that the chance could not be taken.

So, without enthusiasm but at considerable expense, MSC built the Lunar Receiving Laboratory with provisions for two-way biological containment (preventing biological contamination of the samples *and* the escape of contaminants from the samples into the laboratory) and modified recovery procedures to isolate spacecraft and astronauts from the earth's biosphere. As the system was finally implemented, one serious gap remained: the crew would leave the command module by opening the hatch and clambering into the recovery raft. MSC refused to consider hoisting the returned spacecraft aboard ship with the astronauts inside or to add a biological filter to the post-landing ventilation system. The Interagency Committee on Back Contamination objected, but MSC

---

*Carbon compounds, for example—possible relics of extinct life or precursors of life on the moon—could be detected if present to the extent of a few parts per billion.

held firm, on the grounds that crew safety would be imperiled. The committee settled for biological isolation garments and application of a biocide to the spacecraft.

In the event, the precautions proved unnecessary and quarantine was abandoned after the third lunar landing. Apart from the expense, quarantine was irksome to engineers and scientists alike, for it impeded postflight debriefings and delayed the release of samples to principal investigators. In any case, quarantine was not intended to be absolute: the second guideline governing procedures stated, "the preservation of human life should take precedence over the maintenance of quarantine." If a command module had begun to sink during recovery operations, or if a major fire had broken out in the crew quarters of the receiving laboratory, or if a quarantined astronaut had suffered a medical emergency that could not be handled within the LRL, quarantine would have been broken.[41]

The results of biological and chemical examination of the returned lunar samples indicated that life never has existed on the moon—or at least that it left no traces at the sites examined. As did the data returned from Mars by the Viking landers a few years later, the Apollo results showed that the only conclusion that can be reached at present concerning the existence of life elsewhere in our solar system is, "not proven."

Project Apollo: Stunt or Portent?

A decade and a half after Eugene Cernan left the last human footprint on the moon, the value and the wisdom of the Apollo project can still be debated. As an engineering accomplishment it is unparalleled in history. As a scientific project it enabled researchers on earth to study documented specimens from another body in the solar system, perhaps the only such specimens that will be available in this century. As an exercise in the management of an unprecedentedly large and complex effort it stands alone in human experience.

Yet all these achievements can be read two ways. Magnificent as they were, the launch vehicles that carried men to the moon turned out to be too expensive for other missions. The choice of lunar-orbit rendezvous as the mission mode—largely dictated by the end-of-the-decade challenge—produced two spacecraft ideally adapted to their function but without sufficient margin for growth to advance the exploration of the moon as far as scientists wanted. Apollo's scientific results were of vital interest to only a very small fraction of the scientific community and did not authoritatively answer the questions scientists hoped they would answer before the first landing. (As one critic caustically commented, the scientists were able to obtain "a neater fix, so to speak, on the number of angels who can dance on the point of a pin."[42]) The methods devised for managing Apollo were impressive; yet the question remains,

> Has the whole operation represented but another highly successful one-shot exercise in crisis management, or has it represented incorporation into American society of a new way to organize, systematically and purposefully, the development and use of scientific and technological resources to the furtherance of national goals?[43]

James Webb, under whose direction NASA developed those methods, believed them to be the lasting contribution of Apollo, but in the 18 years since the first lunar landing they have not been applied to any undertaking of comparable size.

It was unfortunate, in a sense, for the United States's space program that Kennedy's challenge called for NASA to proceed directly to the most difficult goal that seemed achievable in 1961, for it doomed whatever followed to be an anticlimax:

> The first Apollo landing was, in one sense, a triumph that failed, not because the achievement was anything short of magnificent but because of misdirected expectations and a general misinterpretation of its real meaning. The public was encouraged to view it only as the grand climax of the space program, a geopolitical horse race and extraterrestrial entertainment—not as a dramatic means to the greater end of developing a far-ranging spacefaring capability [italics in the original]. [44]

Americans had plenty to occupy their attention in 1969—civil rights, the plight of the poor, an increasingly unpopular war in southeast Asia, rising federal deficits, and growing concern for the preservation of a livable environment—and plenty of advocates for every cause clamoring for action. If the nation had not set its space goals as high as possible at the outset, the concerns of the late 1960s could easily have stopped a more measured approach well short of the moon.

Future generations may look back on Apollo as a costly technological stunt or the portent of man's destiny in the universe. Scientists studying the origin and evolution of the solar system may devise other means of acquiring data from the moon and the planets, but they will surely be thankful for the samples gathered by the twelve men who made the first lunar voyages.

SOURCE NOTES

CHAPTER 1

1. U.S., Congress, House, *Toward the Endless Frontier: History of the Committee on Science and Technology, 1959–1979* (Washington: U.S. Government Printing Office, 1980), p. 269. This history was written by Ken Hechler, a Ph.D. historian and author of books on political and military history, who served on the House Committee on Science and Technology for 18 years beginning in 1959.
2. Homer E. Newell, *Beyond the Atmosphere: Early Years of Space Science*, NASA SP-4211 (Washington, 1980), pp. 87–99. Newell joined NASA in 1958 as director of the space science program, became Associate Administrator for Space Science (later Space Science and Applications) in 1961 and Associate Administrator in 1967. He retired in 1973 and died in 1984. See also Arnold S. Levine, *Managing NASA in the Apollo Era*, NASA SP-4102 (Washington, 1982), pp. 9–17, and Robert L. Rosholt, *An Administrative History of NASA, 1958–1963*, NASA SP-4101 (Washington, 1966), pp. 37–47.
3. Rosholt, *Administrative History*, pp. 45–47. For a history of JPL's Ranger project, see R. Cargill Hall, *Lunar Impact: A History of Project Ranger*, NASA SP-4210 (Washington, 1977).
4. Newell, *Beyond the Atmosphere*, pp. 104–105.
5. Loyd S. Swenson, Jr., James M. Grimwood, and Charles C. Alexander, *This New Ocean: A History of Project Mercury*, NASA SP-4201 (Washington, 1966), pp. 55–65. The NACA-NASA research in high-speed flight is treated in Richard Hallion's *On the Frontier: A History of the Dryden Flight Research Facility*, NASA SP-4303 (Washington, 1984).
6. Swenson, Grimwood, and Alexander, *This New Ocean*, p. 73.
7. Ibid., pp. 99–102.
8. Courtney G. Brooks, James M. Grimwood, and Loyd S. Swenson, Jr., *Chariots for Apollo: A History of Manned Lunar Spacecraft*, NASA SP-4205 (Washington, 1979), pp. 4, 52–53.
9. Roger E. Bilstein, *Stages to Saturn: A Technological History of the Apollo/Saturn Launch Vehicles*, NASA SP-4206 (Washington, 1980), pp. 26–42.
10. *Documents on International Aspects of the Exploration and Use of Outer Space, 1954–1962*, staff report prepared for the Senate Committee on Aeronautical and Space Sciences (Washington, 1963), p. 188; the quote is from Eisenhower's annual budget message to Congress, Jan. 18, 1961.
11. ''Report to the President-Elect of the Ad Hoc Committee on Space,'' Jerome B. Weisner, chmn., (classified version), Jan. 12, 1961, p. 10.
12. Brooks, Grimwood, and Swenson, *Chariots*, pp. 7–17.
13. John M. Logsdon, *The Decision to Go to the Moon: Project Apollo and the National Interest* (Cambridge, Mass.: MIT Press, 1970), pp. l08–l09, 112–115.
14. President's Science Advisory Committee, *Introduction to Outer Space* (Washington, 1968), p. 6.
15. James R. Killian, address to the M.I.T. Club of New York, Dec. 13, 1960; quoted in Logsdon, *Decision*, p. 20.
16. Space Science Board of the National Academy of Sciences-National Research Council, ''Man's Role in the National Space Program,'' reprinted in Senate, Committee on Aeronautical and Space Sciences, *National Space Goals for the Post-Apollo Period*, 89th Cong., 1st sess. (henceforth 89/1), 1965, pp. 242–43.
17. Logsdon, *Decision*, p. 86.
18. James E. Webb and Robert S. McNamara, memo to the President, ''Recommendations for our National Space Program: Changes, Policies, Goals,'' May 8, 1961.
19. Logsdon, *Decision*, pp. 127–28.
20. Ibid., p. 129.
21. Ibid.
22. Brooks, Grimwood, and Swenson, *Chariots*, p. 31.
23. Robert Colby Nelson, ''Full Moon Debate,'' *Christian Science Monitor*, June 14, 1961.

24. Hall, *Lunar Impact*, pp. 10-12.
25. House, Subcommittee on Space Sciences and Applications of the Committee on Science and Astronautics, *1965 NASA Authorization*, Hearings on H.R. 9641, 88/2, part 3, pp. 1897-98. See also Hechler, *Toward the Endless Frontier*, pp. 541-42.
26. For a study of the changing relationships between science and government before and after World War II, see Daniel S. Greenberg, *The Politics of Pure Science* (New York: The New American Library, Inc., 1967).
27. Ibid., p. 288.
28. National Academy of Sciences-National Research Council, *A Review of Space Research*, NAS/NRC Publication 1079 (Washington, 1962), pp. 1–21 to 1–22; see also Newell, *Beyond the Atmosphere*, pp. 208–209.
29. Newell, *Beyond the Atmosphere*, p. 386.
30. See, for example: Webb's address to the Greater Hartford (Conn.) Chamber of Commerce, Oct. 1, 1962, cited in *Astronautical and Aeronautical Events of 1962*, NASA Report to the House Committee on Science and Astronautics, 88/1 (Washington, 1963), p. 206; Webb's address to the Northeast Commerce and Industry Exposition, Boston, Oct. 2, 1962, ibid., p. 207; Hugh L. Dryden, letter to Sen. Robert S. Kerr, chmn., Senate Committee on Aeronautical and Space Sciences, cited in *Astronautical and Aeronautical Events of 1961*, NASA Report to the House Committee on Science and Astronautics, 87/2 (Washington, 1962), p. 28. While program office officials covered details of the current programs' budgets and plans with congressional subcommittees, Webb and Dryden consistently stressed the broader aims of the programs before the full committees. Consult the various House and Senate committee hearings on NASA authorization bills, fiscal years 1961–1969, for more examples.
31. House, Committee on Science and Astronautics, *1964 NASA Authorization*, hearings on H.R. 5466, 88/1, part 1, pp. 2–3; idem, *1963 NASA Authorization*, hearings on H.R. 10100, 87/2, part 1, pp. 3–9.
32. *Endless Frontier*, p. 171.
33. "A Matter of Priority: An examination of the budget and benefits of the moon shot in relation to other problems," report prepared by the staff of the Senate Republican Policy Committee, May 10, 1963.
34. Logsdon, *Decision*, p. 128.
35. P. H. A[belson]., "Manned Lunar Landing," *Science* 140 (1963): 267.
36. See, for example: Frederick D. Hibben, "NASA, Scientists Divided on Space Goals," *Aviation Week & Space Technology*, Apr. 29, 1963, pp.24–25; Albert Eisele, "Nobel Winners Criticize Moon Project," *Washington Post*, May 6, 1963; Robert Hotz, "Apollo and Its Critics," *Aviation Week & Space Technology*, Apr. 29, 1963; William J. Perkinson, "Engineers vs. Scientists," *Baltimore Sun*, Apr. 30, 1963; Howard Simons's 3-part series in the *Washington Post*, "Scientists Divided on Apollo," May 12-14, 1963.
37. "U.S. Official Answers Critics of Manned Lunar Program," *Baltimore Sun*, Apr. 30, 1963.
38. "Man-on-Moon Project Backed by 8 Scientists," Washington *Evening Star*, May 27, 1963.
39. Senate Committee on Aeronautical and Space Sciences, *Scientists' Testimony on Space Goals*, Hearings, 88/1, June 10–11, 1963.
40. Ibid., pp. 242–43.
41. *Endless Frontier*, pp. 169–74.
42. *Astronautics and Aeronautics, 1963*, pp. 440, 464, 474.
43. Newell, *Beyond the Atmosphere*, p. 385.
44. Ibid., p. 384.
45. Swenson, Grimwood, and Alexander, *This New Ocean*, p. 643.
46. Ibid., pp. 384–86.
47. House Committee on Science and Astronautics, *1965 NASA Authorization*, Hearings on H.R. 9641, 88/2, part 1, p. 10.

CHAPTER 2

1. R. Cargill Hall, *Lunar Impact: A History of Project Ranger*, NASA SP-4210 (Washington 1977), pp. 15–16.
2. Stephen G. Brush, "Nickel for Your Thoughts: Urey and the Origin of the Moon," *Science* 217 (1982):891–98.
3. Senate Committee on Aeronautical and Space Sciences, *Scientists' Testimony on Space Goals*, Hearings, 88/1, June 10, 1963, pp. 51, 52–53.
4. Homer E. Newell, *Beyond the Atmosphere: Early Years of Space Science*, NASA SP-4211 (Washington, 1980), pp. 212–13.
5. Hall, *Lunar Impact*, p. 15.
6. Ibid., pp. 5, 17; Clayton R. Koppes, *JPL and the American Space Program: A History of the Jet Propulsion Laboratory* (New Haven and London: Yale University Press, 1982), pp. 99–100.
7. Hall, *Lunar Impact*, pp. 18, 20–24.
8. Ibid., p. 38.
9. NASA, *Fifth Semiannual Report to Congress, October 1, 1960, Through June 30, 1961* (Washington, 1962), pp. 49–50.
10. For a discussion of some of the problems faced by the engineers in ensuring reliability, see Loyd S. Swenson, Jr., James M. Grimwood, and Charles C. Alexander, *This New Ocean: A History of Project Mercury*, NASA SP-4201 (Washington, 1966), pp. 167–213.
11. Newell, *Beyond the Atmosphere*, p. 163.
12. *Scientists' Testimony on Space Goals*, pp. 110, 244; Newell, "The Mission of Man in Space," address to Symposium on Protection Against Radiation Hazards in Space, Gatlinburg, Tenn., Nov. 5, 1962, text.
13. R. L. F. Boyd, "In Space: Instruments or Man?" *International Science and Technology*, May 1965, pp. 64–75. Boyd, a British astronomer with substantial experience in unmanned space science projects, presents the archetypal sky scientist's view—supremely confident of the potential of computerized systems and condescendingly contemptuous of the capability of man.
14. Hall, *Lunar Impact*, p. 114.
15. Ibid., pp. 289–96.
16. Swenson, Grimwood, and Alexander, *This New Ocean*, pp. 414–15.
17. Joseph F. Shea to Dir., Aerospace Medicine and Dir., Spacecraft & Flight Missions, "Selection and Training of Apollo Crew Members," Mar. 29, 1962.
18. Courtney G. Brooks, James M. Grimwood, and Loyd S. Swenson, Jr., *Chariots for Apollo: A History of Manned Lunar Spacecraft*, NASA SP-4205 (Washington, 1979), chap. 3.
19. Ivan D. Ertel and Mary Louise Morse, *The Apollo Spacecraft: A Chronology*, vol. I, NASA SP-4009 (Washington, 1969), p. 108.
20. "Draft Statement on Scientific Training of the Astronaut for Consideration of the Lunar Exploration Committee," signed by G. McDonald, M. Ewing, T. Gold, E. Stuhlinger, H. Hess, H. Brown, and R. Jastrow (members or consultants of the Lunar Sciences Subcommittee of the Space Sciences Steering Committee), no date [ca. Sept. 1961].
21. House Committee on Science and Astronautics, Subcommittee on Space Sciences and Applications, *1963 NASA Authorization*, Hearings on H.R. 10100, 87/2, pt. 4, pp. 1783–88, 1879–80, 1928–31.
22. Hall, *Lunar Impact*, p. 157.
23. Newell, *Beyond the Atmosphere*, chap. 12, discusses the relations of OSS to the external scientific community, including the scientists' insistence on noninterference. For additional examples, see Daniel S. Greenberg, *The Politics of Pure Science* (New York: The New American Library, 1967), pp. 85, 112, 114, 132, 135, 137.
24. Hall, *Lunar Impact*, p. 162.
25. OMSF, "Requirements for Data in Support of Project Apollo," issue no. 1, June 15, 1962.
26. Hall, *Lunar Impact*, p. 162; Koppes, *JPL and the American Space Program*, p. 116.
27. Shea, "Selection and Training of Apollo Crew Members," Mar. 29, 1962; W. A. Lee to Dr. J. F. Shea, "*Ad Hoc* Working Group on Apollo Scientific Experiments and Training (sic)," Apr. 13, 1962.

28. L. D. Jaffe, "Minutes: Ad Hoc Working Group on Apollo Scientific Experiments and Training, 27 March 1962," Mar. 30, 1962.
29. Jaffe, "Minutes: Ad Hoc Working Group on Apollo Scientific Experiments and Training, 17 April 1962," Apr. 20, 1962, with attachment: Lee, "Guidelines from the Office of Manned Space Flight."
30. Lee, "Guidelines From the Office of Manned Space Flight."
31. Jaffe, "Minutes: Ad Hoc Working Group on Apollo Scientific Experiments and Training, 23 April 1962," no date.
32. Ibid.
33. Homer E. Newell and D. Brainerd Holmes, memo for the Assoc. Adm., "Establishment of a Joint OSS/OMSF Working Group," Oct. 22, 1962.
34. Newell, *Beyond the Atmosphere*, p. 401.
35. NASA Hq., News Release 62-251, "Unit to Coordinate Manned and Unmanned Space Flight," Nov. 27, 1962.
36. Hall, *Lunar Impact*, p. 79.
37. Shoemaker interview, Houston, Mar. 17, 1984.
38. Shoemaker interview; Shoemaker, "Report for week of November 12," Dec. 13, 1962.
39. Lee to Shea, "OSS Participation in Apollo," Sept. 22, 1962.
40. Swenson, Grimwood, and Alexander, *This New Ocean*, pp. 414, 419, 443–44; W. David Compton and Charles D. Benson, *Living and Working in Space: A History of Skylab*, NASA SP-4208 (Washington, 1983), pp. 59–60.
41. Shoemaker to Dir., Lunar and Planetary Programs and Dir., Systems Studies, "Report for Week of November 12," "Report for Week of November 19," and "Report for Week of November 26," all dated Dec. 13, 1962.
42. Shoemaker to Dir., Office of Space Sciences, "Recommended Structure for Manned Space Science Planning Group," Dec. 13, 1962; Shoemaker to Dir., OMSF and Dir., OSS, "Panel Chairmen and Members Recommended for the Manned Space Science Planning Group," Dec. 26, 1962.
43. Shoemaker to Dir., Office of Space Sciences, "Recommended Structure . . . ," Dec. 13, 1962; Shoemaker interview; Shoemaker, "Report for Week of November 19," Dec. 13, 1962.
44. Newell to Robert Gilruth, Feb. 15, 1963; Newell, "Memorandum to A/Mr. Webb," Mar. 26, 1963.
45. Shoemaker interview.
46. Newell, *Beyond the Atmosphere*, p. 292; Nolan to Gilruth, Apr. 24, 1963.
47. "Memorandum of Agreement between Office of Manned Space Flight [and] Office of Space Sciences, Scientific Interfaces," no date; signed by E. M. Cortright July 25, 1963, and J. F. Shea July 26, 1963.
48. NASA Release 63-242, "NASA Office of Space Science and Applications Organization Detailed," Nov. 18, 1963; Newell, *Beyond the Atmosphere*, p. 284.
49. Robert L. Rosholt, *An Administrative History of NASA, 1958–1963*, NASA SP-4101 (Washington, 1966), pp. 289–302; Arnold S. Levine, *Managing NASA in the Apollo Era*, NASA SP-4102 (Washington, 1982), pp. 5–6, 38–46.
50. Levine, *Managing NASA*, p. 19.
51. House Committee on Science and Astronautics, *1965 NASA Authorization*, Hearings on H.R. 9641, 88/2, pt. 1, pp. 113–14.
52. Levine, *Managing NASA*, p. 119.
53. Brooks, Grimwood, and Swenson, *Chariots*, pp. 128–31; "Apollo Schedule and Cost Evaluation," presentation to George E. Mueller, Sept. 28, 1963, copy in JSC History Office files, box 063–65.
54. NASA Management Instruction 9000.002, "Establishment of a Manned Space Flight Experiments Board," Jan. 14, 1964; Compton and Benson, *Living and Working in Space*, pp. 61–62.
55. Willis B. Foster to Chief, Lunar & Planetary Branch [OSSA], "Establishment of Manned Space Flight Experiments Board," Jan. 9, 1964.
56. Barton C. Hacker and James M. Grimwood, *On the Shoulders of Titans: A History of Project Gemini*, NASA SP-4203 (Washington, 1977). Gemini is, to some degree, the forgotten manned space flight program, yet it was essential in determining that men could survive for as long as two weeks in zero gravity with no serious aftereffects, in proving the techniques of orbital rendezvous and controlled earth landing, and in developing the fuel cell as a power source, besides providing many hours of operational experience for mission control.
57. President's Science Advisory Committee, report by the Space Science and Space Vehicle Panels (Donald F. Hornig, chmn.), "Objectives and Means in Lunar Surface Exploration," Oct. 15, 1963.

CHAPTER 3

1. See Harold C. Urey, "The Contending Moons," *Astronautics & Aeronautics,* vol. 7, no. 1 (Jan. 1969), pp. 37–41, reproduced in "Lunar Science Prior to Apollo 11," Robert Jastrow, Vivien Gornitz, Paul W. Gast, and Robert A. Phinney, eds. (NASA Goddard Space Flight Center Institute for Space Studies, 1969), pp. I-1 to I-6. This collection, a summary of discussions at a conference on problems in lunar science held in New York on June 5, 1969, contains papers and excerpts from books by the leading researchers in lunar science. Although it includes some results from *Ranger* and *Surveyor* that are not pertinent to the present discussion, this volume is a good single source of pre-Apollo information on the moon.

2. Eugene M. Shoemaker, "Exploration of the Moon's Surface," *American Scientist* 50 (1962):99–129.

3. Urey, "The Contending Moons"; Harold C. Urey and G. J. F. MacDonald, "Origin and History of the Moon," in "Lunar Science Prior to Apollo 11," pp. I-14 to I-135.

4. Urey, "The Contending Moons."

5. "Principal Issues in Lunar Exploration," in "Lunar Science Prior to Apollo 11," pp. 1-19.

6. Ibid., p. 4.

7. "Draft report of the Ad Hoc Working Group on Apollo Experiments and Training on the Scientific Aspects of the Apollo Program," Charles P. Sonett, chmn., July 6, 1962.

8. L. D. Jaffe, "Minutes: Ad Hoc Working Group on Apollo Scientific Experiments and Training, 23 April 1962," no date.

9. Brooks, Grimwood, and Swenson, *Chariots,* pp. 61–67, 83–86.

10. Robert C. Seamans, Jr., to Homer E. Newell, July 25, 1962.

11. Homer E. Newell, *Beyond the Atmosphere: Early Years of Space Science,* NASA SP-4211 (Washington, 1980), pp. 207–208.

12. Ibid.

13. National Academy of Sciences-National Research Council, *A Review of Space Research,* report of the summer study conducted under the auspices of the Space Science Board at the State University of Iowa, June 17–Aug. 10, 1962, NAS-NRC Publication 1079 (Washington, 1962), pp. 11–4 to 11–5, 4 to 10.

14. Ibid., pp. 11–15 to 11–16.

15. Newell, *Beyond the Atmosphere,* p. 209.

16. Verne C. Fryklund, Jr., to Ernst Stuhlinger and Hans Hueter, MSFC, "Scientific Guidelines for LLS and LLV Studies," July 3, 1963; Fryklund to Dir., MSC, "Scientific Guidelines for the Apollo Project," Oct. 8, 1963.

17. Loyd S. Swenson, Jr., James M. Grimwood, and Charles C. Alexander, *This New Ocean: A History of Project Mercury,* NASA SP-4201 (Washington, 1966) , pp. 414–15, 418, 443–45.

18. For a discussion of the integration of experiments with the Mercury flights, see Lewis R. Fisher, William O. Armstrong, and Carlos S. Warren, "Special Inflight Experiments," in *Mercury Project Summary Including Results of the Fourth Manned Orbital Flight, May 15 and 16, 1963,* NASA SP-45 (Washington, 1963), pp. 213–19.

19. The author found this impression to be rarely if ever documented but pervasive among space scientists and others interviewed. It seemed to be most persistent in the minds of scientists who were not associated with the manned programs over long periods of time; those who worked closely with MSC engineers in developing the Apollo science program came to feel otherwise, for the most part. See author's interviews with P. E. Purser, Mar. 10, 1983, E. M. Shoemaker, Mar. 17, 1984, and H. H. Schmitt, May 30, 1984, and Loyd S. Swenson's interviews with A. J. Dessler, May 16, 1971, and E. H. King, Jr., May 27, 1971, tapes in JSC History Office files. Those who understood the engineers' problems and accepted the manned lunar landing program as the driving force of the entire space program tended to be more sympathetic. Scientist-astronaut H. H. Schmitt felt the scientists' expectations of the engineers were unreasonable: "look what they [the engineers] were trying to do, for crying out loud: they were trying to land on the moon! . . . [As late as 1968] there was not an engineer . . . who could prove that the lunar module was going to be able to fly to the moon [land and return]." (Schmitt interview.) During Mercury, the many foreseeable (and unforeseeable but expected) problems of developing spacecraft

and operations caused engineers to be intolerant of *any* exercise not essential to the lunar landing that might in any way imperil the crew or the engineering and operational objectives of a flight.

20. Shoemaker interview.
21. Newell to Gilruth, Feb. 15, 1963; Wendell W. Mendell, interview with Swenson, Feb. 11, 1971, transcript in JSC History Office files; Newell to A/Mr. Webb, Mar. 16, 1963.
22. King interview; Gilruth to Thomas B. Nolan, Mar. 29, 1963.
23. MSC Circular 19, "Establishment of the Mercury Scientific Experiments Panel," Apr. 18, 1962; MSC gen. mgt. inst. 2-3-1, "Manned Spacecraft Center In-Flight Scientific Experiments Coordination Panel," Oct. 15, 1962; MSC tech. mgt. inst. 37-1-1, "In-Flight Experimental Programs," July 18, 1963; William O. Armstrong, interview with James M. Grimwood and Ivan D. Ertel, Jan. 24, 1967, transcript in JSC History Office files.
24. Fryklund to SD/Deputy Director, "Memo from J. M. Eggleston About a Facility at MSC to House the Space Environment Division," July 24, 1963; Newell to M/Assoc. Adm., "Facility at Manned Spacecraft Center for the Space Environment Division," July 31, 1963.
25. Newell to Gilruth, "Manned Space Science Division Supporting Research and Technology Tasks," Nov. 21, 1963; Fryklund, "Discussions at Manned Spacecraft Center on September 19, 1963," memo for record, Sept. 27, 1963; Fryklund to Dr. Robert Voas, Oct. 11, 1963; Paul R. Brockman to SM/Acting Director, "Manned Space Science Project Management," Nov. 8, 1963; Brockman to SP/Gutheim, "Weekly report," Nov. 22, 1962; Willis B. Foster to Assoc. Adm. for MSF, "Apollo Scientific Guidelines," Dec. 19, 1963; Brockman to Gutheim, "Weekly Report," Dec. 6, 1963.
26. Foster to Richard Mize, "Weekly Report for Week of December 16," Dec. 20, 1963.
27. Foster to Mize, "Weekly Activities Report," Jan. 10, 1964; "Minutes: Manned Space Science Working Group of the Space Sciences Steering Committee," Jan. 30, 1964; Foster to SS/Dir. of Sciences, "Weekly Activities Report," Feb. 12, 1964.
28. Fryklund to Members, Manned Space Science Working Group, Space Science Steering Committee, "First Group of Suggested Apollo Investigations and Investigators," Feb. 3, 1964; Fryklund to chmn., Space Sciences Steering Committee, "Preliminary definition of Apollo investigations," Feb. 13, 1964.
29. Fryklund to John M. Eggleston, "Integrated Apollo Science Program," Mar. 25, 1964; Fryklund to Assoc. Adm., SSA, "Integration of Apollo Science Program," Mar. 26, 1964; "Minutes, Manned Space Science Working Group of the Space Sciences Steering Committee, 26 March 1984," Apr. 9, 1964.
30. E. L. Durst to Chief, Flight Operations Div., "Project Apollo, Operational Ground Rules for the Lunar Landing Mission," Oct. 23, 1963; MSC, "The Development of an Apollo Lunar Landing Mission Reference Trajectory," Internal Note 64-OM-12, May 1964.
31. MSC, "Contributions of MSC Personnel to the Manned Lunar Exploration Symposium, June 15 and 16, 1964," no date; D. A. Beattie, E. M. Davin, and P. D. Lowman to Dir., Manned Space Science Div. and Chief, Lunar and Planetary Science Branch, "Supplemental Recommendations and Comments on 'Apollo Scientific Investigations,'" July 6, 1964; Fryklund to Dir., MSC, "Action Items and Positive Results from the Apollo Science Meeting," July 6, 1964.
32. Davin to multiple addressees, "Request for comments on the attached lunar surface science programs definition document for the approved Apollo missions, no date [Dec. 1964].
33. OSSA, "Apollo Lunar Science Program: Report of Planning Teams," part I, Summary (Dec. 1964), pp. 4–9.
34. Ibid., part II, Appendix.
35. Eggleston to Staff, "Establishment of interim coordinators for scientific equipment," Jan. 14, 1965.
36. Foster to multiple addressees, "Report of Program Plans and Status," Feb. 1, 1965.
37. Foster to Assoc. Adm., MSF, "Matters to discuss with Joe Shea in regard to science experiments," Apr. 1, 1965.
38. Maxime A. Faget to Foster, "Grants or contracts to potential principal investigators for lunar surface experiments," May 11, 1965, with encl., "Chronology [of background information]."
39. Richard J. Allenby to Mueller and Newell, "Minutes of Newell-Mueller Meeting of 23 February 1965," Apr. 19, 1965, with encl., "Memorandum of Agreement between Office of Manned Space Flight [and] Office of Space Science and Applications, Scientific Interfaces."
40. B. A. Linn to the record, "MSF Procurement Plan for Lunar Surface Experiments Package," June 3, 1965; Mueller to MSC, attn. Dave Lang, "Request for Approval of Procurement Plan for Lunar Surface Experiments Package," June 7, 1965.

41. NASA release 65-260, ''Three Firms Selected to Design Apollo Lunar Surface Package,'' Aug. 4, 1965.
42. National Academy of Sciences-National Research Council, *Space Research: Directions for the Future*, report of a study by the Space Science Board, Woods Hole, Mass., 1965, NAS-NRC Publication 1402 (Washington, 1966), p. iii.
43. Ibid.
44. Ibid., pp. 623-26.
45. Ibid., pp. 628-29.
46. Ibid., pp. 5-34.
47. Ibid., pp. 486-93.
48. Ibid., p. 21.
49. *NASA 1965 Summer Conference on Lunar Exploration and Science, Falmouth, Massachusetts, July 19-31, 1965*, NASA SP-88 (Washington, 1965) , pp. 1, 7-19.
50. Ibid., pp. 59-393, 407-17.
51. Neal J. Smith to Allenby, 11 Aug. 1965.
52. *NASA 1965 Summer Conference*, pp. 7-12, 16-19.
53. Shoemaker interview.
54. Harold C. Urey to Newell, June 19, 1961.
55. Shoemaker, ''Exploration of the Moon's Surface,'' *American Scientist* 50 (1962): 121.
56. Ted H. Skopinski to Chief, Systems Integration Div., ''Selection of Lunar Landing Site for the Early Apollo Lunar Missions,'' Mar. 21, 1962.
57. Bruce K. Byers, *Destination Moon: A History of the Lunar Orbiter Program*, NASA TM X-3487 (Washington, 1977), pp. 9-15.
58. Ibid., pp. 19-47, 67-69, 75-78, 227.
59. Gilruth to Mueller, Aug. 5, 1965.
60. T. H. Thompson to G. E. Mueller and Gen. S. C. Phillips, Dec. 23, 1964.
61. Mueller to multiple addressees, ''Establishment of Apollo Site Selection Board,'' July 1, 1965.

CHAPTER 4

1. National Academy of Sciences–National Research Council, *A Review of Space Research*, report of the summer study conducted under the auspices of the Space Science Board of the National Academy of Sciences, NAS-NRC Publication 1079 (Washington, 1962), pp. 4-7, 4-33.
2. John M. Eggleston to M. A. Faget, ''Initial Handling of Geological and Biological Samples Returned from the Apollo Missions,'' Feb. 24, 1964.
3. Aleck C. Bond to Chief, Off. of Technical and Engineering Services, ''Sample Transfer Facility,'' with encl. , ''Functional Description and Tentative Performance Requirements,'' Apr. 14, 1964.
4. OSSA, ''Apollo Lunar Science Program, Report of Planning Teams,'' part II, Appendix: sec. III, ''Preliminary Report on the Sampling and Examination of Lunar Surface Materials,'' pp. 8-9; sec. IV, ''First Report—Geochemistry Planning Team,'' pp. 8-13.
5. E. A. King and D. A. Flory to Asst. Dir. for Engineering and Development, ''Requirements for a Facility to Receive and Accomplish Initial Lunar Sample Investigation at MSC,'' July 7, 1964.
6. Faget to Willis B. Foster, ''Apollo Sample Handling Facility'' (draft) , Aug. 26, 1964.
7. Foster to Eggleston, ''Proposal for Cataloging of Lunar Samples by Elbert A. King and Donald A. Flory,'' Aug. 17, 1964; Foster to Faget, ''Requirement for Laboratory Facilities for Receiving, Unpacking and Preliminary Examination of Lunar Samples,'' Aug. 17, 1964.
8. Foster to Eggleston, ''Lunar Sample Receiving Laboratory,'' Oct. 23, 1964.
9. James C. McLane, Jr., to Mgr., Systems Tests and Evaluation, ''Lunar Sample Receiving Laboratory,'' Nov. 13, 1964.
10. McLane to Mgr., System Tests and Evaluation, ''Lunar Sample Receiving Laboratory,'' Dec. 18, 1964.
11. Homer E. Newell to Harry H. Hess, Dec. 8, 1964.

12. Hess to Newell, with encl., "Report of Ad Hoc Committee on Lunar Sample Handling Facility," Feb. 2, 1965.
13. Ibid.
14. McLane to Chief, Facilities Div., "FY 67 C of F Program," with encl., "Lunar Sample Receiving Laboratory Project Description and Project Justification," Jan. 20, 1965.
15. Foster to Faget, "Lunar Sample Receiving Laboratory," Feb. 24, 1965.
16. Faget to Foster, "MSC Lunar Sample Receiving Laboratory," Mar. 22, 1965.
17. Q. G. Robb to Chief, Test Facilities Branch, "Background Material on the Lunar Sample Facility," Apr. 6, 1965.
18. Hess to Newell, Feb. 2, 1965.
19. See the testimony of Dr. Colin S. Pittendrigh in Senate Committee on Aeronautical and Space Sciences, *Scientists' Testimony on Space Goals*, 88/1, June 10–11, 1963, pp. 73–80. Pittendrigh cited the same estimate of the probability of life in the universe as that given by Su-Shu Huang ("Occurrence of Life in the Universe," *American Scientist* 47 (1959): 397–403), who estimated that one billion billion (10^{18}) planets in the observable universe might be sites for the evolution of life. See also the report of the biology working group in *A Review of Space Research*, pp. 9-1 to 9-4, 9-6. For a discussion of the arguments concerning life in the solar system, see Edward Clinton Ezell and Linda Neumann Ezell, *On Mars: Exploration of the Red Planet, 1958–1978*, NASA SP-4212 (Washington, 1984), pp. 51–66; see also their bibliography, p. 482, and survey of preliminary results of the life-detecting experiments on the Viking landing missions, pp. 400–414.
20. Minutes, meeting of the Exobiology Committee of the Space Science Board, Feb. 20, 1960, cited in Space Science Board, "Conference on Potential Hazards of Back Contamination from the Planets, July 29–30, 1964" (advance copy), no date [Aug. 1964].
21. *A Review of Space Research*, p. 9-13.
22. Homer E. Newell, *Beyond the Atmosphere: Early Years of Space Science*, NASA SP-4211 (Washington, 1980), pp. 274–75.
23. Ibid.
24. Space Science Board, "Conference on Potential Hazards of Back Contamination."
25. Eggleston to multiple addressees, "Sterilization precautions and quarantine of astronauts and equipment following Apollo missions," Feb. 5, 1965.
26. Orr E. Reynolds to Assoc. Adm. for Space Science and Applications, "Responsibility for Space Quarantine," July 2, 1965.
27. Eggleston to multiple addressees, "Sterilization precautions . . .," Feb. 5, 1965; Eggleston to multiple addressees, "Recommendations on NASA Position on Sterilization and Quarantine of Apollo Astronauts and Equipment," Feb. 19, 1965.
28. Reynolds to Assoc. Adm. for SSA, "Status of the Public Health Service–National Aeronautics and Space Administration negotiations on back contamination," May 10, 1965.
29. W. E. Stoney, Jr., to Chief, Engineering Div., "Support Information for FY 67 C of F Project— Lunar Sample Receiving Laboratory " July 30, 1965; Hall to Ed Chao, "Engineering Study and Preliminary Engineering Report for Lunar Sample Receiving Laboratory," July 1, 1965.
30. Reynolds to the record, "Summary of meeting between representatives of the National Aeronautics and Space Administration and the Public Health Service, July 31, 1965," Aug. 17, 1965.
31. W. W. Kemmerer, Jr., and E. A. King, Jr., to Faget, "Proposed MSC Quarantine Policy for the Apollo Program," with encl., "Proposed Quarantine Policy for Apollo Program," Sept. 23, 1965.
32. Elbert A. King, Jr., interview with Loyd S. Swenson, May 27, 1971, tape in JSC History Office files; Kemmerer and King to the record, "Summary of a meeting between representatives of the National Aeronautics and Space Administration, Public Health Service and the Department of Agriculture, MSC, Houston, Texas, September 27, 1965," Sept. 30, 1965.
33. Lawrence B. Hall to Deputy Adm., "Informal Conference on Back Contamination Problems," Oct. 15, 1965.
34. Owen E. Maynard to PS Branches, "Earth contamination from lunar surface organisms," Oct. 29, 1965.
35. King interview.
36. Hugh L. Dryden to William H. Stewart, Nov. 15, 1966.
37. Stewart to Webb, Dec. 22, 1966.
38. Deputy Dir., Space Medicine (OMSF), TWX to McLane, Nov. 22, 1965; McLane to A. C. Bond, "Hqs. Request for Quarantine Fac[ility]. Study," Nov. 22, 1965.

39. Faget to Col. Jack Bollerud, "Study of Existing Personnel Quarantine Facilities," with encl., "Apollo Personnel Quarantine Facilities Review," Dec. 8, 1965.

40. W. David Compton and Charles D. Benson, *Living and Working in Space: A History of Skylab*, NASA SP-4208 (Washington, 1983), pp. 40, 99–100.

41. House Committee on Science and Astronautics, *1967 NASA Authorization*, Hearings on H.R. 12718, 89/2, pt. 1, Mar. 10, 1966, p. 6.

42. Ibid., p. 36.

43. Idem, *1967 NASA Authorization*, Hearings before the Subcommittee on Manned Space Flight on H. R. 12718, 89/2, pt. 2, Mar. 1, 1966, p. 465.

44. Ibid., pp. 417–21.

45. Ibid., p. 476.

46. Ibid., pp. 477–79.

47. Arthur Hill, "Lab Delay May Slow Mission To the Moon," *Houston Chronicle*, Mar. 10, 1966.

48. Robert F. Freitag to George M. Low, Mar. 14, 1966.

49. MSC, "Site Investigation Study, Lunar Receiving Laboratory," draft report, Mar. 24, 1966, unpaginated. The final version was distributed on Apr. 7.

50. Ibid.

51. Paul E. Purser to Gilruth and Low, Mar. 29, 1966; Purser to Freitag, "Supplementary Information on the Lunar Receiving Laboratory," Mar. 30, 1966.

52. House, *1967 NASA Authorization*, pt. 2, Mar. 31, 1966, pp. 1207–59.

53. Charles Culhane, "Panel for Restoring Funds for Moon Lab," *Houston Post*, Apr. 1, 1966.

54. House Committee on Science and Astronautics, *Authorizing Appropriations to the National Aeronautics and Space Administration*, House Rept. 1441, 89/2, Apr. 20, 1966, pp. 121–24.

55. "$4.9-Billion Is Voted For NASA By House," *New York Times*, May 5, 1966.

56. House Committee on Appropriations, *Independent Offices Appropriation Bill, 1967*, House Rept. 1477, 89/2, May 5, 1966, pp. 13–14.

57. Senate Committee on Aeronautical and Space Sciences, *NASA Authorization for Fiscal Year 1967*, Senate Rept. 1184, 89/2, May 23, 1966, pp. 81–83.

58. NASA Off. of Legislative Affairs,, "Legislative Activity Report," vol. V. , no. 141, Aug. 24, 1966.

59. Jane Van Nimmen and Leonard C. Bruno, with Robert L. Rosholt, *NASA Historical Data Book, 1958–1968, vol. I: NASA Resources*, NASA SP-4012 (Washington, 1976), p. 117.

60. Leo T. Zbanek to Chief, Facilities Div., "Lunar Sample Handling Facility," Mar. 5, 1965; Faget to Foster, "MSC Lunar Sample Receiving Laboratory," Mar. 22, 1965; OSSA Ad Hoc Committee on the Lunar Sample Receiving Laboratory, "Concepts, Functional Requirements, Specifications and Recommended Plan of Operation for the Lunar Sample Receiving Laboratory," Mar. 15, 1965; McLane to J. G. Griffith, "Lunar Sample Receiving Laboratory," Apr. 8, 1965; Gilruth to Chief, Engineering Div., "Formation of a Technical Working Committee for the Design of a Lunar Sample Receiving Laboratory and Designation as Consultants to Assist in the Selection of an Architect-Engineer Firm," June 14, 1965; Foster to Faget, "Membership of the Headquarters Advisory Committee on Lunar Sample Receiving Laboratory," July 14, 1965; Foster to Dir., Facilities Programming and Construction, "Lunar Sample Receiving Laboratory," Aug. 19, 1965; Foster to George M. Low, "Lunar Sample Receiving Laboratory," Aug. 25, 1965; OSSA Standing Committee on LSRL, "Review of the Preliminary Engineering Report (PER) of the Lunar Sample Receiving Laboratory (LSRL)," Nov. 10, 1965; Low to Foster through Dr. George E. Mueller, "Manned Space Science Standing Committee for the Lunar Sample Receiving Laboratory," Dec. 9, 1965.

61. House, *1967 NASA Authorization*, pt. 2, Mar. 1, 1966, p. 472.

62. Dave W. Lang to George J. Vecchietti, "Schedule of Procurement Actions—Lunar Receiving Laboratory," Apr. 30, 1966.

63. MSC Announcement 66-57, "Establishment of a Lunar Receiving Laboratory Policy Board and a Lunar Receiving Laboratory Program Office," May 9 , 1966

64. D. D. Wyatt, "Lunar Receiving Facility, memo for record, Manned Spacecraft Center," June 14, 1965.

65. Mueller to MSC, TWX, subj.: "Lunar Receiving Laboratory," July 28, 1966.

66. NASA Releases 66-200, Aug. 1, 1966, and 66-222, Aug. 19, 1966.

67. Space Science Board, "Report of Ad Hoc Committee on Lunar Sample Handling Facility," Feb. 2, 1965.

68. King interview.

69. OSSA Ad Hoc Committee on the Lunar Sample Receiving Laboratory, "Concepts . . . for the Lunar Sample Receiving Laboratory," Mar. 15, 1965, p. 80.

70. McLane to Mgr., Systems Tests and Evaluation, "Staffing for the Lunar Sample Receiving Laboratory," Sept. 2, 1965.

71. Urner Liddell to Chmn., Space Science Steering Committee, "Recommendations of Planetology Subcommittee, Meeting 3–66, February 23–25, 1966," no date.

72. "Summary Minutes, Lunar Receiving Laboratory Working Group of the Planetology Subcommittee, Space Sciences Steering Committee (Meeting No. 1–66)," May 5, 1966.

73. McLane to Dir., Engineering and Development, "Estimate of scientist staffing requirements for the Lunar Receiving Laboratory," May 25, 1966.

74. House Committee on Science and Astronautics, *1968 NASA Authorization*, hearings before the Subcommittee on Manned Space Flight, 90/1, pt. 2, p. 599.

75. "Summary Minutes, Lunar Receiving Laboratory Working Group of the Planetology Subcommittee, Space Sciences Steering Committee (Meeting No. 1–67)," July 11, 1966.

76. Joseph R. Crump to Purser, with encl., "Concerning a Contract for Lunar Receiving Laboratory," July 27, 1966.

77. Draft memo, Gilruth to Mueller, "Procurement Plan for Operational Support Services Contract for the Manned Spacecraft Center Lunar Receiving Laboratory," with encl., procurement plan, no date [c. Sept. 5, 1966].

78. Gilruth to Lt. Gen. Frank A. Bogart, Oct. 12, 1966.

79. J. E. Riley to Mr. Cariski, "MSC Procurement Plan for Operational Support of Lunar Receiving Laboratory," Oct. 18, 1966; Bogart to Mr. Webb, "Meeting with Dr. Seitz on the Lunar Receiving Laboratory," Nov. 7 , 1966; Purser, "Lunar Receiving Laboratory Operations Planning," memo for the record, Nov. 7, 1966; Robert O. Piland to Deputy Dir., "Lunar Receiving Laboratory," Nov. 8, 1966; Wesley L. Hjornevik, TWX to Bogart, Dec. 15, 1966; Francis B. Smith to Webb and Seamans, "December 1966 meeting to discuss plans for the Lunar Receiving Laboratory," Dec. 19, 1966.

80. Smith to Webb and Seamans, "December 1966 meeting . . .," Dec. 19, 1966.

81. Willis M. Shapley to Mueller and Newell, "Lunar Receiving Laboratory," Dec. 21, 1966.

82. "Summary Minutes, Lunar Receiving Laboratory Working Group of the Planetology Subcommittee, Space Science Steering Committee (Meeting No. 2–67)," Sept. 26–27, 1966; Lunar Receiving Laboratory Working Group to Newell through Planetology Subcommittee, Sept. 29, 1966; Bogart to Low, Sept. 22, 1966; "Minutes, Interagency Committee on Back Contamination," Oct. 3, 1966; David J. Sencer to Seamans, Nov. 14, 1966; G. Briggs Phillips to Bogart, with encl., "Status Report on the Lunar Receiving Laboratory," Dec. 13, 1966; Melvin Calvin to Newell, Dec. 29, 1966; Richard J. Allenby to Assoc. Adm. for Space Science and Applications, "Lunar Receiving Laboratory Problems," Dec. 29, 1966.

83. Hjornevik to Bogart, TWX, Dec. 15, 1966; MSC Announcement 67-7, "Organization and Personnel Assignments for the Science and Applications Directorate," Jan. 10, 1967.

84. Low to Bogart, Jan. 13, 1967, with encls.: Gilruth to Chief, Procurement and Contracts Div., "Justification (under NPC 401) for nonpersonal service contract for operation of the Manned Spacecraft Center Lunar Receiving Laboratory," Jan. 13, 1967; Piland to Chief, Procurement and Contracts Div., "Justification for noncompetitive procurement for operational support of the Lunar Receiving Laboratory," Jan. 13, 1967.

CHAPTER 5

1. MSC's philosophy of astronaut selection and training was well summarized in a memo by Christopher C. Kraft, Jr., of MSC's Flight Operations Division in late 1962. Commenting on an article in a national magazine, wherein it was speculated that future astronauts might be computer experts rather than test pilots, Kraft wrote, "It is our feeling that you must use people who are adapted to taking action should emergencies develop and who are well trained in this category. We feel that you can then take this type of individual . . . and give him the necessary training

in the other fields such as navigation and guidance, geology, etc. *The primary need is for spacecraft control, at least in the programs presently planned (Gemini and Apollo), and these people can be trained to accomplish the explorations presently envisioned* [emphasis added]. When you get to the point of conducting experiments in space, such as you would do in the Space Station, . . . we would probably use engineers and scientists who were experts in a given field. However, the people responsible for control of the vehicle ferrying these types of experts to and from the Space Station, and the people on board the Space Station responsible for emergency action will probably still be in the same category as test pilots." C. C. Kraft, Jr., to A. M. Chop, "Article on Astronaut Selection in November 7 issue of Time Magazine," Nov. 8, 1962. This basic philosophy has been followed for 20 years, right down to the operational flights of the Shuttle orbiter.

2. Loyd S. Swenson, Jr., James M. Grimwood, and Charles C. Alexander, *This New Ocean: A History of Project Mercury*, NASA SP-4201 (Washington, 1966), p. 174.

3. Ibid., pp. 131, 160–63.

4. Ibid., p. 174.

5. Ibid., p. 235; John H. Glenn, Jr., interview with Robert B. Merrifield, Mar. 15, 1968, transcript in JSC History Office files. For a colorful account of the beginning of the manned space flight program, including the choice of test pilots as the Original Seven astronauts and their experiences through the Mercury program, see Tom Wolfe, *The Right Stuff* (New York: Farrar, Straus and Giroux, 1979). Wolfe's literary style is objectionable to many, but his treatment of Mercury is (to the present author's knowledge) well researched and is considered by many Mercury participants to accurately convey the spirit of the early days of manned space flight.

6. Swenson, Grimwood, and Alexander, *This New Ocean*, pp. 235–37.

7. National Academy of Sciences–National Research Council, *A Review of Space Research*, report of the summer study conducted under the auspices of the Space Science Board, NAS–NRC Publication 1079 (Washington, 1962), p. 1–22.

8. Ibid., pp. 11-1 to 11-16.

9. Ibid., pp. 11-8 to 11-12.

10. Ibid., pp. 11-17 to 11-19.

11. Swenson, Grimwood, and Alexander, *This New Ocean*, pp. 440–42.

12. Ibid., pp. 177, 442.

13. Donald K. Slayton interview, Oct. 15, 1984. Slayton's comments on this case: "There were two situations where we had a strong input from Headquarters. One was on Apollo 13, where for some reason George Mueller didn't like the crew I had assigned to fly 13 and insisted on turning them around. . . . We wound up switching the 13 and 14 crews around. . . . The other thing, . . . on 13, Mattingly had that medical problem [exposure to rubella]. . . ." Alan Shepard was commander of the crew that flew Apollo 14, so the original assignment would have given him command of Apollo 13. At the time those two crews were publicly named (August 1969) Shepard had been back on flight duty less than six months after spending almost five years off the active list for medical reasons.

14. Slayton interview by Robert B. Merrifield, Oct. 17, 1967, transcript in JSC History Office files; MSC *Space News Roundup*, Sept. 19, 1962; MSC Announcements 190, Apr. 29, 1963, and 268, Nov. 5, 1963.

15. Alan L. Bean interview, Apr. 10, 1984.

16. Barton C. Hacker and James M. Grimwood, *On the Shoulders of Titans: A History of Project Gemini*, NASA SP-4203 (Washington, 1977), pp. 49–73.

17. House Committee on Science and Technology, *Astronauts and Cosmonauts: Biographical and Statistical Data* [Revised May 31, 1978], report prepared by the Congressional Research Service, Library of Congress, July 1978, pp. 6–7; idem, *Astronautical and Aeronautical Events of 1962*, report prepared by the NASA Historical Staff, June 12, 1963, p. 56.

18. *Astronautical and Aeronautical Events of 1962*, pp. 146, 191.

19. Slayton interview, Oct. 17, 1967.

20. Ibid.; MSC, "Flight Crew Training Report No. 1," Oct. 20, 1962.

21. MSC, "Flight Crew Training Reports," nos. 1–16, Oct. 15, 1962, to Feb. 9, 1963; no. 16 includes a summary of topics covered in the basic science course.

22. MSC, "Apollo Spacecraft Project Status Report No. 1 for Period Ending September 31, 1962," p. 48.

23. NASA, *Astronautics and Aeronautics, Chronology on Science, Technology, and Policy, 1963*, NASA SP-4004, (Washington, 1964) p. 197.

24. MSC Release 63–102, June 18, 1963; Slayton interview, Oct. 17, 1967; *Astronauts and Cosmonauts*, p. 7.
25. House Subcommittee on Manned Space Flight of the Committee on Science and Astronautics, *1964 NASA Authorization*, hearings on H.R. 5466, 88/1, pt. 2(a), p. 235, Mar. 7, 1963.
26. *Astronautics and Aeronautics, 1963*, pp. 322, 392.
27. *Astronauts and Cosmonauts*, p. 7.
28. *Astronautics and Aeronautics, 1963*, p. 495.
29. Hacker and Grimwood, *On the Shoulders of Titans*, p. 191.
30. MSC, Flight Crew Training Reports nos. 66–83, Feb. 3–June 1, 1964.
31. Elbert A. King, Jr., interview by Loyd S. Swenson, Jr., May 27, 1971, tape in JSC History Office files.
32. Ibid.; interviews, Alan L. Bean, Apr. 10, 1984, and Eugene A. Cernan, Apr. 6, 1984; Michael Collins, *Carrying the Fire: An Astronaut's Journeys* (New York: Farrar, Straus and Giroux, 1974), pp. 72–75.
33. Hacker and Grimwood, *On the Shoulders of Titans*, pp. 194–200.
34. Ibid., 219.
35. Ibid. , 220–24; Donald K. Slayton, Warren J. North, and C. H. Woodling, "Flight Crew Procedures and Training," in *Gemini Midprogram Conference Including Experimental Results*, NASA SP-121 (Washington, 1966) , pp. 201–11.
36. Hacker and Grimwood, *On the Shoulders of Titans*, pp. 239–40.
37. Ibid., 255.
38. All the astronauts this author has interviewed have stated that they never understood exactly what determined their first assignment to a crew; see author's interviews with Alan L. Bean, Harrison H. Schmitt, Eugene A. Cernan, and Joseph P. Kerwin, Jr., transcripts in JSC History Office files; with Paul J. Weitz, transcript in Skylab files, Fondren Library, Rice Univ.; also Collins, *Carrying the Fire*, p. 141, and Slayton interview, Oct. 15, 1984. A special scientific committee convened at OSSA's request by Rice University to recommend ways to provide more opportunities for scientific training of astronauts noted that criteria for crew selection were a mystery to the astronauts; see n. 61, below.
39. See W. David Compton and Charles D. Benson, *Living and Working in Space: A History of Skylab*, NASA SP-4208 (Washington, 1983) pp. 12–14, 19–20; also House, Subcommittee on Manned Space Flight of the Committee on Science and Astronautics, *1965 NASA Authorization*, hearings on H.R. 9641, 88/2, pt. 2, pp. 446–48, 587–609, and Arnold S. Levine, *Managing NASA in the Apollo Era*, NASA SP-4102 (Washington, 1982) , pp. 239–47.
40. Homer E. Newell to W. N. Hess, Apr. 16, 1964.
41. "NASA to Select Scientist-Astronauts for Future Missions," NASA Release 64–248, Oct. 19, 1964.
42. Homer E. Newell, *Beyond the Atmosphere: Early Years of Space Science*, NASA SP-4211 (Washington, 1980), p. 209.
43. Slayton interview, Oct. 15, 1984.
44. Eugene M. Shoemaker interview, Mar. 17, 1984. According to Jack Schmitt, who worked with Shoemaker before and after he was selected as an astronaut, the committee set the standards too high. He recalled telling Shoemaker that they should have picked more scientists, "because we need[ed] more visibility down here." Shoemaker argued that the Academy did not want to risk choosing people who would not work out as astronauts, so the committee applied very strict standards. H. H. Schmitt interview, July 6, 1984.
45. MSC News Release 65–63, June 29, 1965.
46. "NASA Picks 6 Scientist-Astronauts To Make Field Trips to the Moon, *Washington Post*, June 27, 1965.
47. Associated Press, "Dr. Graveline Quits Project as Astronaut," *Chicago Tribune*, Aug. 19, 1965. Graveline's wife sued him for divorce shortly after he was selected. While no documentation has been found to confirm it, the general impression among the astronauts was that divorce was considered incompatible with the established image of the astronaut, apart from the fact that involvement in a divorce case might create psychological problems that would impair the trainee's (or pilot's) efficiency.
48. Compton and Benson, *Living and Working in Space*, pp. 40–56, 83–86.
49. George E. Mueller to Robert R. Gilruth, Jan. 25, 1964.
50. "NASA To Select Additional Pilot-Astronauts," NASA Release 65–288, Sept. 10, 1965; "U.S. Seeking Astronauts for Apollo," *Baltimore Sun*, Sept. 11, 1965.

51. "Nineteen Pilots Join United States Astronaut Team," NASA Release 66–77, Apr. 4, 1966; "19 Chosen as Spacemen, Some for Moon Missions," *Washington Post*, Apr. 5, 1966.
52. "Scientists Invited to Become Astronauts, Do Research in Space," NASA Release 66–255, Sept. 26, 1966.
53. Russell L. Schweickart, interview with Peter Vorzimmer, May 1, 1967, transcript in JSC History Office files.
54. *NASA 1965 Summer Conference on Lunar Exploration and Science*, NASA SP–88 (Washington, 1965), pp. 407–17.
55. Alan B. Shepard, Jr., to multiple addressees, "Astronaut Technical Assignments," Jan. 6, 1967.
56. Harrison H. Schmitt interview, May 30, 1984.
57. Slayton interview, Oct. 15, 1984.
58. Ibid.; Bean interview.
59. Joseph P. Kerwin, Jr., interview, Mar. 29, 1985.
60. Bean interview; Cernan interview.
61. "Space Science Training for Astronauts Involved in NASA Manned Space Flight Missions," report under contract NSR 44–006–031, Nov. 1965, p. 4.
62. Ibid., pp. 2–4.
63. Cernan interview.
64. Gilruth to Dr. A. J. Dessler, Dec. 22, 1965.
65. Slayton interview, Oct. 15, 1984. When the debate over sending a scientist to the moon intensified in 1969, several wire-service stories quoted Slayton to this effect; see Paul Recer (Associated Press), "They Feud Over Moon Flights," *Miami Herald*, Oct. 12, 1969. Gilruth's position was expressed in a letter to George Mueller, Sept. 2, 1969, responding to Mueller's concern that the science community was growing restive because no scientist-astronauts were assigned to Apollo missions.

CHAPTER 6

1. Harold C. Urey to Homer E. Newell, June 19, 1961.
2. Eugene M. Shoemaker, "Exploration of the Moon's Surface," *American Scientist* 50 (1962):99–130.
3. MSC, "Environmental Factors Involved in the Choice of Lunar Operational Dates and the Choice of Lunar Landing Sites," NASA Project Apollo Working Paper (AWP) No. 1100, Nov. 22, 1963. In 1972 an entire number of *The Bell System Technical Journal* (vol. 51, no. 5, pp. 955–1127) was devoted to a single article, "Where on the Moon? An Apollo Systems Engineering Problem," J.O. Cappellari, Jr., ed. This is a narrative description of the many factors that entered into the choice of an Apollo landing site, how the factors were weighted, and how the interaction of those factors changed as experience with Apollo systems accumulated. It is not so technical as to be incomprehensible to most readers. Not surprisingly, it lays considerable emphasis on the role of Bellcomm, Inc., in the site selection process.
4. AWP 1100, p. 23.
5. Ibid., p. 32.
6. Ibid., pp. 7–13.
7. Ibid., pp. 6–7.
8. Ibid., p. 33.
9. W. E. Thompson (Bellcomm, Inc.), "Lunar Landing Site Constraints: The Arguments for and Against One Preselected Site Versus Several Sites," Jan. 31, 1964.
10. Benjamine J. Garland, memo for Apollo proj. off., "Characteristics of the Lunar Surface," Aug. 14, 1961; Thomas Gold, "Structure of the Moon's Surface," in J.W. Salisbury and P.E. Glaser, eds., *The Lunar Surface Layer* (London: Academic Press, 1964), pp. 345–53. Gold interpreted radar, optical, and thermal properties of the lunar surface as indicating a layer of fine particles up to several meters thick whose mechanical properties would be difficult to predict. He suggested that at the very least a lunar landing would be compromised by blinding clouds of dust raised by the exhaust from the descent engine. In the popular press, Gold's hypothesis was taken to

mean that a lunar module could sink out of sight. The pictures returned by *Ranger 7* (July 1964) gave Gold no reason to change his thinking; see R. Cargill Hall, *Lunar Impact: A History of Project Ranger*, NASA SP-4210 (Washington, 1977), p. 285; also Gold, "Ranger Moon Pictures; Implications," *Science* 145 (1964):1046–48. Most Ranger scientists agreed that the TV pictures from *Ranger 7* gave no basis for drawing conclusions about the load-bearing characteristics of the surface. They did give some useful information about the distribution and size of craters and the slopes of lunar terrain on a smaller scale than had been possible. Gold remarked after the conclusion of Ranger that its pictures were like mirrors: "everyone sees his own theories reflected in them." Hall, *Lunar Impact*, p. 309. For a bibliography of studies on the lunar surface up to the first successful Ranger mission, see J.W. Salisbury, ed., *Bibliography of Lunar and Planetary Research — 1960–1964*, AFCRL-66-52, U.S. Air Force Office of Aerospace Research, Cambridge Research Laboratories, Jan. 1966.

11. Hall, *Lunar Impact*, pp. 156–82.

12. Courtney G. Brooks, James M. Grimwood, and Loyd S. Swenson, Jr., *Chariots for Apollo: A History of Manned Lunar Spacecraft*, NASA SP-4205 (Washington, 1979), pp. 150–54.

13. National Academy of Sciences-National Research Council, *A Review of Space Research*, report of the summer study conducted under the auspices of the Space Science Board at the State University of Iowa, June 17–Aug. 10, 1962, NAS-NRC Publication 1079 (Washington, 1962); idem, *Space Research: Directions for the Future*, report of a study by the Space Science Board, Woods Hole, Mass., 1965, NAS-NRC Publication 1402 (Washington, 1966).

14. *NASA 1965 Summer Conference on Lunar Exploration and Science, Falmouth, Massachusetts, July 19–31, 1965*, NASA SP-88 (Washington, 1965), passim.

15. House, Subcommittee on Manned Space Flight of the Committee on Science and Astronautics, *1966 NASA Authorization*, hearings on H.R. 3730, 89/1, pt. 2, pp. 278–85.

16. John M. Eggleston to Dr. W.A. Lee, "Target Selection for Future Ranger Flights," Feb. 4, 1965.

17. C.J. Byrne, "Ranger VII Photo Analysis—Preliminary Measurements of Apollo Landing Hazards," Bellcomm, Inc., Tech. Memo. TM-65-1012-2, Mar. 17, 1965.

18. Newell to multiple addressees, "Establishment of Ad Hoc Surveyor/Orbiter Utilization Committee," June 22, 1965.

19. George E. Mueller to multiple addressees, "Establishment of Apollo Site Selection Board," Management Instruction (NMI) 1152.20, Aug. 6, 1965.

20. John P. Mayer to Chief, Mission Planning and Analysis Div., MSC, "Lunar landing site selection," Oct. 6, 1964; Robert R. Gilruth to Mueller, "Establishment of Apollo Site Selection Board," July 29, 1965; Gilruth to Newell, "Members of Ad Hoc Surveyor/Orbiter Utilization Committee," July 29, 1965.

21. OSSA, "Ad Hoc Surveyor/Orbiter Utilization Committee, Minutes First Meeting, Washington, D.C., August 20, 1965," Sept. 16, 1965; I.M. Ross (Bellcomm, Inc.) to Maj. Gen. S.C. Phillips, "MSF Position for Surveyor/Orbiter Utilization Committee Meeting August 20, 1965," Aug. 20, 1965. For a more detailed description of Lunar Orbiter mission planning, especially in support of Apollo, see Bruce K. Byers, *Destination Moon: A History of the Lunar Orbiter Program*, NASA TM X-3487 (Washington, 1977), pp. 177–99, 248–53, 261; Orbiter's contributions to Apollo site selection are summarized on pp. 308–14. No useful single source for Surveyor mission planning and its relation to Apollo is available; the project's results (almost entirely scientific) are summarized in *Surveyor Program Results*, NASA SP-184 (Washington, 1969).

22. Willis E. Foster, "Draft Minutes of Second Meeting of Ad Hoc Surveyor/Orbiter Utilization Committee," Oct. 1, 1965, with encl., ltr., R.J. Parks to Benjamin Milwitzky, subj.: Surveyor Mission A Landing Site Selection, Sept. 28, 1965.

23. Albert Sehlstedt, Jr., "Surveyor Makes Soft Landing On Moon," Baltimore *Sun*, June 2, 1966.

24. William Hines, "Man Could Explore the Moon on Foot," Washington *Evening Star*, June 17, 1966; Evert Clark, "Surveyor Found 'Gritty' Moon That Can Support a Spacecraft," *New York Times*, June 17, 1966.

25. Byers, *Destination Moon*, pp. 228–45.

26. Lunar Orbiter Photo Data Screening Group, "Preliminary Terrain Evaluation and Apollo Landing Site Analysis Based on Lunar Orbiter I Photography," Langley Working Paper LWP-323, Nov. 1966.

27. Foster, "Minutes of the Surveyor/Orbiter Utilization Committee, Washington, D.C., September 29, 1966"; Byers, *Destination Moon*, p. 251.

28. Byers, *Destination Moon*, p. 258. See L.J. Kosofsky and Farouk El-Baz, *The Moon as Viewed by Lunar Orbiter*, NASA SP-200 (Washington, 1970), for an annotated collection of many of the best photographs from the Lunar Orbiter missions, along with brief descriptions of the spacecraft and its systems and a complete listing of the sites photographed on each of the five missions. The detail in these photographs, even in halftone reproduction, must be seen to be believed. Much of the geologic interpretation of the moon has been accomplished using Lunar Orbiter photographs and the information they contain is far from exhausted, according to Eugene Shoemaker (interview, Mar. 17, 1984).

29. Foster, "Minutes of the Surveyor/Orbiter Utilization Committee, Washington, D.C., January 5, 1967."

30. S.C. Phillips to multiple addressees, "Minutes of the Apollo Site Selection Board Meeting, December 15, 1966," Mar. 3, 1967.

31. John M. Eggleston to Dir., MSC, "Utilization of Orbiter and Surveyor in Support of Apollo and Apollo Applications Program Objectives," Jan. 18, 1967; Byers, *Destination Moon*, pp. 312–26, 329–30.

32. MSC, "Project Apollo Lunar Excursion Module Development Statement of Work," July 24, 1962, pp. 2, A-107–109.

33. MSC, "Project Apollo Quarterly Status Report No. 3 for Period Ending March 31, 1963," p. 30.

34. MSC, "Project Apollo Quarterly Status Report No. 6 for Period Ending December 31, 1963," p. 34; contract NAS9-2115, "Survey of Lunar Surface Measurements, Experiments, and Geologic Studies," Sept. 30, 1963.

35. John M. Eggleston to Wayne Young (JSC), letter with comments on draft of Apollo explorations history, Sept. 12, 1984.

36. MSC, "Project Apollo Quarterly Status Report No. 8 for Period Ending June 30, 1964," p. 47.

37. Mueller to MSC, "Request for Approval of Procurement Plan for Lunar Surface Experiments Package," June 7, 1965.

38. "Three Firms Selected to Design Apollo Lunar Surface Package," Hqs. Release 65-260, Aug. 4, 1965.

39. MSC, "Lunar Surface Experiments Package Request for Proposal," May 27, 1965, pp. b-1 to b-29.

40. S.C. Phillips to multiple addressees, Apollo Program Directive No. 3, subj.: Management Assignment for the Lunar Surface Experiments Package (LSEP) Project, M-D 8030.003, June 15, 1965.

41. Newell to Dir., MSC, "Selection of Scientific Investigations for Early Apollo Lunar Landing Missions," Oct. 1, 1965.

42. Newell to Dir., MSC, "Selection of Apollo Lunar Science Magnetic Field Investigations," Dec. 15, 1965.

43. William J. O'Donnell to Ames Res. Ctr., MSC, and JPL, TWX, "Moon Surface Experiments Chosen for Apollo," (NASA Hqs. Release 66-17), Jan. 27, 1966.

44. Newell to Dir., MSC, "Authorization to Procure Space Science and Applications Investigations for Apollo Lunar Missions," Feb. 14, 1966.

45. "Bendix Named to Manufacture Lunar Package," NASA Hqs. release 66-63, Mar. 17, 1966.

46. A.P. Fontaine (Bendix Corp.) to Gilruth, Feb. 18, 1966.

47. Barton C. Hacker and James M. Grimwood, *On The Shoulders of Titans: A History of Project Gemini*, NASA SP-4203 (Washington, 1977), pp. 229–31; see *Gemini Midprogram Conference Including Experimental Results*, NASA SP-121 (Washington, 1966), pp. 305–436, and *Gemini Summary Conference*, NASA SP-138 (Washington, 1967), pp. 221–317, for summaries of the Gemini experiments program and its results.

48. See W. David Compton and Charles D. Benson, *Living and Working in Space: A History of Skylab*, NASA SP-4208 (Washington, 1983), pp. 59–63, for pre-Apollo experience with scientific experiments.

49. MSC Announcement 65-81, "Designation of Manager, Experiments, in E&D, and Establishment of the Experiments Program Office," June 21, 1965.

50. Ibid.

51. Compton and Benson, *Living and Working in Space*, pp. 40–52.

52. Maxime A. Faget to multiple addressees, "Establishment of a Space Science Office within E&D," Mar. 31, 1966.

53. Gilruth to Mueller, "Change in the basic MSC organization," Apr. 4, 1966.

54. D.M. Allison, "Summary Minutes: Planetology Subcommittee of the Space Science Steering Committee (Meeting No. 1-67), 26, 27, 28 July 1966," no date.

55. Robert C. Seamans to Hqs. Assoc. Administrators, "Management Responsibilities for Future Manned Flight Activities," July 26, 1966; see also Compton and Benson, *Living and Working in Space*, pp. 48–52.

56. George M. Low to multiple addressees, "Pending MSC Organizational Change, Nov. 17, 1966."

57. Mueller to Gilruth, Jan. 17, 1967, with encls., MSC Organization Chart approved by Webb and functional statement for MSC Director of Science and Applications.

58. MSC Announcement 67-27, "Director, Science and Applications Directorate," Feb. 17, 1967.

59. Brooks, Grimwood, and Swenson, *Chariots*, pp. 131–201.

60. Roger E. Bilstein, *Stages to Saturn: A Technological History of the Apollo/Saturn Launch Vehicles*, NASA SP-4206 (Washington, 1980), pp. 209–32.

61. Ibid., pp. 225–32; Brooks, Grimwood, and Swenson, *Chariots*, pp. 194–96; Maj. Gen. Samuel C. Phillips to J. Leland Atwood, NAA, Dec. 19, 1965, with encls.

62. Brooks, Grimwood, and Swenson, *Chariots*, pp. 183–85, 208.

63. Eugene M. Shoemaker interview, Mar. 17, 1984.

64. Charles D. Benson and William Barnaby Faherty, *Moonport: A History of Apollo Launch Facilities and Operations*, NASA SP-4204 (Washington, 1978), pp. 384–87.

CHAPTER 7

1. Courtney G. Brooks, James M. Grimwood, and Loyd S. Swenson, Jr., *Chariots for Apollo: A History of Manned Lunar Spacecraft*, NASA SP-4205 (Washington, 1979), pp. 214–17; Charles D. Benson and William Barnaby Faherty, *Moonport: A History of Apollo Launch Facilities and Operations*, NASA SP-4204 (Washington, 1978), pp. 390–94.

2. One author has characterized the media coverage of the early period as based on the unspoken view (based on several widely publicized failures) that *"our boys always botch it and our rockets always blow up,"* and suggests that journalists uncritically emphasized the risks of manned space flight. Tom Wolfe, *The Right Stuff* (New York: Farrar, Straus and Giroux, 1979). Wolfe's conclusion—supported by the (admittedly subjective) recollections of the present author, who watched many of the launches, from Mercury through Apollo, on television—gains credibility after viewing a television documentary entitled "Spaceflight," broadcast on the Public Broadcasting System in May 1985. In an interview on that program Walter Cronkite, who was a principal figure in the Columbia Broadcasting System's coverage of manned space flight, recalled that the history of failure in rocket tests up to 1961 heightened his perception of the danger to the astronaut. The astronauts did not share this apprehension; their day-to-day participation in hardware development gave them an informed estimate of the danger they faced.

3. "Entire Crew of Mission Set for Feb. 21 Is Lost; Grissom, White, Chaffee," *Washington Post*, Jan. 18, 1967.

4. "Is Moon Travel Worth the Cost?" *Chicago Tribune*, Jan. 29, 1967; "The Space 'Success Schedule,' " *Philadelphia Inquirer*, Jan. 30, 1967; "A Time for Deliberation," Washington *Evening Star*, Feb. 1, 1967; "Haste in Space," *Wall Street Journal*, Feb. 20, 1967; "The Legacy of Counterfeit Crisis," ibid., Apr. 13, 1967; Jerry E. Bishop, "Moon Race: Can't Machines Do It?" ibid., Apr. 16, 1967.

5. Brooks, Grimwood, and Swenson, *Chariots*, pp. 219–21.

6. Ibid.; *Report of Apollo 204 Review Board to the Administrator, National Aeronautics and Space Administration* (Washington, 1967), Floyd L. Thompson, chmn., Apr. 5, 1967. See also House Subcommittee on NASA Oversight of the Committee on Science and Astronautics, *Investigation into Apollo 204 Accident*, hearings, 90/1, Apr. 10 to May 10, 1967; Senate Committee on Aeronautical and Space Sciences, *Apollo Accident*, hearings, 901 and 902, pts. 1–8, 1967–68; idem, *Apollo 204 Accident: Report*, S. Rept. 956, 90/2, Jan. 30, 1968.

7. Brooks, Grimwood, and Swenson, *Chariots*, pp. 222–27; House, *Investigation*, passim.

8. Brooks, Grimwood, and Swenson, *Chariots*, p. 225.

9. Oscar Griffin, "Rep. Rumsfeld Criticizes NASA For Lack of Reports Before Fire," *Houston Chronicle*, July 19, 1967; "NASA Losing Its Support, Casey Says," ibid.; W. David Compton and Charles

D. Benson, *Living and Working in Space; A History of Skylab,* NASA SP-4208 (Washington, 1983), p.84–91.

10. House, *Investigation,* pp. 10–11; J.V. Reistrup, "Webb Defends Judgment on Apollo," *Washington Post,* Apr. 11, 1967. Considerable misunderstanding was generated when members of the committees demanded to see the "Phillips Report," which both NASA officials and North American executives professed not to recognize by that name. The members of Congress were referring to a hard-hitting critique of NAA's management, which Gen. Sam Phillips had submitted to NASA and to the contractor in late 1965 (see Chap. 6), but because it took the form of notes enclosed with a letter it was not called a "report." When the document was finally identified, however, Webb, Mueller, and Phillips provoked charges of a cover-up by refusing to furnish it to the committees on the grounds that public disclosure could undermine NASA-contractor relations; William Hines, "Ryan Says NASA Knew of Apollo 'Incompetence,' " Washington *Evening Star,* Apr. 26, 1967; Thomas O'Toole, "NASA Accused of Covering Up Troubles," *Washington Post,* May 11, 1967. On this point S. Rept. 956 (see note 6, above) stresses the effect of NASA officials' behavior on the legislators' confidence in the agency; see "Additional Views of Mr. Brooke and Mr. Percy," pp. 12–14, and "Additional Views of Mr. Mondale," pp. 15–16. For more extensive discussion of the fire and its aftermath, see Brooks, Grimwood, and Swenson, *Chariots;* Benson and Faherty, *Moonport;* and the records of the Senate and House committee investigations cited earlier.

As the investigations proceeded, the press raised allegations of questionable political deals that gave North American the contract for the command and service module; see Harry Schwartz, "Unanswered Questions on the Apollo Tragedy," *New York Times,* Apr. 15, 1967; William Hines, "Apollo: A Shining Vision in Trouble," *Washington Star,* May 21, 1967. These allegations were never put in the committee records and remained unsubstantiated. Three years later, however, they were revived in *Journey to Tranquility* (Garden City, N.Ỳ.: Doubleday and Co., 1970), by three English journalists, Hugo Young, Bryan Silcock, and Peter Dunn.

11. Brooks, Grimwood, and Swenson, *Chariots,* pp. 225–32; Benson and Faherty, *Moonport,* pp. 401–402.

12. OMSF, "Manned Space Flight Schedules, vol. I, Level 1 Schedules and Resources Summary," Dec. 1966; House Subcommittee on Manned Space Flight of the Committee on Science and Astronautics, *1967 NASA Authorization,* hearings on H.R. 12718, 89/2, pt. 2, pp. 72–76.

13. OMSF, "Manned Space Flight Schedules, vol. I, Level 1 Schedules and Resources Summary," May 1967.

14. Ibid., February 1966.

15. John T. Holloway to Dir., MSC, "Development of Experiments for the Apollo Lunar Surface Experiments Package (ALSEP)," Apr. 14, 1966.

16. OMSF, "Manned Space Flight Schedules, vol. I, Level 1 Schedules and Resources Summary," Nov. 1966.

17. William T. O'Bryant to MSC, attn. Robert O. Piland, "Guidelines for Possible Substitution of Other Instruments for the Lunar Surface Magnetometer in ALSEP," Dec. 23, 1966.; D.K. Slayton to Mgr., Experiments Program Off., "Possible Replacement of the Lunar Surface Magnetometer on ALSEP," Dec. 22, 1966.

18. Anon., "ALSEP OSSA Review, October 3, 1966."

19. Slayton to Director, Engineering and Development, "Comments on the ALSEP Delta Preliminary Design Review," Dec. 15, 1966.

20. OMSF, "Manned Space Flight Schedules, vol. I, Level 1 Schedules and Resources Summary," July 1967, pp. 49, 33.

21. Leonard Reiffel to Gen. S.C. Phillips, "Flight Schedule for ALSEP and Related Matters," June 20, 1967.

22. Harrison H. Schmitt interview, May 30, 1984.

23. Phillips to Administrator, "Lunar Surface Magnetometer," June 6, 1967; Piland to Mgr., Apollo Spacecraft Program Office, "ALSEP Program Review with General Phillips, June 1, 1967," June 7, 1967.

24. Phillips to Dep. Adm.,"Lunar Surface Magnetometer," July 5, 1967, with encl., preliminary report of review team; Phillips to Dep. Adm., same subj., Aug. 30, 1967; John F. Parsons (Ames Res. Ctr.) to NASA Hqs., attn. Dr. James H. Turnock, "Lunar Surface Magnetometer Program," Aug. 31, 1967.

25. OMSF, ''Manned Space Flight Schedules, vol. I, Level 1 Schedules and Resources Summary,'' monthly issues, Jan. 9 through Oct. 8, 1968.

26. Arnold S. Levine, *Managing NASA in the Apollo Era*, NASA SP-4102 (Washington, 1982), pp. 239–53; Compton and Benson, *Living and Working in Space*, pp. 40–44, 46–48, 79–82.

27. President's Science Advisory Committee, *The Space Program in the Post-Apollo Period* (The White House, February 1967), pp. 13–16.

28. Brooks, Grimwood, and Swenson, *Chariots*, pp. 362–63.

29. *1967 Summer Study of Lunar Science and Exploration*, NASA SP-157 (Washington, 1967), pp. 3–6.

30. Anon., ''Falmouth plus two years or how much nearer is the whale to the water?'' no date [c. Aug. 1966].

31. *1967 Summer Study of Lunar Science and Exploration*, pp. 9–11.

32. Ibid., pp. 12–13.

33. Ibid., pp. 18–19.

34. Ibid., p. 19.

35. Ibid., pp. 19–29.

36. Ibid., p. 4.

37. Minutes of the Lunar Mission Planning Board, Sept. 28, 1967.

38. Robert R. Gilruth, TWX to Hqs., subj. ''OSSA Activities—Weekly Report,'' Nov. 30, 1967.

39. Homer E. Newell to Gilruth, Nov. 20, 1967.

40. Compton and Benson, *Living and Working in Space*, pp. 83–102.

41. Weekly Compilation of Presidential Documents, vol. 3, no. 34, p. 1193.

42. Senate, Committee on Aeronautical and Space Sciences, *NASA's Proposed Operating Plan for Fiscal Year 1968*, hearing, 90/1, Nov. 8, 1967. For the tribulations and demise of Voyager, see Edward Clinton Ezell and Linda Neuman Ezell, *On Mars: Exploration of the Red Planet 1958–1978*, NASA SP-4212 (Washington, 1984), pp. 85–119; for a general discussion of NASA's budget process and the ''phasing down of the space program'' in the middle to late l960s, see Arnold S. Levine, *Managing NASA in the Apollo Era*, NASA SP-4102 (Washington, 1982), pp. 179–209.

43. Compton and Benson, *Living and Working in Space*, chapters 3 and 5, deal with AAP's fluctuating plans and fortunes during the period before Apollo 11.

44. Robert C. Seamans, Jr., memorandum for the record, subj.: ''Apollo Program Decisions—Manned Apollo Flights and Apollo Applications Program Plans,'' May 17, 1967.

45. ''Unified Lunar Exploration Office,'' NASA Release 68-5, Jan. 4, 1968; Philip E. Culbertson to Wilmot N. Hess, Jan. 3, 1968.

46. Gilruth to Newell, ''Principal Investigator requirements for the Lunar Receiving Laboratory (LRL),'' Jan. 11, 1967.

47. Joseph V. Piland, ''Lunar Receiving Laboratory Bi-weekly Status Reports,'' Jan. 9, Jan. 23, Feb. 6, Feb. 20, Mar. 6, Mar. 20, Apr. 3, Apr. 17, 1967; Piland to Dir., MSC, ''Phase-out of Lunar Receiving Laboratory Program Office,'' Apr. 14, 1967.

48. Clark Goodman to members of LRL Working Group, ''Science and Applications Directorate at MSC,'' no date [c. Jan. 14, 1967].

49. MSC Announcements no. 67-7, ''Organization and Personnel Assignments for the Science and Applications Directorate,'' Jan. 10, 1967, and 67-4, ''Interim Plan for the Management and Operation of the Lunar Receiving Laboratory,'' Jan. 13, 1967.

50. Melvin Calvin to Newell, Dec. 29, 1966; Paul E. Purser to Deputy Dir., MSC, ''Planetary Biology Subcommittee Meeting at MSC on January 10–11, 1967,'' Jan. 17, 1967.

51. George M. Low, ''Resume of Meeting of January 16, 1967, at the Communicable Disease Center, Atlanta, Georgia,'' Jan. 16, 1967.

52. ''First Lunar Samples Set for Experiment by 110 Scientists,'' Hqs. Release 67-55, Mar. 16, 1967.

53. Low to John E. Naugle, Mar. 9, 1967; Newell to MSC, ''Lunar Sample Analysis Planning,'' June 16, 1967.

54. Gilruth to John E. Naugle, ''Lunar Sample Analysis Planning,'' Nov. 15, 1967.

55. Ibid.

56. MSC Announcement 67-140, ''Designation of Chief, Lunar and Earth Sciences Division,'' Oct. 11, 1967.

57. Low to Earle Young, ''Various documents concerning back-contamination and lunar sample release,'' Jan. 23, 1967.

58. Robert L. Tweedie, memo for record, "NASA/DOD Interface Conference, Recovery Quarantine Equipment," Aug. 25, 1966; James C. McLane, Jr., to C.R. Haines, "Lunar Receiving Laboratory (LRL) functional interfaces with the Apollo Command Module and returned astronauts," Aug. 24, 1966.

59. John E. Pickering, "Minutes, Interagency Committee on Back Contamination," Oct. 3, 1966.

60. Collection of related documents, "Interagency Committee on Back Contamination," no date [approx. Aug. 1966], box 076-12, JSC History Office files.

61. Aleck C. Bond to record, "Interagency Committee Meeting, October 3, 1966," Oct. 6, 1966.

62. W.W. Kemmerer, "Lunar Receiving Laboratory Sample Protocol Briefing December 16, 1966," Dec. 16, 1966; Minutes, Interagency Committee on Back Contamination, Dec. 16, 1966.

63. Wolf Vishniac to Dr. Harry S. Lipscomb (Baylor Coll. of Med.), Feb. 2, 1967.

64. Bond to Deputy Dir., MSC, "Biological protocol for LRL," Feb. 9. 1967.

65. Pickering, "Minutes, Interagency Committee on Back Contamination," Mar. 2, 1967.

66. Pickering, "Minutes, Interagency Committee on Back Contamination," June 7, 1967.

67. Interagency Committee on Back Contamination. "Quarantine Schemes for Manned Lunar Missions," no date [Aug. 1967].

68. George E. Mueller to Gilruth, June 2, 1967.

69. Joseph V. Piland to Dir., Science and Applications, "Simulated handling of astronauts and returned equipment for Apollo missions," Aug. 9, 1967.

70. Brooks, Grimwood, and Swenson, Chariots, pp. 228–35.

71. Bilstein, Stages to Saturn, pp. 211–33.

72. Ibid., pp. 355–60; Brooks, Grimwood, and Swenson, Chariots, pp. 232–34; Benson and Faherty, Moonport, pp. 403–29; MSC, "Apollo 4 Mission Report," MSC-PA-R-68-1, Jan., 1968, pp. 1-1 through 1-4; MSFC, "Saturn V AS-501 Flight Evaluation," MPR-SAT-FE-68-1, Jan. 15, 1968, pp. xxxviii–xlii.

73. Brooks, Grimwood, and Swenson, Chariots, pp. 234–35.

74. James E. Webb to Frank B. Smith, Jan. 7, 1967; Webb to Frederick Seitz, NAS, Feb. 2, 1967; Philip H. Whitbeck to Deputy Dir., Administration, MSC, "Operation of the Lunar Receiving Laboratory," Feb. 14, 1967, with encl., "Report on meeting with Mr. Webb on organizational location and operation of the lunar receiving laboratory"; Newell, draft memo to Seamans, "Manned Spacecraft Center Science and Applications Directorate and the National Lunar Research Laboratory," Mar. 2, 1967; Smith, "Comments Made at MSC at 3/16/67 Meeting with Dr. Seitz, George Low, et al., Relative to Management of the LRL," Mar. 27, 1967; anon., viewgraph for presentations on functions of Center for Lunar and Earth Sciences, Dec. 20, 1967, JSC History Office files, box 076-41.

75. OMSF, "Manned Space Flight Schedules, vol. I, Level 1 Schedules and Resources Summary," Jan. 9, 1968, pp. ii, 27–29.

CHAPTER 8

1. Courtney G. Brooks, James M. Grimwood, and Loyd S. Swenson, Jr., Chariots for Apollo: A History of Manned Lunar Spacecraft, NASA SP–4205 (Washington, 1979), pp. 228–29, 237–41.

2. Ibid., pp. 244–47.

3. Ibid., pp. 241–53; Roger E. Bilstein, Stages to Saturn: A Technological History of the Apollo/Saturn Launch Vehicles, NASA SP–4206 (Washington, 1980), pp. 360–63.

4. Brooks, Grimwood, and Swenson, Chariots, p. 253.

5. Leonard Reiffel to Gen. S. C. Phillips, "Flight Schedule for ALSEP and Related Matters," June 10, 1967.

6. Donald K. Slayton to Dir., Science and Applications (MSC), "Apollo Lunar Surface Operations Planning," Feb. 28, 1968.

7. Verl R. Wilmarth, "Summary Minutes, Planetology Subcommittee of the Space Science and Applications Steering Committee (Meeting No. 3–FY68), May 15, 16, 17, 1968"; Edward M. Davin to Mgr., Apollo Surface Expts. Program, "Contingency Science Payloads," July 3, 1968; Richard J. Green to MSC, attn. : John W. Small, "Science Payload Options, July 3, 1968.

8. George E. Mueller to Robert R. Gilruth, June 5, 1968; Gilruth to Mueller, June 27, 1968.

9. Phillips to Dr. Charles H. Townes, Aug. 31, 1968, with encl., letter to members of OMSF Science and Technology Advisory Committee, same date.

10. George M. Low to G. W. S. Abbey, "Lunar Mission Planning," Sept. 3, 1968; Wilmot N. Hess to Mgr., Apollo Spacecraft Program (MSC), "Changes in Mission G Plans," Sept. 4, 1968.

11 Hess to Dir. (MSC) , "Contingency Apollo Science Program," Sept. 30, 1968, with encl., "Contingency Apollo Science Program Implementation Plan" (preliminary), Sept. 27, 1968; idem, "Plans for Alternate Science Program on First Apollo Lunar Landing," Oct. 4, 1968.

12. Mueller, TWX to MSC attn.: Robert R. Gilruth, Nov. 5, 1968.

13. A. E. Morse, Jr., to Mgr., Apollo program, "KSC support for the LM–5 Early Apollo Scientific Experiment Payload (EASEP)," Nov. 12, 1968; Phillips to Gilruth, Nov. 15, 1968.

14. Phillips to Gilruth, Nov. 15, 1968.

15. Phillips, TWX to MSC attn.: G. M. Low, W. N. Hess, "Experiment Assignments to Lunar Missions," Dec. 5, 1968.

16. Wilmarth, "Summary Minutes."

17. Gilruth to Hqs., attn.: E. Taylor, TWX, "OSSA Activities—Weekly Report," Oct. 24, 1968.

18. MSC, "Lunar Receiving Laboratory Briefing," transcript of presentation, June 29, 1967; MSC, "Annual Report for Calendar Year 1967, The Directorate of Medical Research and Operations, Manned Spacecraft Center," n.d. [Dec. 1967].

19. Edgar M. Cortright to Gilruth, Mar. 27, 1968; Hess to Dir., Medical Research & Operations, "Initiation of monthly review meetings on LRL," Mar. 29, 1968.

20. Hess to multiple addressees, "Preliminary Examination Team Simulation Tests and Full Scale LRL Simulation," Aug. 20, 1968.

21. Gilruth to Charles W. Mathews, May 17, 1968; Minutes, Monthly LRL Reviews, June 10, July 1, Aug. 12, 1968.

22. MSC, "Minutes of Monthly LRL Review," Sept. 16, 1968; John E. Pickering, "Trip Report [to LRL monthly manager's review]," Sept. 18, 1968.

23. P. R. Bell to Martin Favero, June 18, 1968; "[Minutes,] Lunar Science Analysis Planning Team, July 26 [1968]"; Hess to multiple addressees, "Preliminary Examination Team Simulation Tests and Full Scale LRL Simulation," Aug. 20, 1968.

24. Gilruth to Edward H. Levi (Univ. of Chicago), Feb. 25, 1969.

25. Bell to Members of Preliminary Examination Team, "The Preliminary Examination Team simulation and training session, October 22 through November 1, 1968," Oct. 16, 1968.

26. Test dir., vacuum laboratory, "Report on PET Simulation, October 25–30, 1968," Oct. 31, 1968; Bell to Dir., Science and Applications, "LRL Equipment and System Problems, Nov. 7, 1968.

27. Hess to L. R. Scherer, "Changes in the Lunar Receiving Laboratory," Dec. 6, 1968; Hess to LRL Staff, "Configuration Control Board (CCB)," Dec. 6, 1968.

28. Gilruth to multiple addressees, "Operational Readiness Inspection of the Lunar Receiving Laboratory," Oct. 21, 1968.

29. Peter J. Armitage to multiple addressees: "Minutes of the First ORI Committee Meeting for the Lunar Receiving Laboratory," Nov. 5, l968; "Minutes of the Second ORI Committee Meeting for the Lunar Receiving Laboratory," Nov. 22, 1968; "Minutes of the Third ORI Committee Meeting for the Lunar Receiving Laboratory," Dec. 4, 1968; "Minutes of the Fourth ORI Committee Meeting for the Lunar Receiving Laboratory," Dec. 4, 1968; "Status of the Operational Readiness Inspection on the Lunar Receiving Laboratory and the Recovery Quarantine Equipment," Dec. 19, 1968; "Minutes of the Fifth ORI Committee Meeting for the Lunar Receiving Laboratory," Dec. 23, 1968.

30. Armitage, "Final Report, Operational Readiness Inspection of the Lunar Receiving Laboratory and Recovery Quarantine Equipment," May 7, l969.

31. John E. Pickering, "Minutes, Interagency Committee on Back Contamination, 8–9 Feb. 1968."

32. Idem, "Minutes, Interagency Committee on Back Contamination, 11 June 1968."

33. Idem, "Minutes, Interagency Committee on Back Contamination, 17–l8 Oct. 1968."

34. Richard S. Johnston to C. C. Kraft, M. A. Faget, and C. A. Berry, "Apollo Back Contamination," Nov. 8, 1968.

35. Johnston to multiple addressees, "Back-Contamination program review," Jan. 16, 1969; "ICBC Meeting at MSC on March 28 and 19," Mar. 18, 1969.
36. Gilruth to Maj. Gen. J. W. Humphreys, Jr., Mar. 19, 1969.
37. Pickering, "Minutes, Interagency Committee on Back Contamination, 28–29 March 1969."
38. David J. Sencer to Thomas O. Paine, Apr. 7, 1969.
39. S. C. Phillips, TWX to MSC, attn.: G. M. Low, "Biological containment on 'G' mission prior to entry into MQF," Apr. 10, 1969.
40. Ibid.; Low to multiple addressees, "Apollo Back Contamination Action Items," Apr. 14, 1969; Low to Phillips, "Apollo Back Contamination Action Status," Apr. 15, 1969.
41. Low to Phillips, "Transmittal of Summary Report and Actions from Back Contamination LDX Conference Held April 21, 1969," Apr. 24, 1969, with encl., Low to Multiple Addressees, "Back Contamination LDX Conference held April 21, 1969, Between MSC/NASA Headquarters; minutes and action items," Apr. 24, 1969.
42. MSC Announcement 69–60, "Lunar Receiving Laboratory Operations," May 1, 1969.
43. Johnston, draft of MSC position paper on Apollo back contamination program, Apr. 25, 1969.
44. Pickering, "Minutes, Interagency Committee on Back Contamination, 2 May 1969."
45. Phillips to Gilruth, "Lunar Receiving Laboratory Readiness Review," Jan. 16, 1969; Gilruth to Phillips, Jan. 29, 1969, with tentative agenda for proposed review.
46. Mueller to Gilruth, Jan. 13, 1969; Gilruth to Mueller, Feb. 8, 1969.
47. A. B. Park to Pickering, Mar. 7, 1969, with encl., "Report by U.S. Department of Agriculture Team on Lunar Receiving Laboratory Evaluation for Quarantine Requirements," n.d.; Howard H. Eckels to Pickering, Mar. 11, 1969, with encl., "Department of Interior Observations on the Lunar Receiving Laboratory Procedures Relative to Invertebrate and Fish Species, February 12, 13, and 14, 1969"; idem, Feb. 27, 1969, with encl., Kenneth E. Wolf to Eckles, "Technical Review of Lunar Receiving Laboratory," Feb. 25, 1969; Pickering to Dir., Apollo Program, "LRL Bioprotocol Readiness Review and Certification Procedures," Feb. 24, 1969.
48. MSC, "Lunar Receiving Laboratory Simulation Plan, March 3–April 16, 1969," n.d., pp. 1–2.
49. Rudy Trabanino to LTD's [Laboratory Test Directors], "Debriefing of Laboratory Personnel," Mar. 16, 1969; D. White to LTD's, "Critique of 3–13 March Simulation in Vacuum Laboratory—Shift I," Mar. 16, 1969; T. McPherson to LTD's, "Vacuum Laboratory Critique—Second Shift," Mar. 16, 1969; Trabanino to LTD's, "Critical Vacuum Laboratory Problems," Mar. 21, 1969; O. A. Schaeffer and J. G. Funkhauser to Bryan Erb, "Leak Checking of RCL Containers with Gas Analysis System," Mar. 24, 1969; Bell to Deputy Mgr., LRL, "Problems Identified in February and March LRL simulations," Mar. 27, 1969; H. C. Sweet and Charles H. Walkinsbau, Jr., "Summary of Simulation of Botanical Protocol—March 1969," n.d. [Mar. 1969]; "Minutes, Interagency Committee on Back Contamination, 28–29 March 1969—Manned Spacecraft Center—Houston, Texas," Mar. 29, 1969; Trabanino, "Critical Vacuum Laboratory Problems," Apr. 3, 1969.
50. Hess to Bell, "Organic Contamination of F201," Apr. 8, 1969.
51. Johnston to multiple addressees, "Action items from Interagency Committee on Back Contamination (ICBC) Meeting," May 5, 1969; Gilruth to multiple addressees, "Establishment of Apollo Back Contamination Control Panel," May 8, 1969; Johnston to Chief, Crew Systems Div., and Chief, Flight Crew Support Div., "Lunar Module Back Contamination Simulation, May 12, 1969; Johnston to Col. John E. Pickering, May 14, 1969; Johnston to multiple addressees, "Action Items from Interagency Committee on Back Contamination (ICBC) Meeting, May 2, 1969," May 15, 1969; idem, "ICBC Telephone Conference Summary and Action Items," May 21, 1969; MSC, "Back Contamination Mission Rules (Recovery to Receiving Lab)," MSC 00005, May 21, 1969; Low to multiple addressees, "Back Contamination Procedures, May 20, 1969; Robert E. Smylie to Mgr., Apollo Spacecraft Program, "Back Contamination Procedures," May 22, 1969; Johnston to multiple addressees, "Apollo Back Contamination Simulation Meeting Summary," May 23, 1969; idem, "Back Contamination Action Items," May 28, 1969.
52. Mueller to Administrator, "Manned Space Flight Weekly Report," June 9, 1969.
53. Johnston to multiple addressees, "Apollo Back Contamination Simulation Meeting Summary, May 23, 1969.
54. Sencer to Paine, Apr. 7, 1969.
55. W. David Compton and Charles D. Benson, *Living and Working in Space: A History of Skylab*, NASA SP-4208 (Washington, 1983), pp. 99–102.

56. Hess to MSC Dir., "Summary OSSA Senior Council Meeting," Apr. 30, 1968; Hess to John E. Naugle, "Staffing of the LRL," June 6, 1968; Naugle to Hess, same sub., July 9, 1968; M. L. Raines to Assoc. Dir., "Systems and Support Contractor Reduction," July 12, 1968; MSC, "Minutes of Monthly LRL Review, September 16, 1968"; Earle B. Young to LRL ORI Committee, "Proposed LRL Staffing," Nov. 26, 1968; P.R. Bell to Dir., Science and Applications, "Staffing Plan for the Lunar and Receiving Laboratory," Jan. 17, 1969; Wesley L. Hjornevik to Lt. Gen. Frank A. Bogart, "Lunar Receiving Laboratory Staffing, Jan. 27, 1969; MSC, "LRL Staffing Plan Status, April 8, 1969, Science & Applications Directorate and Medical Research & Operations Directorate."

57. John D. Stevenson, TWX to multiple addressees, "MSF mission operations forecast for January 1969," Jan. 3, 1969; Gilruth to Stevenson, "Ability to support launch schedule for CY 1969," Mar. 4, 1969.

58. Bell to Dir., Science and Applications, "Staffing plan for the Lunar and Receiving Laboratory," Jan. 17, 1969.

59. W. W. Kemmerer, Jr., to all LRL personnel, "Initiation of mission operating conditions for the Sample Laboratory secondary biological barrier," June 3, 1969.

60. Johnston to all NASA LRL personnel, "Lunar Receiving Laboratory Apollo 11 Task Group," June 4, 1969.

61. MSC, "Lunar Landing Site Selection Briefing, Compilation of Presentation Material," Mar. 8, 1967; John E. Eggleston to Dir., Science and Applications, "Trip report—Apollo Site Selection Board Meeting; Surveyor/Orbiter Utilization Committee Meeting," Apr. 3, 1967; Owen E. Maynard to Mgr., Apollo Spacecraft Program Office, "Trip report—Apollo Site Selection Board and Surveyor/Orbiter Utilization Committee Meetings," Apr. 20, 1967.

62. Low to Hqs., attn. Dr. James A. Turnock, "Selection of Lunar Landing Sites for First Lunar Landing Mission," Sept. 3, 1967.

63. Phillips to multiple addressees, "Minutes of the Apollo Site Selection Board meeting of December 15, 1967," Jan. 29, 1968; MSC, "Apollo Site Selection Board Briefing, Compilation of Presentation Material," Dec. 15, 1967.

64. Ibid.

65. Ibid.

66. Compton and Benson, *Living and Working in Space*, p. 87; Robert C. Seamans, Jr., to Mueller, "Termination of the Lunar Mapping and Survey System," July 25, 1967; U.S., Congress, House, Subcommittee on Manned Space Flight of the Committee on Science and Astronautics, *1968 NASA Authorization*, Hearings on H.R. 4450, H.R. 6470, 90/1, pt. 2, pp. 127–28, 568–69; idem, *1969 NASA Authorization*, Hearings on H.R. 15086, 90/2, pt. 2, p. 64.

67. Phillips to Gilruth, "Lunar Photography from the CSM," Mar. 29, 1968.

68. John R. Brinkmann to Chief, Syst. Eng. Div., "Lunar photography from the CSM," May 13, 1968; Warren J. North to Chief, Syst. Eng. Div., same subj., May 15, 1968.

69. MSC, "MSC Lunar Scientific Exploration Plan," Apr. 14, 1967.

70. Hess to multiple addressees, "Pre-GLEP meeting January 10 in Washington, D.C.," Jan. 3, 1968.

71. NASA Hqs., "The Plan for Lunar Exploration," Feb. 1968.

72. Ibid.

73. Ibid.

74. Gilruth to Mueller, Mar. 4, 1968.

75. Gilruth to Mueller, Apr. 1, 1968.

76. House, Subcommittee on Manned Space Flight of the Committee on Science and Astronautics, *1969 NASA, Authorization*, Hearings on H.R. 15086, 90/2, pt. 2, pp. 26–35. The same plan was presented by John E. Naugle to the subcommittee on science and applications, also without stimulating discussion; see idem, Subcommittee on Space Science and Applications of the Committee on Science and Astronautics, *1969 NASA Authorization*, Hearings on H.R. 15086, 90/2, pt. 3, pp. 146–58.

77. Andre J. Meyer, Jr., "Minutes of the Lunar Mission Planning Board," Feb. 13, 1968.

78. Brooks, Grimwood, and Swenson, *Chariots*, pp. 237–53.

79. Compton and Benson, *Living and Working in Space*, pp. 99–104.

80. Phillips, "Minutes of the Apollo Site Selection Board Meeting of March 26, 1968," May 6, 1968.

81. Brooks, Grimwood, and Swenson, *Chariots*, pp. 265–72.

82. Ibid., pp. 256–60, 272–74.

83. Ibid.
84. Ibid., pp. 274–84.
85. MSC PA–R–69–1, "Apollo 8 Mission Report," Feb. 1969, p. 3-2.
86. Ibid., pp. 4–1 to 4–9.
87. Ibid., pp. 4–8 to 4–9, 7-19 to 7-20.
88. John D. Stevenson, TWX, "MSF Mission Operations Forecast for January 1969," Jan. 3, 1969.

CHAPTER 9

1. Courtney G. Brooks, James M. Grimwood, and Loyd S. Swenson, Jr., *Chariots for Apollo: A History of Manned Lunar Spacecraft*, NASA SP–4205 (Washington, 1979), p. 261.
2. Ibid., pp. 261–62.
3. Interview with Joseph P. Kerwin, Jr., Mar. 29, 1985; interview with John R. Sevier, Apr. 24, 1986.
4. NASA Release 67–211, Aug. 4, 1967.
5. W. David Compton and Charles D. Benson, *Living and Working in Space: A History of Skylab*, NASA SP–4208 (Washington, 1983), pp. 84–86.
6. Interview with Donald K. Slayton, Oct. 15, 1984.
7. Brooks, Grimwood, and Swenson, *Chariots*, pp. 373–77.
8. Ibid., pp. 234–35.
9. Ibid., pp. 290–91.
10. Ibid., pp. 292–99; "Apollo 9 Mission Report," MSC PA–R–69–2, May 1969, pp. 1–1 to 1–2.
11. "Apollo 10 Mission Report," MSC–00126, Aug. 1969, p. 3–1.
12. Charles D. Benson and William Barnaby Faherty, *Moonport: A History of Apollo Launch Facilities and Operations*, NASA SP–4204 (Washington, 1978), p. 474.
13. Brooks, Grimwood, and Swenson, *Chariots*, pp. 303–12.
14. Ibid., pp. 318–26; MSC, "Apollo 11 Crew Training Summaries (Jan. 31–July 15, 1969)," folder in box 081–13, JSC History Office Apollo files.
15. Brooks, Grimwood, and Swenson, *Chariots*, pp. 322–25; Langley Research Center, "Lunar Landing Research Facility and Landing Loads Track," presentation at Field Inspection of Advanced Research and Technology, Hampton, Va., May 18–22, 1964, copy in box 081–45, JSC History Office Apollo files; George E. Mueller to Robert R. Gilruth, Feb. 19, 1969; Gilruth to Mueller, Apr. 1, 1969.
16. Bell Aerosystems Co. release, Jan. 20, 1967, in box 081–45, JSC History Office Apollo files; several good photographs of the LLTV in flight are found in "Series of Lunar Landings Simulated," *Aviation Week & Space Technology*, June 30, 1969. See also David F. Gebhard, "Factors in the Choice of LEM Piloting and Testing Techniques," SAE Preprint 866F, April 1964; Gene J. Matringa, Donald L. Mallick, and Emil E. Kleuver, "An Assessment of Ground and Flight Simulators for the Examination of Manned Lunar Landing," AIAA Paper 67–238, presented at AIAA Flight Test, Simulation and Support Conference, Cocoa Beach, Fla., Feb. 6–8, 1967; and James P. Bigham, Jr. , "The lunar landing training vehicle," *Simulation*, July 1970, pp. 17-18, copies in box 081-46, JSC History Office Apollo files; Richard H. Holzapfel, TWX to NASA, attn.: B. P. Helgeson, May 7, 1968; T. O. Paine to LLRV-1 Review Board, "Investigation and Review of Crash of Lunar Landing Training Vehicle #1," Dec. 11, 1968; idem, "Investigation and Review of Crash of Lunar Landing Research Vehicle #1," Dec. 31, 1968.
17. Bigham, "The Lunar Landing Training Vehicle."
18. "Apollo 11 Crew Training Summary."
19. Elbert A. King, Jr., interview with Loyd S. Swenson, Jr., May 17, 1971, tape in JSC History Office Apollo files.
20. Mueller to Administrator, "Manned Space Flight Report—May 26, 1969."
21. Idem, Manned Space Flight Reports: May 26, June 2, 9, 16, 23, 30, 1969; "Apollo 11 Crew Training Summary"; Slayton, Flight Crew Operations Directorate Weekly Activity Reports: May 30, June 6, 13, 20, July 3, 11, 18, 1969.
22. Julian Scheer to Slayton, June 3, 1969; Slayton to Scheer, "Attendance of Apollo 10 and 11 crews at White House dinner," June 9, 1969.

23. Mueller to Administrator, "Manned Space Flight Weekly Report—July 7, 1969."
24. Slayton, "Weekly Activity Report—June 14–20, 1969, Flight Crew Operations Directorate."
25. Mueller to Administrator, "Manned Space Flight Weekly Report—June 23, 1969"; ibid. , June 9, 1969; Brooks, Grimwood, and Swenson, *Chariots*, pp. 318–26, 332–34.
26. Brooks, Grimwood, and Swenson, *Chariots*, pp. 337–44; MSC, "Apollo 11 Technical Air-to-Ground Voice Transcription (GOSS Net 1)," July 1969 (hereinafter cited as "Air-to-Ground"), p. 317.
27. Air-to-Ground, p. 319.
28. Ibid., p. 321.
29. Ibid., p. 324.
30. Ibid., p. 335.
31. Ibid., pp. 322, 332–33, 336, 343–46, 349–50, 412–15.
32. Air-to-Ground, p. 377; Brooks, Grimwood, and Swenson, *Chariots*, p. 346. After the crew returned to Houston, press representatives repeatedly asked what Armstrong had actually said. The Apollo news center at MSC issued the following release (copy in box 078-56, JSC History Office files) on July 30, 1969: "Armstrong said that his words when he first stepped onto the moon were: "That's one small step for a man, one giant leap for mankind" not "That's one small step for man, one giant leap for mankind" as originally transcribed."
33. Air-to-Ground, pp. 377–79.
34. Ibid., pp. 379–400.
35. King interview.
36. Air-to-Ground, pp. 400–406.
37. Ibid., pp. 407–30.
38. Ibid., pp. 434–51.
39. Ibid., pp. 460–88; MSC, "Apollo 11 Technical Crew Debriefing, July 31, 1969," pp. 12–41 to 12–43.
40. Air-to-Ground, pp. 488–505.
41. Brooks, Grimwood, and Swenson, *Chariots*, pp. 355–57; "Technical Crew Debriefing," pp. 16–7 to 16–12; Neil A. Armstrong, Michael Collins, and Edwin E. Aldrin, Jr., to Dir., Flight Crew Operations, "Suitability of the Biological Isolation Garments," Oct. 2, 1969.
42. Brooks, Grimwood, and Swenson, *Chariots*, p. 357.
43. "Apollo 11 Mission Report," MSC-00171, pp. 13–3 to 13–5.
44. "Apollo 11 Mission Report," pp. 13–3 to 13–5; Kenneth L. Suit, "LRL Apollo 11 Daily Summary Report (1200 July 24 to 1200 July 25)," July 25, 1969.
45. Suit, "Apollo 11 LRL Daily Summary Report, 1200 July 25 to 1200 July 26."
46. Victor Cohn, "Old Moon Game Taunts Players," *Washington Post*, Sept. 21, 1969.
47. MSC, "LRL Daily Summary Report No. 3," July 26, 1969.
48. Suit, "Apollo 11 LRL Daily Summary Report, 1200 July 25 to 1200 July 26."
49. [Suit], "Apollo 11 LTD Daily Summary Report, 1200 July 26 to 1200 July 27"; Richard S. Johnston, "LRL Daily Summary Report No. 4," July 27, 1969.
50. Cohn, "Old Moon Game"; "Scientists Get First Look at Moon Rocks," Washington *Sunday Star*, July 27, 1969.
51. Johnston, "LRL Daily Summary Report No. 4."
52. Lunar Sample Preliminary Examination Team, "Preliminary Examination of Lunar Samples from Apollo 11," *Science* 165 (1969): 1211–17 (19 Sept. 1969).
53. Johnston, "LRL Daily Summary Report No. 4"; Cohn, "Old Moon Game."
54. Lunar Sample Analysis Planning Team, "Sample Information Summary #1," July 29, 1969, "Sample Information Summary #2," July 31, 1969.
55. "Sample Information Summary #1."
56. "LRL Summary Report No. 7," July 31, 1969.
57. "LRL Summary Report No. 8," Aug. 1, 1969.
58. Ibid.
59. "LRL Summary Report No. 9," Aug. 2, 1969; "Sample Information Summary #3," Aug. 2, 1969.
60. Ibid.; Johnston, "Lunar Receiving Laboratory Sample Flow Directive, Revision A," MSC-00002, July 1, 1969, p. 6.
61. "Apollo 11 Mission Report," p. 13–4.
62. Ibid.
63. MSC Public Affairs Office, "Lunar Receiving Laboratory— Status Report No. 1," July 21, 1969.

64. Idem, "Status Report, John McLeaish Comments on Crew, 7/28/69, CDT 08:05 AM," July 28, 1969; Richard S. Johnston, Lawrence F. Dietlein, M.D., and Charles A. Berry, M.D., eds., *Biomedical Results of Apollo*, NASA SP-368 (Washington, 1975), p. 418. If indeed there was a serious risk of contaminating the earth with alien organisms, the decision to breach quarantine in the event of a life-threatening emergency would have meant taking that risk rather than sacrificing an astronaut. No one seems to have contested that decision; perhaps this indicates the real perception of risk from "moon bugs."

65. MSC, "Lunar Receiving Laboratory, MSC Building 37, Facility Description," September 1968, pp. 8–19; "Apollo 11 Technical Crew Debriefing," vol. II, p. 26–4.

66. MSC Public Affairs Office, Crew Status Reports, July 28-Aug. 10, 1969.

67. George M. Low to NASA Hqs., attn.: Lt. Gen. S. C. Phillips, "Special Lunar Samples for Museum Display or Presentation to VIPs," July 7, 1969.

68. Phillips to Low, "Lunar Sample Accounting," Aug. 5, 1969.

69. Willis H. Shapley to multiple addressees, "Lunar Sample Material," Aug. 19, 1969. It was reported that Nixon offered the President of Indonesia and other chiefs of state "a piece of the moon as a souvenir"; *New York Times*, July 28, 1969, cited in *Astronautics and Aeronautics, 1969: Chronology on Science, Technology, and Policy*, NASA SP-4014 (Washington, 1970), p. 249.

70. Phillips, TWX to Low, Aug. 15, 1969; Robert R. Gilruth to Rocco A. Petrone, "Non-Scientific Use of Lunar Samples," Sept. 5, 1969.

Forwarding the latter memo to George Low, Wilmot Hess (whose deputy, Anthony Calio, had drafted it for Gilruth's signature) stressed his own distaste for the plan to present large samples to VIPs—a distaste that was shared by the lunar scientists. Elbert King, who had been appointed first curator of lunar samples, did everything he could to prevent distribution of *any* lunar samples as souvenirs. When he first heard of the plan to present souvenirs to VIPs, his impression was that "they were talking about big polished pieces of rock for paperweights. . . . Here was priceless scientific material that was going to be cut up as political trophies." The White House proposal was even more offensive to King, and when a Washington paper called him to ask for his reaction, he commented that it would be unthinkable—a reaction which, he recalled later, "caused much unhappiness within the center, because here was someone [at MSC] disagreeing with the President, and that's very poor form." Elbert A. King, Jr., interview with L. S. Swenson, Jr., May 27, 1971, tape in JSC History Office files; Marti Mueller, "Trouble at NASA: Space Scientists Resign," *Science* 165 (1969): 776–79.

Nixon had already drawn considerable adverse criticism from some newspapers, which accused him of trying to turn Apollo to his own political advantage in spite of his negligible record of supporting the space program. His telephone call to the astronauts on the moon, the addition of his signature to those of the astronauts on the commemorative plaque on the lunar module *Eagle*, and his visit to the *Hornet* (on his way to a tour of Asia), were all cited as attempts to get unmerited credit for the Apollo accomplishment. See the editorial in the *Washington Post*, "Our Mark on the Moon," July 3, 1969; Herblock's cartoon "Lunar Hitchhiker," *Washington Post*, July 6, 1969; "Plaque Pique," Washington *Evening Star*, July 9, 1969; Marianne Means, "President Nixon and the Astronauts," *Los Angeles Herald-Examiner*, July 13, 1969; "Hate the President," *Richmond* (Va.) *News Leader*, July 15, 1969; William Hines, "Nixon Skims Off the Cream," Washington *Sunday Star*, July 27, 1969.

71. "8,200 View Moon Rock," Washington *Evening Star*, Sept. 18, 1969, " 'It looks like any rock,' " *Washington Daily News*, Sept. 18, 1969.

72. Johnston et al., *Biomedical Results of Apollo*, p. 419.

73. "Minutes, Interagency Committee on Back Contamination," Aug. 10, 1969.

74. Don Kirkman, "Germ-Free Apollos Sent Home," *Washington Daily News*, Aug. 11, 1969; Albert Sehlstedt, Jr., "Astronauts Brace For Celebrations," Baltimore *Sun*, Aug. 12, 1969.

75. "Apollo 11 LTD Daily Summary Report, 1200 July 26 to 1200 July 27"; Wilmot N. Hess, "LRL Summary Report No. 7," July 31, 1969.

76. A summary of the investigations is found in Johnston et al., *Biomedical Results of Apollo*, pp. 425–34.

77. MSC, "Sample Information Summary #5 Final," Aug. 27, 1969, p. 13; see also LRL Summary Reports for Aug. 1 through Aug. 31.

78. J. W. Humphreys, Jr., to Dir., MSC, "Release of all lunar material and lunar exposed material in Building 37, Manned Spacecraft Center, Houston, Texas," Sept. 11, 1969.

79. Rocco A. Petrone, TWX to Gilruth, "Approval of MSC proposed allocation of Apollo 11 lunar sample material for distribution to approved principal investigators for scientific analysis," Sept. 9, 1969; Petrone, TWX to Gilruth, "Release of lunar samples," Sept. 12, 1969; NASA Hqs., "Moon Surface Samples Distributed," Release No. 69-130, Sept. 12, 1969.
80. Richard A. Wright, "Lunar and Earth Sciences weekly activity report," Oct. 17, 1969.
81. Paine to Gilruth, Sept. 8, 1969. Some investigators did not conform to the ban. In November, a scientist at the U.S. Geological Survey wrote indignantly to MSC's director of science and applications, enclosing clippings that indicated some findings had been released prematurely and outside of normal scientific channels. According to one, Harold Urey had stated in a talk in San Diego that the Apollo 11 samples had been found to be 4.6 billion years old, which supported his theory of the moon's origin. G. Brent Dalrymple to Anthony J. Calio, Nov. 13, 1969, with encl., copy of article, "Moon Rocks' Age Is Now Firmly Fixed," *San Francisco Chronicle*, Nov. 8, 1969.
82. E. F. Bartell (Cornell Univ.), TWX to Space Sciences Procurement Branch, MSC, Sept. 16, 1969.
83. William L. Green, TWX to MSC, attn. Brian Duff, Oct. 3, 1969.
84. Donald U. Wise, TWX to MSC, attn. Dr. W. Hess, Aug. 11, 1969.
85. Philip H. Abelson, "The Moon Issue," *Science* 167 (1969):447.
86. The Lunar Sample Preliminary Examination Team, "Preliminary Examination of Lunar Samples from Apollo 11," *Science* 165 (1969):1211-27.
87. Ibid., p. 1226.
88. Cohn, "Seismometer on Moon Registers Disturbance," *Washington Post*, July 23, 1969, "Moon Tremor Recorded by Apollo Device," ibid., July 24, 1969.
89. Idem, "Moon Tremor Recorded."
90. Idem, "Old Moon Game Taunts Players."
91. Mueller, "Manned Space Flight Weekly Report—July 28, 1969."

CHAPTER 10

1. Stuart Auerbach, "Mariner 6 Relays 33 Pictures of Mars," *Washington Post*, July 30, 1969; idem, "Mariner 7 Photographs Mysterious Mars Canals," ibid., Aug. 5, 1969; Edward Clinton Ezell and Linda Neuman Ezell, *On Mars: Exploration of the Red Planet 1958-1978*, NASA SP-4212 (Washington, 1984), pp. 74-80, 175-80.
2. W. David Compton and Charles D. Benson, *Living and Working in Space: A History of Skylab*, NASA SP-4208 (Washington, 1983), pp. 104-11.
3. Ibid., pp. 115-18; Homer E. Newell, *Beyond the Atmosphere: Early Years of Space Science*, NASA SP-4211 (Washington, 1980), pp. 286-89.
4. John Noble Wilford, "A Spacefaring People: Keynote Address," in *A Spacefaring People: Perspectives on Early Spaceflight*, Alex Roland, ed., NASA SP-4405 (Washington, 1985), p. 72. Wilford, a veteran reporter and one of the best on space, covered the space program for the *New York Times* for many years. He was the keynote speaker at this conference on manned space flight history at Yale University, Feb. 6-7, 1981. Wilford asserted (p. 71) as one of the major themes of space history that *"The first Apollo landing was, in one sense, a triumph that failed, not because the achievement was anything short of magnificent but because of misdirected expectations and a general misperception of its real meaning. The public was encouraged to view it only as the grand climax of the space program, a geopolitical horse race and extraterrestrial entertainment—not as a dramatic means to the greater end of developing a far-ranging spacefaring capability. This led to the space program's post-Apollo slump* [emphasis in the original]."
5. John D. Stevenson, TWX to MSF centers and support elements, "MSF Mission Operations Forecast for June 1969," June 5, 1969; Courtney G. Brooks, James M. Grimwood, and Loyd S. Swenson, Jr., *Chariots for Apollo: A History of Manned Lunar Spacecraft*, NASA SP-4205 (Washington, 1979), p. 285; Mueller, "Manned Space Flight Weekly Report—July 1, 1969."
6. Stevenson to multiple addressees, TWX, "MSF mission operations forecast for August 1969," July 27, 1969; Mueller, "Manned Space Flight Weekly Report—July 29, 1969."

7. *1967 Summer Study of Lunar Science and Exploration*, NASA SP–157 (Washington, 1967), pp. 163–73, 229–33, 324, 358–59.

8. Andre J. Meyer, handwritten notebooks III–VI (late 1967 to early 1969), in JSC History Office files. Besides being involved in other advanced project planning, Meyer headed the Apollo Explorations Project Office at MSC.

9. Mueller, "Manned Space Flight Weekly Report—January 6, 1969," "Manned Space Flight Weekly Report—April 21, 1969."

10. Wilmot N. Hess to AA/Director, "Summary OSSA Senior Council Meeting," Apr. 30, 1968; John D. Hodge to PA/Office of Program Manager, MSC, "Lunar Exploration Support," June 4, 1968, with encl., "ELM [extended lunar module]—Development."

11. MSC, draft statement of work, "LM Modification Study for Extended Lunar Staytime," rev. 28, Feb. 1969; Mueller, "Manned Space Flight Weekly Report—March 4, 1969."

12. Lee R. Scherer to MSC, attn. Wilmot N. Hess, "Subsatellite on Apollo Lunar Missions," May 3, 1968; Hess to Mgr., Apollo Spacecraft Program Off., "CSM Lunar Orbiting Science Study," May 10, 1968.

13. J. J. Eslinger to Apollo CSM Engineering Supervision, North American Rockwell Corp., Project Memorandum #3, "Apollo Scientific Experiments Program (ASEP)," Feb. 27, 1969.

14. MSC, "Weekly Activities Report, Science and Applications Directorate, January 20–27, 1969."

15. Robert R. Gilruth to Mueller, "Lunar Exploration CSM Orbital Science, Phase I," Mar. 27, 1969.

16. John A. Naugle to Gilruth, TWX, Apr. 4, 1969.

17. S. C. Phillips to Gilruth, TWX, subj.: CSM Phase I Orbital Science, Apr. 21, 1969.

18. William E. Stoney, Jr., to MSC, attn.: J. P. Loftus, "CSM Orbital Science Missions," May 15, 1969.

19. Phillips to MSC, attn.: R. Gilruth, TWX, subj.: Authority to proceed with SM orbital experiments for Apollo 16 through 20, June 27, 1969.

20. J. O. Cappelari, Jr., ed., "Where on the Moon? An Apollo Systems Engineering Problem," *The Bell System Technical Journal* 51 (1972): 976–84.

21. John R. Sevier interview, Apr. 24, 1986.

22. MA/Apollo Program Dir. to multiple addressees, "Minutes of the Apollo Site Selection Board Meeting of June 3, 1969."

23. Ibid.

24. Ibid.; Benjamin Milwitzky to Dir., Apollo Lunar Exploration Office, "Biasing Apollo Missions to Land Near Surveyor Spacecraft on the Moon," Jan. 10, 1969.

25. Minutes of the Apollo Site Selection Board Meeting, June 3, 1969; Owen E. Maynard to Mgr., Apollo Spacecraft Program, "Apollo Site Selection Board trip report—June 3, 1969," with encl., "Lunar Landing Site Recommendations for Apollo 12 as Presented to Apollo Site Selection Board June 3, 1969," June 10, 1969.

26. Minutes of the Apollo Site Selection Board Meeting, June 3, 1969.

27. George M. Low to NASA Hqs. , attn.: S. C. Phillips, TWX, "Lunar Landing Sites for H-1 Mission," June 12, 1969.

28. Phillips to MSC, attn.: G. Low, TWX, "Lunar Landing Sites for H-1 Mission," June 16, 1969.

29. N. W. Hinners to Captain L. R. Scherer, "Second Mission Landing Sites," June 18, 1969; Hinners to Group for Lunar Exploration Planning and Site Selection Subgroup, "Fourth GLEP Site Selection Subgroup Meeting—June 17, 1969," June 23, 1969; Hinners to file, "GLEP Site Selection Subgroup, Fourth Meeting, June 17, 1969," Aug. 4, 1969.

30. Hinners, "Fourth GLEP Site Selection Subgroup Meeting—June 17, 1969."

31. Minutes of the Apollo Site Selection Board Meeting, July 10, 1969.

32. NASA Release 69–53, Apr. 10, 1969; "Crew to Make 2d Landing On Moon Set," *Washington Post*, Apr. 11, 1969.

33. Ivan D. Ertel and Roland W. Newkirk, with Courtney G. Brooks, *The Apollo Spacecraft: A Chronology*, vol. IV, NASA SP-4009 (Washington, 1978), pp. 408–9.

34. Sevier interview.

35. "Apollo 12 Crew Training Summaries (Mar. 28-Nov. 14, 1969)," folder in box 081-14, JSC History Office files.

36. MSC News Release, "Apollo 12 Landing Sites," July 25, 1969; John G. Zarcaro to multiple addressees, "Apollo 12 Mission H$_1$ planning,"July 28, 1969.

37. Sevier interview.

38. B. E. Oldfield (Hughes Aircraft Co.) to John Hodge, July 10, 1969; Hodge to mgr., Apollo space-craft program, "Science and Technological Items Available Upon Landing at a Surveyor Site," July 31, 1969.

39. Howard W. Tindall, Jr., "A Lengthy Status Report on Lunar Point Landing Including Some Remarks about CSM DOI [Descent Orbit Insertion]," Aug. 29, 1969. Tindall's clear and breezy style in writing memos ("Tindallgrams," as they came to be called), usually prefaced with an attention-grabbing title, gained him a well-deserved local reputation. He had the gift, rare among engineers and still rarer among managers, of compacting large amounts of information into a few clear and understandable paragraphs, leavened with considerable humor.

40. Tindall, "How to Land Next to a Surveyor—a Short Novel for Do-It-Yourselfers," Aug. 1, 1969.

41. *Surveyor III: A Preliminary Report*, NASA SP-146 (Washington, 1967), pp. 12–16; Walter Sullivan, "Surveyor Tracked Down By a Photographic Sleuth," *New York Times*, Nov. 19, 1969.

42. John P. Mayer, "The Search for Surveyor III," MSC Internal Note 69-FM-329, Jan. 16, 1970. Its language shows that the report was written before the launch of Apollo 12, although it is dated two months later.

43. Tindall, "How to Land next to a Surveyor," Aug. 1, 1969.

44. Tindall, "A Lengthy Status Report," Aug. 29, 1969.

45. Tindall to multiple addressees, "Apollo 12 Descent—Final Comments," Nov. 4, 1969.

46. MSC, "Mission Requirements, SA-507/CSM-108/LM-6, H-1 Type Mission, Lunar Landing," MSC SPD9-R-051, July 18, 1969, pp. 2–1 to 2–3.

47. MSC, Apollo 12 Mission Reviews, Aug. 15 and Oct. 2, 1969.

48. Andre J. Meyer to PA/Mgr., Apollo Spacecraft Program, "Surveyor 3 Activities on Apollo 12 Mission," Aug. 28, 1969.

49. Hinners to L. R. Scherer, "Second Mission Landing Sites," June 18, 1969; James H. Sasser to multiple addressees, "Highlights of GLEP Site Selection Subgroup Meeting in the Mapping Sciences Laboratory on June 17, 1969," June 22, 1969.

50. Thomas O'Toole, "Veteran Astronauts Lovell, Shepard to Lead '70 Moon Flights," *Washington Post*, Aug. 7, 1969. A month earlier a Houston paper had run an unconfirmed story that Shepard, Roosa, and Mitchell would be the crew for Apollo 13, with Lovell, Mattingly, and Haise manning 14 (Arthur Hill, "Alan Shepard Will Command 3rd Moon Team," *Houston Chronicle*, July 3, 1969). This may have been the cause for George Mueller's overriding Slayton's choice for these crews—the only instance in which Headquarters reversed Slayton's selections (see note 13, Chap. 5). Shepard had been off the active list until May 1969, serving meanwhile as head of the Astronaut Office ("Mercury's Shepard rejoins flight group," MSC *Roundup*, May 16, 1969). The report that he had been named to the first mission scheduled after he returned to active status would have considerably upset the scientist astronauts. In the course of an interview unrelated to this book, one of them commented off the record to the present author, "the test pilots ran the astronaut office and flew their friends." True or not, such a perception would have been hard to refute after this crew selection.

51. Marti Mueller, "Trouble at NASA: Space Scientists Resign," *Science* 165 (1969):776–79.

52. Ibid.

53. Wilmot N. Hess, interview with Robert Merrifield, Nov. 7, 1968, transcript in JSC History Office files; John Noble Wilford, "Moon Scientists Seek Place in the Sun," *New York Times*, Aug. 10, 1969; Gilruth to Mueller, Sept. 2, 1969.

54. Eugene Shoemaker, "Space—Where Now, and Why?" *Engineering and Science* 33(1) (1969): 9–12.

55. "Geologist to Quit Apollo Project; Weak Scientific Effort Charged," *New York Times* (UPI), Oct. 9, 1969.

56. Harold Urey, who during this period continually criticized NASA's choice of lunar landing sites, wrote off geologists as "mostly . . . a second-rate lot. . . . [Geology] is descriptive, and very often [geologists] do not learn more than the most elementary things about chemistry and physics." Urey to Mueller, Oct. 7, 1969. Of Shoemaker, Urey said, "Gene is one of the few capable people who has had a prominent part in advising NASA, though he has had a very rigid geological point-of-view. . . . I do not agree with him in some ways, but his general criticism of the lack of good scientific advice is correct." Urey to Paine, Oct. 9, 1969. Urey and Shoemaker had both worked on Project Ranger.

Shoemaker's criticism deeply affronted some in NASA, who felt it could only do the program harm; see Homer E. Newell, *Beyond the Atmosphere: Early Days of Space Science*, NASA SP-4211

(Washington, 1980), pp. 292–93. To this day Shoemaker insists that his intent was grossly misunderstood: "What I was trying to do," he told the author, "was to get people to focus on something they were losing on this thing, but I never got the point across." He and a few others thought the point was, "let's make the astronaut himself an instrument of scientific discovery," which could not be done by slavishly following preplanned operations. "The sample scientists didn't give a damn, frankly, whether the astronauts discovered something or not; *they* were going to discover something with the samples that came back, you see. * * * As far as I'm concerned, the kinds of things you could discover, by human observer, under the constraints of Apollo, . . . [have not] been touched. If you could get me a spacecraft tomorrow, I've got the whole program . . . in my head, . . . and it's still there [on the moon]. All you need to get me is a hand lens and a shovel. . . . There's three and a half billion years of history in three meters [10 feet] of dirt . . . on the lunar surface, and . . . we haven't even touched it yet." Shoemaker interview, Mar. 17, 1984. Since Shoemaker did not make this point central to his public remarks, it is not surprising that his former colleagues in NASA were affronted.

57. "Scientists vs. NASA," unsigned editorial in the *New York Times*, Aug. 12, 1969.
58. "Before We Start to Mars," unsigned editorial in the *Washington Post*, Aug. 8, 1969.
59. Fred L. Whipple to George E. Mueller, Aug. 1, 1969.
60. Alex J. Dessler, "Discontent of Space-Science Community," Oct. 30, 1969, paper written for unknown purpose, copy in box 075-36, JSC History Office files.
61. Newell, *Beyond the Atmosphere*, pp. 213, 221–22. The fact was that no eminent research scientist was willing to give up research for a career (or even a year's tenure) as a research administrator; yet Whipple, Dessler, and other scientists continued to urge NASA to find such a person to manage the science programs. See Newell, chap. 12 ("Who Decides?"). A similar problem arose in the life sciences program; see John A. Pitts, *The Human Factor: Biomedicine in the Manned Space Program to 1980*, NASA SP-4213 (Washington, 1985), passim.
62. John Lannan, "Money, Space, Environment Dominate Scientists' Parley," Washington *Evening Star*, Dec. 27, 1969; "Scientists Hit Threat to Cut 4 Apollo Flights," *Boston Sunday Globe*, Dec. 28, 1969; "Technology Misused, Radicals Charge," *Washington Post*, Dec. 29, 1969. An expression of the plight of science in 1969 and the evil effects of military-sponsored research is found in "Support of Science on the University's Own Terms," by Gerard Piel (publisher of *Scientific American*), *Science* 166 (1969): 1101.
63. Mueller to Gilruth, Sept. 3, 1969.
64. James R. Arnold to A. J. Calio, "Apollo 12 Recommendations," Sept. 5, 1969; Arnold to Calio, "Display of Sample Information in the LRL," Sept. 6, 1969.
65. Paul W. Gast to Calio, "Dry N_2 Facility as a primary ALSRC Receiving and Handling System," Sept. 8, 1969; LSAPT to W. N. Hess, "Minutes of LSAPT dated March 13, 1969," Mar. 21, 1969; Gast to LRL Management and Mission Review Board, no subject, Aug. 8, 1969.
66. E. C. T. Chao and R. L. Smith to Hess, "Recommendations and Suggestions for Preliminary Examination of Apollo 12 Returned Lunar Samples," Sept. 8, 1969.
67. R. S. Johnston to P. R. Bell and W. W. Kemmerer, Jr., "Apollo 12 Sterile Nitrogen Processing," Sept. 22, 1969.
68. Johnston to NASA Hqs., attn.: Maj. Gen. J. W. Humphreys, "Apollo 12 Back Contamination Program Plan," Oct. 7, 1969; Johnston to Mgr., Command and Service Modules, Apollo Spacecraft Program, "Requirement for Installation of a Postlanding Vent Valve Back Contamination Filter on CM's 108 and Subsequent," Sept. 18, 1969.
69. Johnston to the record, "Apollo 12 Back Contamination Program," Sept. 17, 1969.
70. Apollo 12 Crew Training Summaries, Aug. 15, Sept. 15, Nov. 14, 1969, folder in box 081-14, JSC History Office.
71. Harrison H. Schmitt interview, May 30, 1984.
72. Manned Space Flight Weekly Reports, Apr. 1, July 1, Sept. 15, Nov. 10, 1969; Charles D. Benson and William Barnaby Faherty, *Moonport: A History of Apollo Launch Facilities and Operations*, NASA SP-4204 (Washington, 1978), pp. 480–81.

CHAPTER 11

1. Charles D. Benson and William Barnaby Faherty, *Moonport: A History of Apollo Launch Facilities and Operations*, NASA SF-204 (Washington, 1978), pp. 481–82.
2. Ibid.
3. MSC, "Apollo 12 Technical Air-to-Ground Voice Transcription," prepared for Data Logistics Office, Test Div., Apollo Spacecraft Program Office, Nov. 1969 (cited hereinafter as "12 Air-to-Ground"), p. 17; MSC, "Apollo 12 Mission Report," MSC-01855, March 1970, pp. 14-2 to 14-4.
   Postflight analysis of available data supported the first tentative conclusion. Electric-field measurements at the launch site indicated that the clouds were charged, but not enough to produce lightning. The passage of a conductor (such as the launch vehicle) is known to produce lightning under such conditions. Witnesses disagreed on the question of lightning strikes, but photographs showed two bolts from cloud to ground 36.5 seconds after launch, neither of which produced damage. The second discharge, 52 seconds after launch, was probably between clouds. See MSC-01540, "Analysis of Apollo 12 Lightning Incident," Feb. 1970. Damage to the launch vehicle was negligible (several telemetry sensors on the service module were knocked out), but launch rules were subsequently tightened to forbid launch under weather conditions favorable to lightning strikes.
4. 12 Air-to-Ground, p. 24.
5. Apollo 12 Mission Report, p. 5-4.
6. Ibid., p. 5-5.
7. 12 Air-to-Ground, p. 292.
8. Ibid., pp. 281–92.
9. Ibid., pp. 292–322.
10. Apollo 12 Mission Report, pp. 5–7 to 5–8.
11. 12 Air-to-Ground, pp. 335–45; Apollo 12 Mission Report, p. 5–9.
12. 12 Air-to-Ground, pp. 346–68; Apollo 12 Mission Report, p. 6–3.
13. 12 Air-to-Ground, pp. 348–68.
14. Apollo 12 Mission Report, pp. 4–25 to 4–27.
15. 12 Air-to-Ground, pp. 370–73.
16. Ibid., pp. 406–11.
17. Ibid., pp. 418–26; Apollo 12 Mission Report, p. 14–50.
18. 12 Air-to-Ground, pp. 429–60.
19. Apollo 12 Mission Report, p. 3–26; 12 Air-to-Ground, pp. 438–73.
20. 12 Air-to-Ground, pp. 473–521. Neil Armstrong and Buzz Aldrin had difficulty finding a way to sleep comfortably in the cramped LM. Hammocks were added on *Apollo 12* and Bean and Conrad found them quite comfortable in lunar gravity.
21. 12 Air-to-Ground, pp. 553–89.
22. Ibid., pp. 597–716.
23. The Lunar and Planetary Missions Board recommended in early 1968 that astronauts should talk as much as possible ("almost incessant talking," as the Board put it) while on the lunar surface, describing what they were doing, what they could see and how they interpreted it, and what they intended to do next and why, to provide a complete record of their surface activity. Homer E. Newell to Robert R. Gilruth, Feb. 5, 1968. MSC agreed in principle only, suggesting that the astronauts observe the tried and proven rules of aircraft communications, transmitting only necessary information while at the same time making an effort to provide "a running commentary of significant events and actions so preplanned that normal communications discipline is not violated but yet is complete enough to assure a maximum of scientific return from the exercise." That kind of commentary would be "comparable to the usual field notes of the practicing geologist on a field trip." Gilruth to Newell, Feb. 20, 1968; W. N. Hess to Special Asst. to the Dir., "Astronaut Activity on Lunar Surface," Feb. 19, 1968.

24. 12 Air-to-Ground, pp. 602, 603, 614-52.

    After Buzz Aldrin drew criticism from scientists for his misidentification of biotite (see footnote, chap. 9, p. 282), Bean and Conrad may have been more sensitive to what they were saying. Geologist-astronaut Jack Schmitt noted that in postmission discussions Conrad admitted he was fairly sure the sample was olivine, but "he wasn't about to say so," for fear of making a mistake. H. H. Schmitt interview, May 30, 1984.

25. 12 Air-to-Ground, pp. 661, 666, 670.

26. Ibid., pp. 671-95.

27. Ibid., p. 731.

28. Apollo 12 Mission Report, pp. 10-7, 8-20.

29. Ibid., p. 9-10; 12 Air-to-Ground, p. 740.

30. 12 Air-to-Ground, pp. 757-770; MSC, "Apollo 12 Technical Crew Debriefing," Dec. 1, 1969, pp. 12-21 to 12-23; Apollo 12 Mission Report, pp. 6-5 to 6-6.

31. Apollo 12 Mission Report, pp. 3-10 to 3-11, 9-39 to 9-40; Victor Cohn, "Moon Quake Caused by Lem Called 'Unlike Any' on Earth," *Washington Post*, Nov. 21, 1969; Gary V. Latham, Maurice Ewing, Frank Press, George Sutton, James Dorman, Hosio Nakamura, Nafi Toksoz, Ralph Wiggins, and Robert Kovach, "Passive Seismic Experiment," in *Apollo 12 Preliminary Science Report*, NASA SP-235 (Washington, 1970) , pp. 39-53.

32. 12 Air-to-Ground, pp. 863-1066, 944-57; Apollo 12 Mission Report, p. 9-28.

33. 12 Air-to-Ground, pp. 924-25.

34. Alan L. Bean interview, Apr. 10, 1984.

35. 12 Air-to-Ground, p. 1018.

36. Apollo 12 Technical Crew Debriefing, pp.16-1, 16-4.

37. Apollo 12 Mission Report, pp. 11-6 to 11-7.

38. Cohn, "Moon Quake Caused by Lem."

39. Leo T. Zbanek, "Weekly Activity Report, Engineering Division, November 13, 1969"; Richard A. Wright to TA/Special Asst., "Weekly Activity Report," Nov. 14, 1969.

40. Wright to TA/Dir., Science and Applications, "Weekly Activities Report," Nov. 27, 1969.

41. "Lunar Receiving Laboratory Daily Report, 1400 hours 11-28-69 to 1400 hours 11-29-69," Nov. 29, 1969; J. L. Warner, "Apollo 12 Sample Inventory," Dec. 5, 1969.

42. Bryan Erb to DC/Chief, Preventive Medicine Div., "Selection of lunar samples for biological testing," Nov. 29, 1969.

43. "Lunar Receiving Laboratory Daily Report, 1400 hours 12-1-60 to 1400 hours 12-2-69," Dec. 2, 1969; Charles A. Berry, "Findings and Determinations of Extraterrestrial Exposure and Quarantine of Particular Person(s), Property, Animal(s), or Other Form(s) of Life or Matter," Dec. 1, 1969.

44. MSC, "Containment Fault Press Conference," Dec. 1, 1969, transcript; "Apollo 12 Ward Status Report," transcript of press conf., Dec. 2, 1969.

45. MSC, "LRL Operational Summary, Apollo 12, November 24 to December 6, 1969," Dec. 6, 1969.

46. Ross Taylor, "Comparison Between Apollo 11 Samples from Mare Tranquillitatis and Apollo 12 Samples from Oceanus Procellarum," Dec. 8, 1969; MSC, "Lunar Receiving Laboratory Daily Report, 1400 hours 12-10-69 to 1400 hours 12-11-69," Dec. 11, 1969.

47. MSC, "LRL Operational Summary, Apollo 12, Dec. 6 to Dec. 13, 1969," Dec. 13, 1969.

48. MSC, "Lunar Receiving Laboratory Daily [*sic*] Report, 1400 hours 12-12-69 to 1400 hours, 12-18-69," Dec. 18, 1969; John Noble Wilford, "Apollo 12 Samples Appear to Be Younger," *New York Times*, Dec. 13, 1969.

49. MSC, "Lunar Receiving Laboratory Reports," transcripts of press briefings, Nov. 25–Dec. 10, 1969, box 079-16, JSC History Office Apollo files.

50. Berry to Maj. Gen. J. W. Humphreys, Jr., TWX, Dec 8, 1969; Berry to Chairman, Interagency Committee on Back Contamination, "Recommendation for Release of Apollo XII Crew and Crew Reception Area from Personnel Quarantine," Dec. 9, 1969; Humphreys, TWX to MSC, Dec. 9, 1969; W. Carter Alexander, "Quarantine Release Report No. 3," n.d. [Dec. 8, 1969]; Humphreys to MSC, attn.: Director and Dir. of Medical Research and Operations, "Release of Personnel from CRA/LRL," Dec. 10, 1969.

51. Victor Cohn, "Same Chemical Origin Seen for Earth, Moon," *Washington Post*, Jan. 7, 1970; "Summary of Apollo 11 Lunar Science Conference" (prepared by the Lunar Sample Preliminary Examination Team), *Science* 167 (1970): 449–51. The entire January 30 issue of *Science* was given over to the papers resulting from studies on the Apollo 11 samples (see Chap. 9), requiring some extraordinary effort in reviewing and editing. Manuscripts were received the first week in January, refereed and revised during the next two weeks, and the issue was printed three weeks later. See Philip H. Abelson, "The Moon Issue," ibid. , p. 447, and "Acknowledgements," ibid., p 781. Revised and expanded versions of the papers presented at the conference were published in *Proceedings of the Apollo 11 Lunar Science Conference* (3 vols.), A. A. Levinson, ed. (New York and London: Pergamon Press, 1970).

52. "'Summary of Apollo 11 Lunar Science Conference"; Wilford, "Lunar Churning Surmised From Samples of Rock," *New York Times*, Jan. 7, 1970.

53. "Summary of Apollo 11 Lunar Science Conference."

54. James C. Tanner, "Moon Rocks' Age Put at 4.6 Billion Years, Much Older Than Any Found on the Earth," *Wall Street Journal*, Jan. 6, 1970.

55. Wilford, "Apollo Conference Ends With a Wealth of Data but an Unclear Picture of Lunar Origin," *New York Times*, Jan. 9, 1970; Tanner, "Moon Scientists End Their Session Today, But Many Questions Remain Unanswered," *Wall Street Journal*, Jan. 8, 1970.

56. "Summary of Apollo 11 Lunar Science Conference."

57. George E. Mueller to Administrator, "Manned Space Flight Weekly Report," Dec. 15, 1969.

58. James A. McDivitt to multiple addressees, "Apollo Experiments Review Group," Jan. 6, 1970.

59. Homer E. Newell, "Conference Report, February 5, 1970, Lunar Science Institute, Houston, Texas, Subject: Critique of Apollo Lunar Missions and the Maximization of Scientific Returns for the Remaining Apollo Flights," Feb. 6, 1970.

60. Ibid.

61. Newell to Gilruth, Feb. 6, 1970.

62. Calio to PA/Mgr., Apollo Spacecraft Program, "Lunar Surface Science Requirements," Mar. 27, 1970; Calio to multiple addressees, "Science Mission Manager and Mission Scientist Assignments for Apollo 13 and 14," Apr. 6, 1970; Calio to TJ/James H. Sasser [and others], "Science Working Panel," June 4, 1970.

63. Eugene F. Kranz to TM/Acting Mgr., Lunar Missions Office, "Science Briefings for Flight Crews," Apr. 23, 1970.

64. Newell, *Beyond the Atmosphere: Early Years of Space Science*, NASA SP-4211 (Washington, 1980), p. 293.

65. National Archives and Records Service, *Weekly Compilation of Presidential Documents*, Nov. 17, 1969, p. 1597; NASA Releases 69-151, Nov. 10, 1969, and 70-4, Jan. 8, 1970; NASA Announcement, Aug. 31, 1969; MSC Releases 69-66, Sept. 25, 1969, and 69-70, Nov. 26, 1969; Thomas O'Toole, "Petrone, Launch Director at Cape, Promoted to Head Apollo Project," *Washington Post*, Aug. 23, 1969; "Mueller Leaves NASA Dec. 10; Paine Lauds His Apollo Role," ibid., Nov. 11, 1969.

66. NASA Release, Jan. 15, 1969.

67. *Presidential Documents*, Mar. 21, 1969; NASA Budget Briefing, Apr. 15, 1969, transcript.

68. *Presidential Documents*, Dec. 1, 1969.

69. Space Task Group, *The Post-Apollo Space Program: Directions for the Future*, report to the President, Sept. 15, 1969.

70. "Nixon Backs Mars Flight but Rejects All-Out Drive," *New York Times*, Sept. 16, 1969.

71. "Before We Start to Mars," unsigned editorial in the *Washington Post*, Aug. 8, 1969.

72. Victor K. McElheny, "NASA Adviser: Sending Men to Mars 'Utmost Folly'," *Boston Globe*, Dec. 19, 1969; Cohn, "Cut in Space Program Urged," *Washington Post*, Dec. 29, 1969.

73. U.S., Congress, House, *Toward the Endless Frontier: History of the Committee on Science and Technology, 1959-79* [by Ken Hechler] (Washington: U.S. Govt. Printing Office, 1980), p. 269.

74. "Apollo Missions Extended to '74," *New York Times*, Jan. 5, 1970; "Experts Hail Schedule for Moon Trips," *Washington Post*, Jan. 6, 1970. Rumors that four Apollo missions might be canceled circulated in late 1969; see O'Toole, "Nixon to Give Space Goals; Modest Program Expected," *Washington Post*, Dec. 22, 1969; "Scientists hit threat to cut 4 Apollo flights," *Boston Sunday Globe*, Dec. 28, 1969. For the effects of budget cuts on Skylab, see W. David Compton and Charles D. Benson, *Living and Working in Space: A History of Skylab*, NASA SP-4208 (Washington, 1983), Chap. 4.

75. Richard D. Lyons, "50,000 NASA Jobs to Be Eliminated," *New York Times*, Jan. 14, 1970; O'Toole, "Cutbacks Planned by NASA," *Washington Post*, Jan. 14, 1970.

76. *Astronauts and Aeronautics, 1970: Chronology on Science, Technology, and Policy*, NASA SP-4015 (Washington, 1972) p. 203.

77. *Toward the Endless Frontier*, pp. 313-14.

78. Mueller to Administrator, "Manned Space Flight Weekly Report," Dec. 15, 1969; Dale D. Myers to Administrator, "Manned Space Flight Weekly Report," Jan. 19, 1970.

79. NASA Release 70-5, Jan. 8, 1970.

80. Calio to Mgr., Apollo Spacecraft Program, "Recommendations for Science Sites for Apollo Missions 13-20," Oct. 20, 1969.

81. MSC, "Apollo 13 Mission Report," MSC-02680, Sept. 1970, p. 12-2 and App. A, p. A-2.

82. The best account of Apollo 13 is that of Henry S. F. Cooper, *13: The Flight That Failed* (New York: Dial Press, 1973), a taut narrative that captures the flavor of the entire mission. For a first-person account by a participant, see mission commander James A. Lovell's chapter, "Houston, We've Had A Problem," in *Apollo Expeditions to the Moon*, Edgar M. Cortright, ed., NASA SP-350 (Washington, 1975). A brief summary of the mission and the investigation is given in Benson and Faherty, *Moonport*, pp. 489-94. Headquarters' Public Affairs Office prepared a 25-page booklet on the flight, "Houston, we've got a problem," NASA EP-76 (Washington, 1970). A concise summary of the constraints faced by flight controllers and the rationale for the real-time decisions made during Apollo 13 was presented by Glynn S. Lunney, one of 13's flight directors, as AIAA Paper No. 70-1260, "Discussion of Several Problem Areas During the Apollo 13 Operation," at the AIAA 7th Annual Meeting and Technical Display, Houston, Oct. 19-22, 1970. The *Report of the Apollo 13 Review Board*, a summary volume and eight technical appendices, was released two months after the accident. It contains full technical details of the causes of the accident as deduced from inflight telemetry, the history of the faulty tank, and tests performed to support the investigation. Board chairman Edgar M. Cortright presented the board's findings to the House space committee on June 16, 1970; see House, *The Apollo 13 Accident*, hearings before the Committee on Science and Astronautics, 91/2, June 16, 1970.

83. Apollo 13 Mission Report, pp. 11-9 to 11-10.

CHAPTER 12

1. U.S. Congress, House Committee on Science and Astronautics, *1971 NASA Authorization*, Hearings on H.R. 15695, 91/2, Feb. 1970, vol. 1, pp. 2–38; W. David Compton and Charles D. Benson, *Living and Working in Space: A History of Skylab*, NASA SP-4208 (Washington, 1983), pp. 115–16.

2. Idem, *1971 NASA Authorization*, vol. 1, p. 7. The Electronics Research Center (ERC) in Cambridge, Mass., proposed in 1963, had provoked serious opposition from the start. Nonetheless, ERC was approved and began operation in 1965 while permanent buildings were being constructed. By late 1969, when ERC was abolished, some $80 million had been invested in buildings and equipment. The facility was transferred to the Department of Transportation in March 1970. See House Committee on Science and Technology, *Toward the Endless Frontier: History of the Committee on Science and Technology, 1959–79* (Washington: U.S. Govt. Printing Office, 1980), pp. 219–31.

3. Thomas O'Toole, "3 Apollo Cancellations Weighed for Big Skylab," *Washington Post*, July 9, 1970.

4. National Archives and Records Service, *Weekly Compilation of Presidential Documents*, Aug. 17, 1970, pp. 1056–57; cited in NASA, *Astronautics and Aeronautics, 1970*, p. 263.

5. Thomas O'Toole, "NASA Head Quits, Plans to Rejoin GE," *Washington Post*, July 29, 1970; Richard D. Lyons, "Paine Quits Post in Space Agency," *New York Times*, July 29, 1970.

6. NASA, "FY 1971 Interim Operating Plan News Conference," Sept. 2, 1970, transcript; Thomas O'Toole, "2 Moon Landings Canceled by NASA," *Washington Post*, Sept. 3, 1970.

7. "FY 1971 Interim Operating Plan News Conference." One Saturn V was earmarked for the launch of Skylab; later NASA built a backup Skylab orbital workshop and committed another Saturn V to launch it. Never used, the workshop is now in the National Air and Space Museum and its launch vehicle was put on exhibit at Kennedy Space Center. The third Saturn V is on display

at the Johnson Space Center beside a Mercury-Redstone—relics of the heroic age of manned space flight; John E. Naugle to Dir., JSC, "Disposition of Residual Apollo Hardware," June 17, 1977.

8. "Retreat From the Moon," *New York Times*, Sept. 4, 1970.

9. Richard D. Lyons, "Scientists Decry Moon Flight Cut," *New York Times*, Sept. 4, 1970, and "New Cuts For Apollo: 'No Gas for The Rolls Royce?'," ibid., Sept. 6, 1970.

10. Harold C. Urey, *Washington Post*, Sept. 17, 1970.

11. Gerald J. Wasserburg, *Los Angeles Times*, Sept. 15, 1970. "Return to the dark ages of planetary science" seems to overstate the consequences of canceling two lunar exploration missions, but as a veteran observer of science politics noted, scientists were insecure about financial support and inclined to rhetorical extravagance when lamenting cutbacks in funding: "they recognize no territory between paradise and hell." Daniel S. Greenberg, "Mr. President Meets Mr. Wizard," *Washington Post*, Dec. 6, 1976.

12. Edward Clinton Ezell and Linda Neuman Ezell, *On Mars: Exploration of the Red Planet 1958–1978*, NASA SP-4212, pp. 185–92. See also House, *Toward The Endless Frontier*, pp. 272–76.

13. Compton and Benson, *Living and Working in Space*, pp. 182–85.

14. Letter to George P. Miller, signed by 39 lunar scientists, Sept. 10, 1970, copy included in House Committee on Science and Astronautics press release, Sept. 21, 1970.

15. Miller to multiple addressees, Sept. 21, 1970, House Committee on Science and Astronautics press release, Sept. 21, 1970.

16. Rocco A. Petrone to multiple addressees, "Apollo Site Selection Board Minutes of Meeting," Mar. 16, 1970; James A. McDivitt to multiple addressees, "Apollo Site Selection Board Meeting March 6, 1970"; Myers to Administrator, "Manned Space Flight Weekly Report—March 9, 1970," Mar. 9, 1970.

17. Calio to Mgr., Apollo Spacecraft Program, "Site Selections for Apollo Missions 14 and 15," May 8, 1970.

18. U.S. Congress, Senate, *Apollo 13 Mission Review*, Hearings before the Committee on Aeronautical and Space Sciences, 91/2, June 30, 1970, p. 52; "Apollo 13 Rescheduled," NASA Release 70-5, Jan. 8, 1970; John Noble Wilford, "Apollo 14 to Use Cart on the Moon," *New York Times*, Jan. 20, 1970; Richard D. Lyons, "NASA and 2 Companies Blamed for Apollo Blast," *New York Times*, June 16, 1970.

19. Bendix Corp., Mission Assignments, "Apollo Lunar Surface Experiments Package (ALSEP)," no date [c. Dec. 1970], box 079-51, JSC History Office files.

20. "New Labor-Saving device: the MSC 'Rickshaw,'" MSC *Roundup*, Jan. 30, 1970.

21. MSC-03465-Rev. A, "Flight Crew Health Stabilization Program," Oct. 26, 1970; "MSC Flight Readiness Review, Apollo 14," part IV, Dec. 11, 1970, pp. 13–14; "Crew Health Stabilization Plan Announced for Apollo 14," MSC *Roundup*, Oct. 9, 1970.

22. "MSC Flight Readiness Review, Apollo 14," part IV, pp. 27, 31–34; MSC-04112, "Apollo 14 Mission Report," May 1971, pp. 10–14 to 10–15.

23. Benson and Faherty, *Moonport*, pp. 494–96.

24. Myers to Administrator, "Manned Space Flight Weekly Report—January 11, 1971," Jan. 11, 1971.

25. Benson and Faherty, *Moonport*, p. 499.

26. MSC, "Apollo 14 Mission Report," MSC-04112, May 1971, pp. 1–1, 9–3, 14–1 to 14–5.

27. Ibid., p. 3–19.

28. Ibid., p. 6–2.

29. Ibid., pp. 6–2, 6–7.

30. MSC, "Apollo 14 Technical Air-to-Ground Voice Transcription" (hereinafter cited as "14 Air-to-Ground"), pp. 400–401.

31. 14 Air-to-Ground, p. 449.

32. Ibid., pp. 449–531, 535–54, 575–85.

33. Ibid., pp. 637–68; MSC, "Apollo 14 Technical Crew Debriefing," Feb. 17, 1971, pp. 10–35 to 10–36, 10–50 to 10–57.

34. "Apollo 14 Mission Report," p. 1–2.

35. 14 Air-to-Ground, p. 723; "Moon Trip Prescribed For Golf Ills," *Washington Post*, Feb. 18, 1971. Shepard's golf club was fashioned at his request by technicians in MSC's Technical Services Division. According to Jack Kinzler, chief of Technical Services, it was "bootlegged" through the shops because no one wanted to draw high-level managerial attention to it.

36. "Apollo 14 Mission Report," p. 4–4, 4–5.

37. 14 Air-to-Ground, pp. 757–67.
38. "Apollo 14 Mission Report," pp. 6–4, 6–10.
39. Ibid., pp. 5–1 to 5–6.
40. Ibid., pp. 11–3 to 11–7, 10–15. It later came out that Edgar Mitchell had participated in an experiment of his own, not in the official flight plan. In cooperation with a Midwestern drafting engineer who claimed psychic powers, he attempted some telepathic communication with earth at various times during the flight. See "Apollo ESP test told," *Washington Daily News*, Feb. 22, 1971.
41. MSC to Grumman Aircraft Engineering Co. (GAEC), contract change authorization (CCA) no. 2002, Amendment no. 1, "LM Modifications Study for Extended Lunar Staytime," Mar. 13, 1969; Samuel C. Phillips, TWX to MSC, May 9, 1969; MSC to GAEC, CCA no. 2333, "LM-10 and Subsequent Modification Program," June 9, 1969.
42. GAEC LMA790-2, "Lunar Module Modification Program [LMMP], Vehicle Familiarization Manual," Aug. 28, 1969.
43. Joseph N. Kotanchik to Asst. Dir. for Chemical and Mechanical Systems, MSC, "Report on Attendance at LMMP CDR at GAC, September 12, 1969," Sept. 15, 1969; GAEC, "LMMP Critical Design Review Board Meeting Minutes, September 12, 1969," Sept. 17, 1969.
44. Lee R. Scherer to multiple addressees, TWX, May 27, 1969.
45. Saverio F. Morea, MSFC, to James A. McDivitt, "LRV Weight Growth," Nov. 6, 1969; McDivitt to Roy E. Godfrey, MSFC, Dec. 12, 1969.
46. McDivitt to multiple addressees, "Lunar Roving Vehicle," Nov. 1, 1969; George E. Mueller to Administrator, "Manned Space Flight Weekly Report—November 3, 1969," Nov. 3, 1969.
47. Roger E. Bilstein, *Stages to Saturn: A Technological History of the Apollo/Saturn Launch Vehicles*, NASA SP-4206 (Washington, 1980), pp. 376–77.
48. Donald C. Wade to multiple addressees, "August 12 Saturn V payload improvement meeting," Aug. 15, 1969; John D. Hodge to Phillips, "Increased Saturn V Payload Capacity," Aug. 29, 1969.
49. J. O. Cappellari, Jr., ed., "Where on the Moon? An Apollo Systems Engineering Problem," *The Bell System Technical Journal* 51(5) (May–June 1972): 1028–34; Rocco A. Petrone to multiple addressees, "Saturn V Performance Capability for Apollo 16–20," Nov. 24, 1969.
50. Cappellari, "Where on the Moon?", p. 1030.
51. Richard G. Smith, MSFC, to Petrone, "Use of 510 Launch Vehicle for 107,000 Pounds Payload Mission," July 22, 1970.
52. Scherer, "Minutes of the Apollo Site Selection Board Meeting, Oct. 30, 1969"; Farouk El-Baz and D. B. James, "Minutes of the August 12–14 Meeting of an Ad Hoc Working Group on the Science Objectives of Apollo Missions 12–20," Bellcomm Memo for File, Aug. 18, 1969.
53. Gene Simmons to Dr. Charles H. Townes and Dr. John Findlay, Oct. 21, 1969; Urey to Townes and Findlay, Oct. 27, 1969. Urey felt unable to attend more meetings because "I shall miss three classes this quarter in my thermodynamics course—one to attend the Flagstaff [site evaluation] meeting and two to collect medal awards. . . . I cannot serve on more committees and do my university work."
54. Simmons to Townes and Findlay, Oct. 21, 1969.
55. Scherer, "Minutes of the Apollo Site Selection Board Meeting, Oct. 30, 1969."
56. Ibid.
57. "Group for Lunar Exploration Planning, Minutes of Meeting, February 6–7, 1970"; A. J. Calio to Mgr., Apollo Spacecraft Program, "Site Selections for the Remaining Apollo missions," Mar. 24, 1970.
58. Calio to Mgr., Apollo Spacecraft Program, "Site Selections for the Remaining Apollo missions," Mar. 24, 1970.
59. Idem, "Site Selections for Apollo missions 14 and 15," May 6, 1970.
60. Ronald L. Berry to Chief, Lunar Mission Analysis Branch, "Results of the Sub-GLEP Apollo Site Selection Meeting on June 5, 1970," June 17, 1970; Hinners to multiple addressees, "Subgroup Meeting of June 5, 1970," June 10, 1970.
61. Minutes of the Apollo Site Selection Board Meeting, Sept. 24, 1970; Calio to Mgr., Apollo Spacecraft Program, "Landing Site Recommendation for Apollo 15 Mission," Aug. 17, 1970.
62. Berry to multiple addressees, "Results of the Apollo Lunar Landing Site Selection Board Meeting of September 24, 1970," Sept. 30, 1970; Minutes of the Apollo Site Selection Board Meeting, Sept. 24, 1970.

63. NASA Release 69–115, Aug. 7, 1969; Thomas O'Toole, "Veteran Astronauts Lovell, Shepard to Lead '70 Moon Flights," *Washington Post*, Aug. 7, 1969; "Before We Start to Mars, ibid., Aug. 8, 1969.

64. Paul Recer (Associated Press), "They Feud Over Moon Flights," *The Miami Herald*, Aug. 18, 1969.

65. B. J. Richey, "Lunar Landing Teams Are Unlikely To Include Scientists Until 1971," *St. Louis Globe-Democrat*, Aug. 18, 1969.

66. Richard Witkin, "Scientist Expected to Be Picked for Moon Trip," *New York Times*, Dec. 12, 1969.

67. NASA Release 70–46, "Apollo 15 Crew Selected," Mar. 26, 1970.

68. Harrison H. Schmitt interview, May 30, 1984.

69. Homer E. Newell, *Beyond the Atmosphere: Early Years of Space Science*, NASA SP-4211 (Washington, 1980), p. 210; handwritten notes on meetings with scientist-astronauts, Houston, Jan. 12–13, 1971, in Newell's notebook, box 28 of the Newell files stored in the Federal Records Center, Suitland, Md., accession no. 255-79-0649; Newell to Dr. Fletcher, July 21, 1971.

70. Ibid.

71. Dale D. Myers to Charles H. Townes, Mar. 1, 1971.

72. Compton and Benson, *Living and Working in Space*, p. 220.

73. NASA Release 71–31, "Apollo 16 Crew Selected," Mar. 3, 1971.

74. A. P. Vinogradov, "Preliminary Data on Lunar Ground Brought to Earth by Automatic Probe Luna-16,'" in *Proceedings of the Second Lunar Science Conference*, A. A. Levinson, ed. (Cambridge: MIT Press, 1971), pp. 1–16.

75. Norman J. Hubbard and Paul W. Gast, "Chemical composition and origin of nonmare lunar basalts," ibid., pp. 999–1020; John Noble Wilford, "Exotic Fragments Found In Apollo Lunar Samples," *New York Times*, Jan. 11, 1971.

76. Wilford, "Exotic Fragments"; Ursula B. Marvin, J. A. Wood, G. J. Taylor, J. B. Reid, Jr., B. N. Powell, J. S. Dickey, Jr., and J. F. Bower, "Relative Proportions and Probable Sources of Rock Fragments in the Apollo 12 Samples," *Proceedings of the Second Lunar Science Conference*, pp. 679–99.

77. Arthur Hill, "Primitive Lunar Crust Is Hinted by Fragments," *Houston Chronicle*, Jan. 11, 1971.

78. Cappellari, "Where on the Moon?", pp. 990–94.

79. Earl H. Arnold, "Minutes, Interagency Committee on Back Contamination, January 15, 1970."

80. Space Science Board, National Academy of Sciences, "Report of Meeting on Review of Lunar Quarantine Program, February 17, 1970."

81. J. W. Humphreys, Jr., to Dir., Apollo Program, "Quarantine for Apollo 14 and Subs," July 16, 1970.

82. Humphreys, TWX to multiple addressees, "Discontinuation of Lunar Quarantine," Apr. 28, 1971; Myers to MSC Director, "Decision to Terminate Quarantine," May 10, 1971 with encl., "Decision to Terminate Quarantine Under NMI 1052.90 (Attachment A, Change 1, 2)," signed by George M. Low, Apr. 26, 1971.

CHAPTER 13

1. Rocco A. Petrone to Administrator, "Apollo 15 Mission (AS-510)," Prelaunch Mission Operation Report M-933-71-15, July 17, 1971.

2. Ibid.

3. MSC, "Apollo 15, To the Mountains on the Moon," NASA Fact Sheet MSC-04065, n. d. [c. June 1971].

4. OMSF, "Manned Space Flight Management Council, October 15, 1969, Status Highlights."

5. MSC, "Apollo 15 Press Kit," July 15, 1971; S. F. Morea, "Lunar Rover Briefing, Kennedy Space Center, July 23, 1971," transcript.

6. Morea, "Lunar Rover Briefing"; Dale D. Myers to Administrator, "Manned Space Flight Weekly Report," Feb. 2, 1970; ibid., Sept. 8, 1970.

7. Myers to Adm., "Manned Space Flight Weekly Report, Nov. 23, 1970.

8. Charles D. Benson and William Barnaby Faherty, *Moonport: A History of Apollo Launch Facilities and Operations*, NASA SP-4204 (Washington, 1978), pp. 508–14.

9. Ibid., pp. 507–508.

10. Ibid., pp. 517–18.
11. MSC Public Affairs Office, "Apollo 15 Postlaunch Briefing," July 26, 1971.
12. MSC-05161, "Apollo 15 Mission Report," Dec. 1971, pp. 1-1, 3-2.
13. Ibid., pp. 3-1 to 3-7, 9-1 to 9-6.
14. Ibid., pp. 9-7 to 9-8.
15. MSC-04558, "Apollo 15 Technical Air-to-Ground Voice Transcription" (hereinafter cited as "15 Air-to-Ground"), pp. 337–66; Apollo 15 Mission Report, p. 9-9.
16. Apollo 15 Mission Report, pp. 9-14 to 9-15; 15 Air-to-Ground, p. 453.
17. 15 Air-to-Ground, 485–89.
18. Ibid., pp. 490–506.
19. Ibid., pp. 536–39.
20. Ibid., p. 546.
21. Ibid., p. 544.
22. Ibid., pp. 547–48, 550, 554, 556, 560–61.
23. Ibid., pp. 568–610; Apollo 15 Mission Report, p. 4-1; JSC-09423, "Apollo Program Summary Report," Apr. 1975, p. A-7.
24. 15 Air-to-Ground, pp. 635–77.
25. MSC, Change of Shift Briefing, July 31, 1971, 6:20 p.m. CDT, transcript.
26. 15 Air-to-Ground, pp. 675–76.
27. Ibid., p. 676.
28. David R. Scott, "Finding the Golden Easter Egg," New York Times, Aug. 13, 1971.
29. 15 Air-to-Ground, p. 732; Apollo 15 Mission Report, p. 14-86.
30. 15 Air-to-Ground, pp. 734–824.
31. Ibid., p. 827.
32. William K. Stevens, "Astronauts Fly to Texas Bearing Moon Rock Cargo," New York Times, Aug. 9, 1971.
33. Apollo 15 Mission Report, pp. 4-1 to 4-3, 14-62 to 14-68; 15 Air-to-Ground, pp. 832–952.
34. MSC, "Change of Shift Briefing, August 1, 1971, 2:40 p.m. (CDT)," transcript.
35. Ibid.
36. 15 Air-to-Ground, p. 994.
37. Ibid., pp. 366–70, 460–67, 512–14, and elsewhere; Apollo 15 Mission Report, p. 5-14; Abigail Brett, "Worden Discovers 'Hot Spot' Of Radioactive Deposits," Washington Post, Aug. 3, 1971.
38. Apollo 15 Mission Report, p. 4-3.
39. 15 Air-to-Ground, p. 1140. Another batch of stamps caused acute embarrassment to NASA and resulted in disciplinary action for the three Apollo 15 astronauts. After a stamp dealer in West Germany advertised philatelic covers that had been to the moon on Apollo 15, it came to light that the crew had taken 400 covers with them. Before the flight the covers, which bore 10-cent Apollo 11 commemorative stamps, were canceled at KSC; on return, the astronauts affixed two 8-cent "decade of achievement" commemoratives (the same as the one canceled on the moon for the U.S. Postal Service) and had them canceled aboard ship. These covers were later autographed by the three astronauts and a notarized statement intended to authenticate them was typed on each. One hundred of these covers were given to the German dealer, who reportedly sold them for 4,850 deutschmarks (about $1,500) each. Belmont Faries, "A Lunar Bonanza," Washington Star, June 18, 1972.

    Investigation revealed that the astronauts had been authorized to carry a smaller number of souvenir covers in their personal preference kits—normal practice for manned space flights, it being understood that such souvenirs would not be exploited for personal profit. When the reports became the subject of inquiries to Deke Slayton and John P. Donnelly, assistant administrator for public affairs at Headquarters, NASA impounded the remaining covers and began an official investigation. William Hines, "NASA Probing Moon Stamp Caper," Washington Star, July 2, 1972.

    It developed that the astronauts had given the flown covers to an intermediary in Florida, who passed them on to the German dealer. It was understood that the proceeds from sale of the covers would be put into a trust fund for the astronauts' children; but they later had second thoughts and refused the offer. Nonetheless, after the basic facts were made known, NASA officially reprimanded Scott, Worden, and Irwin with the statement that they had used "poor judgment" and "their actions will be given due consideration in their selection for future assignments." Harold M. Schmeck, Jr., "Apollo 15 Crew Is Reprimanded," *New York Times*, July 12, 1972; "Apollo 15 'Postmen' Officially Reprimanded," *Houston Post*, July 12, 1972; "Postmark: The Moon," *Newsweek*, July 24, 1972; Thomas O'Toole, "Ex-Astronauts Disregarded Warning Against 'Souvenirs'," *Washington Post*, Aug. 1, 1972. The Senate space committee looked into the matter, concluded that no laws had been violated, and took no action. They discovered that unauthorized items had been taken on other lunar landing missions as well. Richard D. Lyons, "Astronauts and Space Officials Heard At Inquiry on Exploitation of Souvenirs," *New York Times*, Aug. 4, 1972. The results of NASA's own investigation were publicized in an 18-page press release, NASA Release 72-189, "Articles Carried on Manned Space Flights," Sept. 15, 1972, which stated that the astronauts' "official Efficiency Reports as military officers reflect a formal finding of lack of judgment," and also noted changes in NASA Regulations that were promulgated to prevent recurrence of such embarrassing incidents.

The incident caused an uproar among American philatelists. The journal of the American Philatelic Society carried two reports of an investigation intended to make known exactly what happened, so that collectors would be aware of the possibilities of fraud. Lester E. Winick, "Report on Apollo 15 Covers—Smuggled and Authorized," *The American Philatelist* 86(10) (1972):887-95, and "'Lookalike' Apollo 15 Covers Prompt Philatelists' Caution," ibid., 86(11) (Nov. 1972):992-98.

40. 15 Air-to-Ground, p. 1142. Scott correctly credited Galileo Galilei (1564-1642), Italian astronomer and physicist, with the deduction that bodies of different weight should fall with the same velocity in the absence of air resistance. He would have been more accurate to assign the credit for getting him to the moon to Sir Isaac Newton (1642-1727), who first formulated the laws that govern the motion of a body moving in a gravitational field and provided a mathematical basis for Galileo's hypothesis. Only with the development of high-speed electronic computers, however, was it possible to solve the equations for the motion of a spacecraft with sufficient accuracy to allow Scott to land at a specific spot on the plain at Hadley. The only notice taken of Scott's demonstration, other than press reports, was in the preliminary science report. Joe Allen, concluding a paper summarizing the scientific results, noted that the result was "predicted by well-established theory, but [was] nonetheless reassuring considering . . . that the homeward journey was based critically on the validity of the particular theory being tested." *Apollo 15 Preliminary Science Report*, NASA SP-289 (Washington, 1972), p. 2-11.

41. Fred Farrar, "Astronauts Leave Moon, Link Up with Mother Ship," *Chicago Tribune*, Aug. 3, 1971; Apollo 15 Mission Report, pp. 14-71 to 14-72.

42. Tom Wicker, "Man and the Moon," *New York Times*, Aug. 3, 1971.

43. Apollo 15 Mission Report, p. 1-3.

44. Ibid., p. 1-4; John Noble Wilford, "Worden Ventures Outside the Cabin of Apollo Craft," *New York Times*, Aug. 6, 1971.

45. O'Toole, "Apollo 15 Lands Hard but Safely," *Washington Post*, Aug. 8, 1971; Apollo 15 Mission Report, pp. 1-4, 14-17 to 14-21.

46. Apollo 15 Preliminary Examination Team, "The Apollo 15 Lunar Samples: A Preliminary Description," *Science* 175 (1972):363-75.

47. Gerald Wasserburg to Robert R. Gilruth, Aug. 9, 1971.

48. Larry A. Haskin to Gilruth, Aug. 5, 1971.

49. "Apollo 15 Science Briefing, Manned Spacecraft Center, August 11, 1971, 3:30 PM (CDT)."

50. "The Apollo 15 Lunar Samples"; Petrone to multiple addressees, "Report of Preliminary Scientific Results of Apollo 15," Sept. 28, 1971, with encl., "Preliminary Scientific Results of Apollo 15 (As of September 24, 1971)."

51. "The Apollo 15 Lunar Samples."

52. I. Adler, J. Trombka, J. Gerard, R. Schmadebeck, P. Lowman, H. Blodgett, L. Yin, E. Eller, R. Lamothe, P. Gorenstein, and P. Bjorkholm, "X-Ray Fluorescence Experiment, Preliminary Report," Goddard Space Flight Center X-641-71-421, October 1971.

53. Gilruth to Myers, TWX, Aug. 31, 1971.

54. "Distribution of Apollo 15 Lunar Samples," NASA Release 71-223, Nov. 9, 1971.

55. NASA Release 71-149, "Apollo 17 Crew Named," Aug. 13, 1971.

56. Donald K. Slayton interview, Oct. 15, 1984.

57. Jim Maloney, "A new goal for dropped astronaut," *Houston Post*, Sept. 8, 1971.

58. MSC, "Apollo 17 Crew Press Conference, Manned Spacecraft Center, August 19, 1971, 10:00 AM (CDT)," transcript; Reuters, "Geologist Defends Selection of Him for Moon Flight," *New York Times*, Aug. 20, 1971.

59. N. W. Hinners, "Apollo 16 Site Selection," in *Apollo 16 Preliminary Science Report*, prepared by NASA MSC, NASA SP-315 (Washington, 1972).

60. Petrone to Mgr., Apollo Spacecraft Program, MSC, "Selection of Landing Sites for Apollo 16 and 17," Mar. 11, 1971; Lee R. Scherer to multiple addressees, "Apollo 16 and 17 Site Selection Discussions," May 5, 1971, with encl., "Minutes of the Ad Hoc Site Selection Committee Meeting on Preliminary Apollo 16 and 17 Site Selection Discussions," Apr. 14, 1971; Harold C. Urey to Hinners, May 11, 1971; James R. Arnold to Scherer, May 13, 1971; Paul W. Gast to Dir., Apollo Lunar Exploration, "Site for Apollo 16," May 14, 1971; John A. O'Keefe to Hinners, May 14, 1971.

61. Petrone, TWX to multiple addressees, subject: Apollo 16 Landing Site, June 10, 1971; James A. McDivitt to multiple addressees, "Apollo 16 Landing Site," June 11, 1971.

62. MSC, "Apollo 16 Mission Report," MSC-07230, Aug. 1972, pp. 12-1, A-7 to A-22; Richard R. Baldwin, "Mission Description," in *Apollo 16 Preliminary Science Report*.

63. Apollo 16 Mission Report, pp. 3-1, 3-2.

64. Baldwin, "Mission Description." The cable incident was the first to knock out an experiment, but every mission had experienced similar problems with cables that would not lie flat on the lunar surface. Furthermore, the space suits did not permit astronauts to see their own feet. Nonetheless, it caused one columnist to splutter, " . . . what we have been watching [is not] science. Those two klutzes up there on the moon, bumping into each other, unable to repair what their clumsiness has damaged, didn't look like scientists or lab technicians even. They looked like . . . a couple of miscast wahoo military officers." Nicholas von Hoffman, "Two Klutzes On the Moon," *Washington Post*, Apr. 24, 1972. Von Hoffman was rather hyperbolically protesting the continuation of Apollo and the amount of time devoted to coverage of the missions by the television networks.

65. Apollo 16 Mission Report, p. 1-3; O'Toole, "90-Lb. Scientific Satellite Crashes On Moon, Ends Study 47 Weeks Early," *Washington Post*, May 31, 1972.

66. Apollo 16 Mission Report, pp. 5-20 to 5-21, 3-9; Baldwin, "Mission Description."

67. O'Toole, "Scientists Think Apollo 16 Rocks Most Important Haul Yet," *Washington Post*, Apr. 30, 1972.

68. Idem, "Metallic Iron in Rocks Baffles Moon Scientists," ibid., May 17, 1972; "Moon Theory Upset By Study of Rocks Astronauts Gathered," *Wall Street Journal*, May 17, 1972.

69. Hinners, "Apollo 16 Site Selection."

70. Allen L. Hammond, "Lunar Research: No Agreement on Evolutionary Models," *Science* 175 (1972):868-70.

71. Wilford, "Apollo 17 Crew to Seek Questions for Answers Found by Previous Moon Flights," *New York Times*, Nov. 12, 1972.

72. Bruce R. Doe to Homer E. Newell, Oct. 18, 1971.

73. Petrone to multiple addressees, "Apollo 17 Site Selection," Oct. 28, 1971.

74. N. W. Hinners to Capt. W. T. O'Bryant, "A.S.S.B. Minutes," March 23, 1972; O'Toole, "Moon Landing Site Chosen For the Last Apollo Flight," *Washington Post*, Feb. 17, 1972.

75. Hinners to O'Bryant, "A.S.S.B. Minutes"; idem, "Apollo 17 Site Selection," in *Apollo 17 Preliminary Science Report*, NASA SP-330 (Washington, 1973), pp. 1-1 to 1-5; John R. Sevier, interview, Apr. 24, 1986.

76. T. McGetchin and J. W. Head, "Lunar Cinder Cones," Bellcomm paper, Aug. 4, 1972; *Apollo 15 Preliminary Science Report*, p. 25-17.

77. OMSF, "Mission Operation Report, Apollo 17 Mission," report no. M-933-72-17, Nov. 8, 1972; "Apollo 17 Site Selection," NASA Release 72-44, Feb. 16, 1972; Harrison H. Schmitt, "Apollo 17 Report on the Valley of Taurus-Littrow," *Science* 182 (1972):681-90; Apollo 17 Preliminary Examination Team, "Apollo 17 Lunar Samples: Chemical and Petrographic Description," ibid., pp. 659-72.

78. McDivitt to multiple addressees, "Apollo launch dates," Feb. 22, 1971.

79. John O. Cappellari, ed., "Where on the Moon? An Apollo Systems Engineering Problem," *Bell System Technical Journal* 51(5) (May-June 1972):1033-45.

80. Hinners to O'Bryant, "A.S.S.B. Minutes," Feb. 11, 1972; Petrone, TWX to McDivitt, subject: "Apollo 13 March Launch Window," Dec. 4, 1969; Gilruth, TWX to Petrone, subject: "Night Launches," Jan. 28, 1970.
81. McDivitt to multiple addressees, "Apollo 17 Landing Site," Feb. 23, 1972.
82. JSC, "Apollo 17 Mission Report," JSC-07904, March 1973, p. 14-1; Benson and Faherty, *Moonport*, pp. 522-26.
83. Apollo 17 Mission Report, p. 1-1.
84. Ibid., p. 3-2.
85. Ibid., pp. 1-1 to 1-2; Schmitt, "Apollo 17 Report on the Valley of Taurus-Littrow."
86. MSC, "Apollo 17 Technical Air-to-Ground Voice Transcription," Dec. 1972 (hereinafter cited as "17 Air-to-Ground"), tape 76A, pp. 4-8, tape 77A, p. 34, 39-41. This transcript is paginated by tape number; tape numbers run sequentially in three series, A, B, and undesignated, corresponding to communication with the lunar module (A) and the command module (B) when the two were separated, and with the command module when all three crewmen were aboard before and after separation (undesignated).
87. Schmitt, "Apollo 17 Report," p. 682.
88. Apollo 17 Mission Report, pp. 4-5 to 4-11.
89. 17 Air-to-Ground, pp. 91A/1-94A/25; Apollo 17 Mission Report, pp. 10-15 to 10-16; Apollo Field Geology Investigation Team, "Geologic Exploration of Taurus-Littrow: Apollo 17 Landing Site," *Science* 182 (1973):672-80.
90. 17 Air-to-Ground, pp. 96A/8-96A/10.
91. Ibid., pp. 96A/11-96A/55.
92. Apollo 17 Mission Report, pp. 10-16 to 10-17.
93. Ibid.; 17 Air-to-Ground, pp. 112A/20-113A/1.
94. 17 Air-to-Ground, pp. 113A/1-3.
95. "Apollo 17 at Taurus-Littrow," NASA EP-102, p. 30.
96. Apollo 17 Mission Report, pp. 3-2, 12-4, 12-5.
97. Ibid., p. 1-3.
98. William Hines, "End of a crazy business," *Chicago Sun-Times*, Dec. 21, 1972. Hines went on to castigate the Apollo project from concept to execution, and concluded by quoting Max Born, Nobel laureate in physics, who had said in 1958 as the "space race" began, "Intellect distinguishes the possible from the impossible; reason distinguishes the sensible from the senseless. Spaceflight is a triumph of intellect and a tragic failure of reason."
99. *Christian Science Monitor*, Dec. 21, 1972, cited in *Astronautics and Aeronautics, 1972: Chronology of Science, Technology and Policy*, NASA SP-4017 (Washington, 1974) , p. 440.
100. *New York Times*, Dec. 14, 1972, cited in *Astronautics and Aeronautics, 1972*, p. 429.
101. *Time*, Dec. 11, 1972, cited in *Astronautics and Aeronautics, 1972*, p. 424.

CHAPTER 14

1. W. David Compton and Charles D. Benson, *Living and Working in Space: A History of Skylab*, NASA SP-4208 (Washington, 1983), p. 231, 247, 220.
2. Edward C. Ezell and Linda N. Ezell, *The Partnership: A History of the Apollo-Soyuz Test Project*, NASA SP-4209 (Washington, 1978), p. 247; "ASTP Crew Named," MSC *Roundup*, Feb. 2, 1973.
3. "Moon Rocks Issued," ibid.
4. "Post-Apollo Lunar Science," Report of a study by the Lunar Science Institute, July 1972.
5. Minutes, Apollo Lunar Sample Analysis Planning Team meeting, Jan. 29-Feb. 2, 1973.
6. Ibid.
7. John Noble Wilford, "Expert Describes Lunar Cataclysm," *New York Times*, Mar. 6, 1973.
8. Robert Gillette and Allen L. Hammond, "Lunar Science: Letting Bygones be Bygones," *Science* 179 (1973):1309; Noel W. Hinners, TWX to A. J. Calio, "Outline of Dr. Low's Talk for 4th Lunar Science Conference," Mar. 2, 1973.

9. Wilford, "Expert Describes Cataclysm"; Boyce Rensberger, "Moon Dead for 3 Billion Years, Experts Say," *New York Times*, Feb. 16, 1973; J. C. Huneke, E. K. Jessberger, F. A. Podosek, and G. J. Wasserburg, "[40]Ar/[39]Ar Measurements in Apollo 16 and 17 Samples and the Chronology of Metamorphic and Volcanic Activity in the Taurus-Littrow Region," *Proceedings of the Fourth Lunar Science Conference* (New York: Pergamon Press, 1973), pp. 1724–56.

10. Minutes, Lunar Sample Analysis Planning Team meeting.

11. Hammond, "Lunar Science: Analyzing the Apollo Legacy," *Science* 179 (1973):1313–15.

12. Gillette and Hammond, "Lunar Science: Letting Bygones be Bygones."

13. Robert Jastrow, "Moon Still Is A Generally Silent Witness," *New York Times*, Mar. 24, 1974.

14. G. J. Wasserburg, "The Moon and Sixpence of Science," *Astronautics and Aeronautics* 10(4) (Apr. 1972):16–21.

15. Material in this and the next paragraph is a synthesis of information from several sources: "Planetology," L. S. Walter, B. M. French, and P. D. Lowman, eds., in *Significant Achievements in Space Science, 1967*, NASA SP-167 (Washington, 1968), pp. 326–52; Harold Urey, "The Contending Moons," *Astronautics and Aeronautics* 7(1) (Jan. 1969):37–41; A. L. Turkevich, W. A. Anderson, T. E. Economou, E. J. Franzgrote, H. E. Griffin, S. L. Grotch, J. H. Patterson, and K. P. Sowinski, "The Alpha-Scattering Chemical Analysis Experiment on the Surveyor Lunar Missions," in *Surveyor Program Results*, NASA SP-184 (Washington, 1969), pp. 271–350; Wasserburg, "The Moon and Sixpence of Science."

16. Gary V. Latham, Maurice Ewing, Frank Press, George Sutton, James Dorman, Yosio Nakamura, Nafi Toksoz, David Lammlein, and Fred Duennebier, "Passive Seismic Experiment," in *Apollo 16 Preliminary Science Report*, NASA SP-315 (Washington, 1972), p. 9–1. This massive chunk of rock struck approximately 145 kilometers (90 miles) north of the Apollo 14 station on May 13, 1972.

17. Hammond, "Lunar Science: Analyzing the Apollo Legacy."

18. Ibid.; Wasserburg, "The Moon and Sixpence of Science."

19. Material in the following paragraphs is based primarily on Harrison H. Schmitt's summary, "Evolution of the Moon: The 1974 Model," in *The Soviet-American Conference on Cosmochemistry of the Moon and Planets*, John H. Pomero and Norman J. Hubbard, eds., NASA SP-370 (Washington, 1977), pp. 63–80, and S. Ross Taylor's *Lunar Science: A Post-Apollo View* (New York: Pergamon Press, 1975).

20. The Lunar Geoscience Working Group, *Status and Future of Lunar Geoscience*, NASA SP-484 (Washington, 1986), pp. 3–32. Members of this working group were: P. D. Spudis, U.S. Geological Survey, chairman; B. R. Hawke, Univ. of Hawaii; L. L. Hood, Univ. of Arizona; P. H. Schultz, Brown Univ.; G. J. Taylor, Univ. of New Mexico; and D. E. Wilhelms, U.S. Geological Survey. This 54-page booklet provides a brief summary of the current understanding of the moon and a prospectus for future lunar exploration (manned and unmanned). It also contains a 5-page scientific bibliography covering recent work in lunar science.

21. Wasserburg, "The Moon and Sixpence of Science."

22. Ibid.

23. Ibid.

24. R. Cargill Hall, *Lunar Impact: A History of Project Ranger*, NASA SP-4210 (Washington, 1977), pp. 308–309.

25. L. D. Jaffe and others, "Principal Scientific Results From the Surveyor Program, in *Surveyor Program Results*, NASA SP-184 (Washington, 1969), pp. 13–17. Wasserburg remarks ("The Moon and Sixpence of Science") that with hindsight, the clues from Surveyor provided broad answers to the question of the moon's internal structure.

26. *Astronautics and Aeronautics, 1970: Chronology on Science, Technology, and Policy*, NASA SP-4015 (Washington, 1972), pp. 52–53; ibid., 1972, NASA SP-4017 (Washington, 1974), pp. 52–53.

27. Wasserburg, "The Moon and Sixpence of Science."

28. Courtney G. Brooks, James M. Grimwood, and Loyd S. Swenson, Jr., *Chariots for Apollo: A History of Manned Lunar Spacecraft*, NASA SP-4205 (Washington, 1979), p. xv.

29. Ibid., p. 31.

30. Homer E. Newell, *Beyond the Atmosphere: Early Years of Space Science*, NASA SP-4211 (Washington, 1980), p. 246. Newell also commented that the Houston center "developed an arrogance born of unbounded self-confidence and possession of a leading role in the nation's number-one space project, Apollo."

31. Ibid., pp. 204–205.

32. Brooks, Grimwood, and Swenson, *Chariots*, p. 127.

33. *NASA 1965 Summer Conference on Lunar Exploration and Science*, NASA SP-88 (Washington, 1965), p. 2.

34. 1967 *Summer Study of Lunar Science and Exploration*, NASA SP-157 (Washington, 1967), p. iii.

35. Michael Collins, *Carrying the Fire: An Astronaut's Journeys* (New York: Farrar, Straus and Giroux, 1974), p. 360.

36. *A Review of Space Research*, report of the Summer Study conducted under the auspices of the Space Science Board of the National Academy of Sciences at the State University of Iowa, June 17–August 10, 1962, National Academy of Sciences-National Research Council Publication 1079 (Washington, 1962), p. 11–19. A questionnaire sent to several prominent scientists before the study sought opinions on the place of scientists in manned space flight. To the question, "How do we develop our astro-scientists?" the consensus of replies included the statement that qualified scientists "should go through astronaut training for part of each year to become familiar with problems of space flight. It is hoped that this would not involve too large a fraction of their time, since emphasis should be on their development as scientists."

37. An unnamed Marshall Space Flight Center engineer paraphrased Slayton's attitude as, "it is easier to teach an astronaut to pick up rocks than to teach a geologist to land on the moon." Minutes of the combined MSFC staff and board meeting, Nov. 10, 1969. This statement should be read alongside the scientists' estimate of the difficulties of space flight training cited in 36.

38. Harrison H. Schmitt interview, May 30, 1984. Elbert King, who was involved with crew training for the earlier missions, confirmed Schmitt's evaluation of Armstrong's ability as an observer: "Armstrong . . . never said much, was kind of quiet on field trips, and we really didn't know how much of this he was soaking up. . . . It turned out that he had really picked up a lot of it and was doing very well with it." Elbert H. King, Jr., interview with Loyd S. Swenson, Jr., May 27, 1971, tape in JSC History Office files.

39. Schmitt interview.

40. See, for example, the report of a series of science workshops sponsored by NASA's Ames Research Center, *The Search for Extraterrestrial Intelligence: SETI*, NASA SP-419 (Washington, 1977). One group stated the central proposition this way: "The conclusion that the origin and evolution of life is inextricably interwoven with the origin and evolution of the cosmos seems inescapable" (p. 41). The question arose again in the Viking mission to Mars, which carried experiments designed to detect life. Some scientists, it seems, desperately wanted to detect life on Mars. See Edward Clinton Ezell and Linda Neuman Ezell, *On Mars: Exploration of the Red Planet 1958–1978*, NASA SP-4212 (Washington, 1984) , pp. 409–14.

41. Richard S. Johnston, Lawrence F. Dietlein, M.D., and Charles A. Berry, M.D., eds., *Biomedical Results of Apollo*, NASA SP-368 (Washington, 1975), pp. 410–11, 418.

42. William Hines, "End of a Crazy Business," *Chicago Sun-Times*, Dec. 21, 1972.

43. Foy D. Kohler and Dodd L. Harvey, "Administering and Managing the U.S. and Soviet Space Programs," *Science* 169 (1970):1049–55.

44. John Noble Wilford, "A Spacefaring People: Keynote Address," in Alex Roland, ed., *A Spacefaring People: Perspectives on Early Spaceflight*, NASA SP-4405 (Washington, 1985), proceedings of a conference on the history of space activity held at Yale University, Feb. 6–7, 1981.

APPENDIX 1

ORGANIZATION CHARTS:
NASA, OMSF, OSSA, MSC

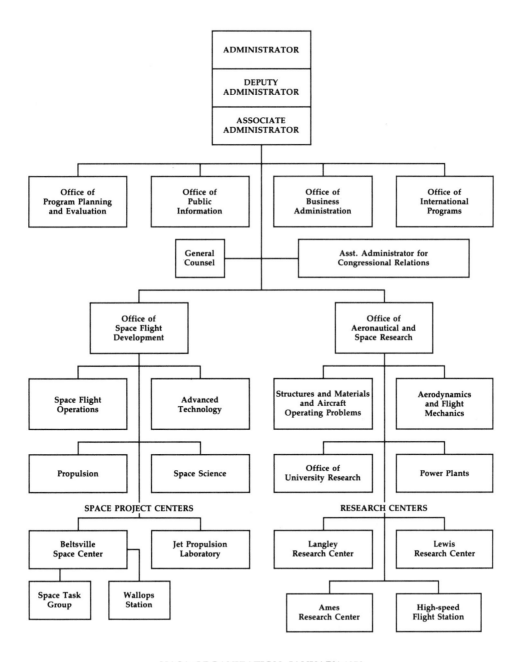

```
                        ┌─────────────────────┐
                        │   ADMINISTRATOR     │
                        ├─────────────────────┤
                        │      DEPUTY         │
                        │   ADMINISTRATOR     │
                        ├─────────────────────┤
                        │    ASSOCIATE        │
                        │   ADMINISTRATOR     │
                        └─────────────────────┘
```

| Office of Program Planning and Evaluation | Office of Public Information | Office of Business Administration | Office of International Programs |

| General Counsel | Asst. Administrator for Congressional Relations |

| Office of Space Flight Development | Office of Aeronautical and Space Research |

| Space Flight Operations | Advanced Technology | Structures and Materials and Aircraft Operating Problems | Aerodynamics and Flight Mechanics |

| Propulsion | Space Science | Office of University Research | Power Plants |

SPACE PROJECT CENTERS   **RESEARCH CENTERS**

| Beltsville Space Center | Jet Propulsion Laboratory | Langley Research Center | Lewis Research Center |

| Space Task Group | Wallops Station | Ames Research Center | High-speed Flight Station |

NASA ORGANIZATION, JANUARY 1959

313

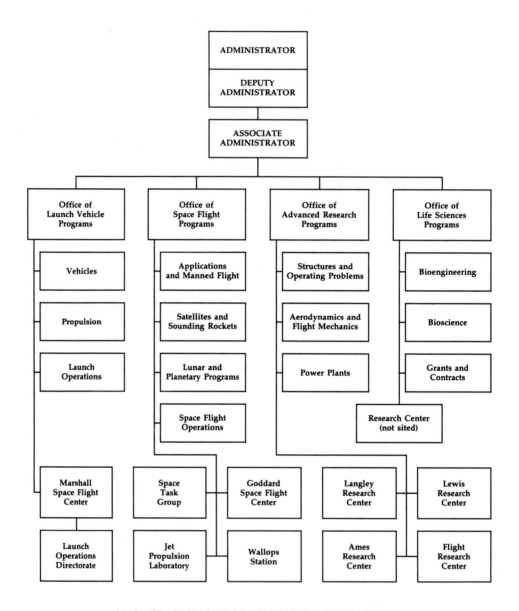

NASA SPACE FLIGHT ORGANIZATION, JANUARY 1961

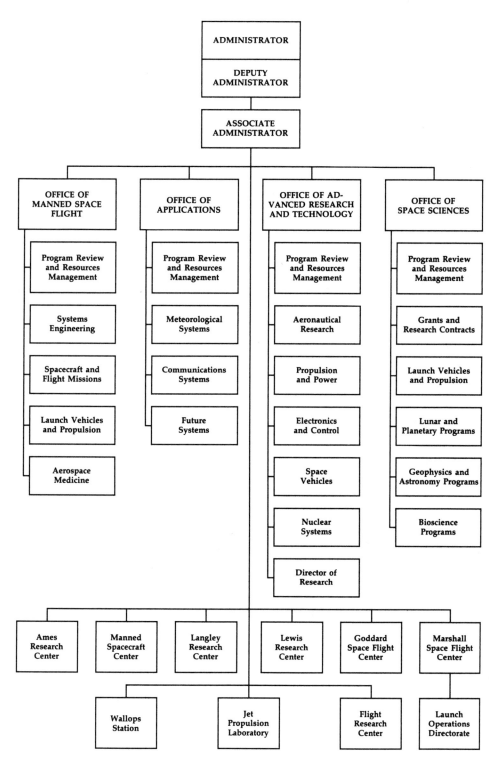

NASA SPACE FLIGHT ORGANIZATION, NOVEMBER 1961

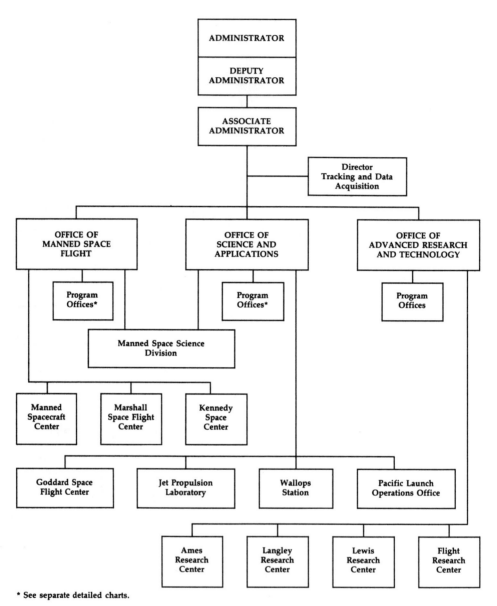

* See separate detailed charts.

NASA SPACE FLIGHT ORGANIZATION, NOVEMBER 1963

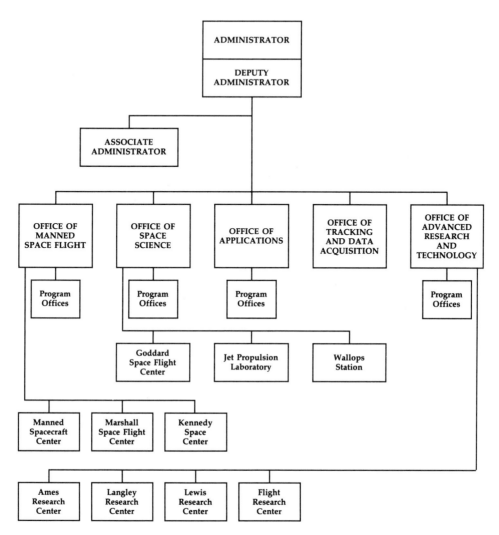

NASA HEADQUARTERS SPACE FLIGHT ORGANIZATION, JANUARY 1972

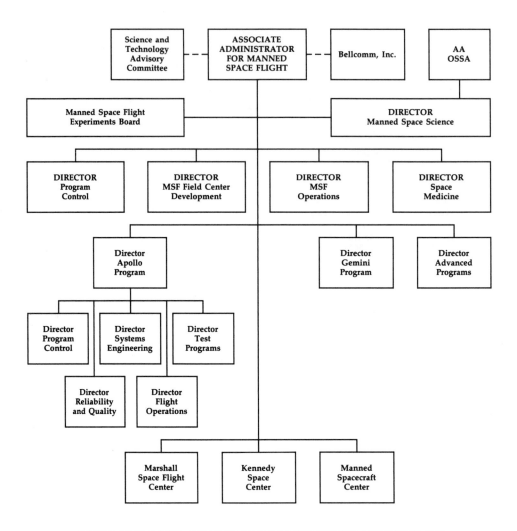

OFFICE OF MANNED SPACE FLIGHT ORGANIZATION, FEBRUARY 1964

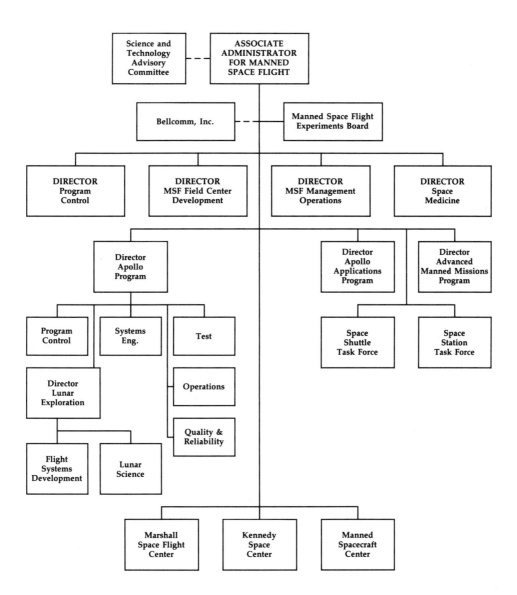

OFFICE OF MANNED SPACE FLIGHT ORGANIZATION, 1969

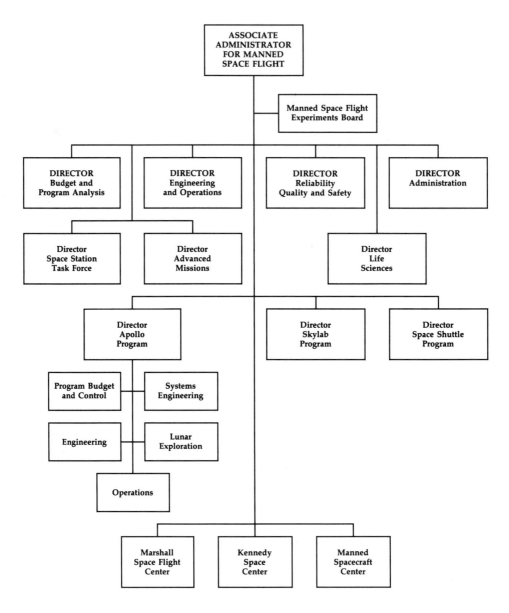

OFFICE OF MANNED SPACE FLIGHT ORGANIZATION, MARCH 1972

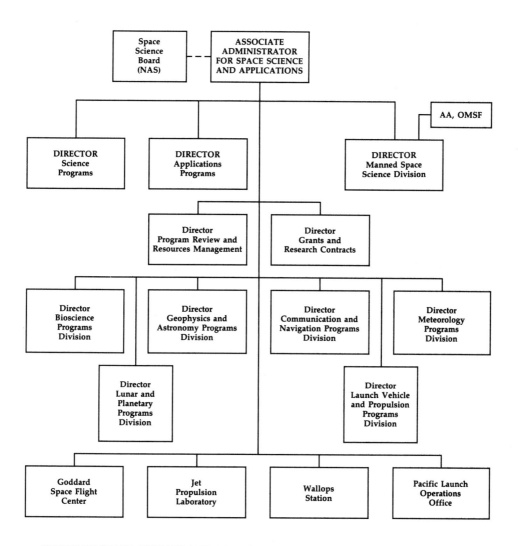

OFFICE OF SPACE SCIENCE AND APPLICATIONS ORGANIZATION, FEBRUARY 1964

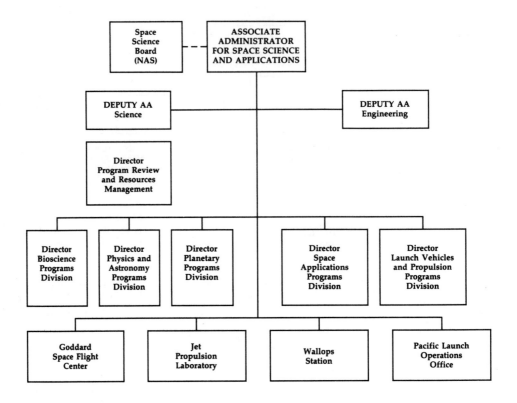

OFFICE OF SPACE SCIENCE AND APPLICATIONS ORGANIZATION, JUNE 1969

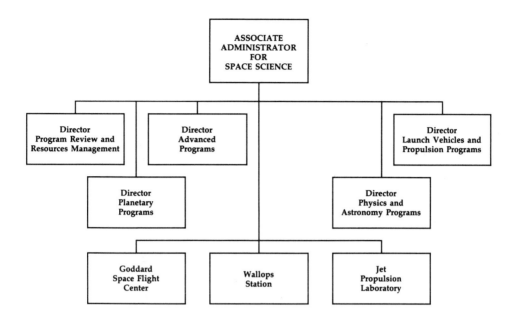

OFFICE OF SPACE SCIENCES ORGANIZATION, MARCH 1972

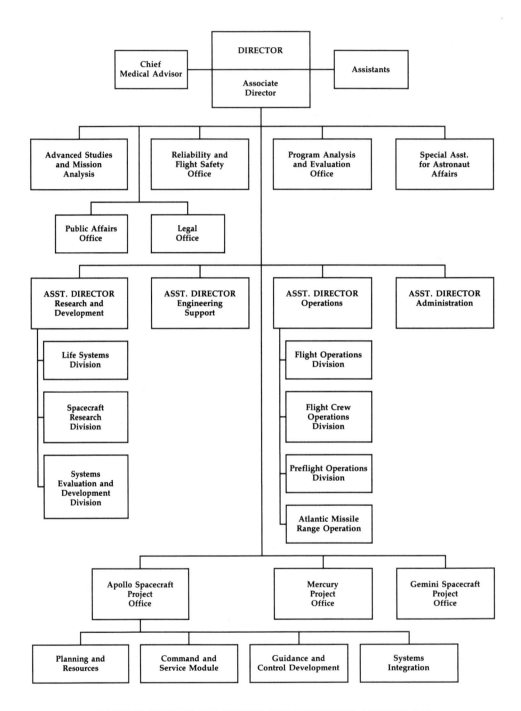

MANNED SPACECRAFT CENTER ORGANIZATION, AUGUST 1962

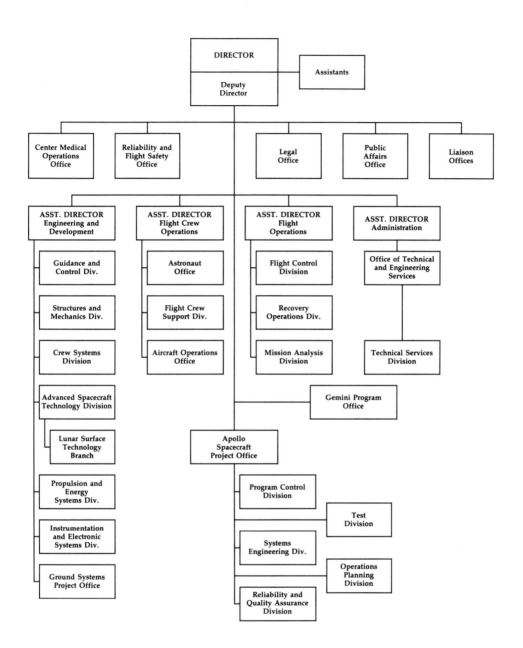

MANNED SPACECRAFT CENTER ORGANIZATION, JANUARY 1964

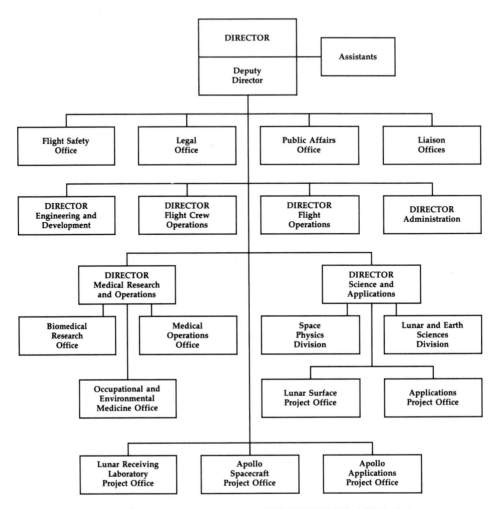

MANNED SPACECRAFT CENTER ORGANIZATION, APRIL 1967

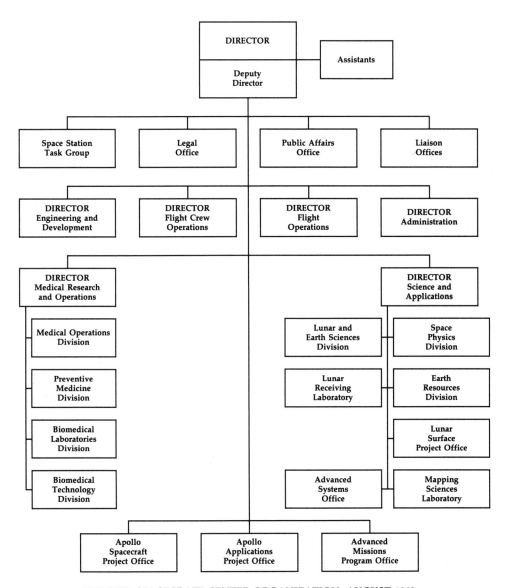

MANNED SPACECRAFT CENTER ORGANIZATION, AUGUST 1969

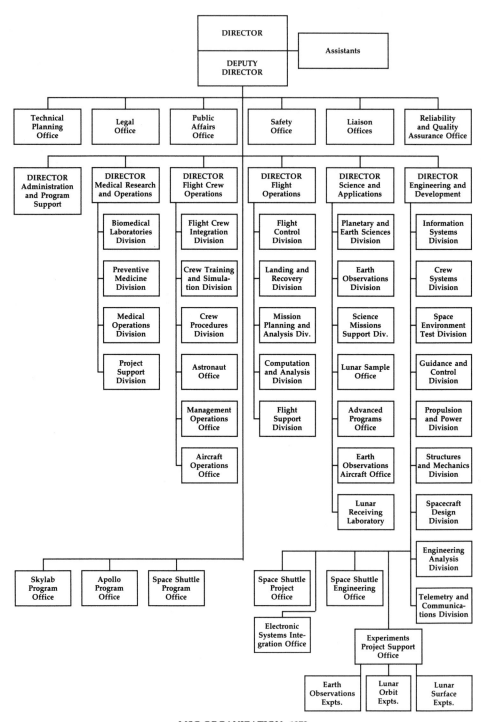

MSC ORGANIZATION, 1972

APPENDIX 2

APOLLO FUNDING HISTORY

Apollo Funding History*

Fiscal Year				Funding Breakdown (thousands of dollars)	
1962					
Original budget request	NASA:	$1,235,300	Oribital flight tests	$	63,900
	Apollo:	29,500	Biomedical flight tests		16,550
			High-speed reentry tests		27,550
Final budget appropriation	NASA:	1,671,750	Spacecraft development		52,000
	Apollo:	160,000			
1963					
Original budget request	NASA:	$3,787,276	Command/service module	$	345,000
including FY 1962 supplemental	Apollo:	617,164	Lunar module		123,100
			Guidance & navigation		32,400
			Instrumentation & scientific equipment		11,500
Final budget appropriation	NASA:	3,674,115	Operational Support		2,500
with FY 1962 supplemental	Apollo:	617,164	Supporting development		3,000
			Little Joe II devl.		8,800
			Saturn I		90,864
1964					
Original budget request	NASA:	$3,926,000	Command/service module	$	545,874
including FY 1963 supplemental	Apollo:	2,243,900	Lunar module		135,000
			Guidance & navigation		91,499
			Integration, reliability, & checkout		60,699
Fiscal budget appropriation	NASA:	3,974,979	Spacecraft support		43,503
with FY 1963 supplemental	Apollo:	2,272,952	Saturn I		187,077
			Saturn IB		146,817
			Saturn V		763,382
			Engine development		166,000
			Apollo mission support		133,101
1965					
Original budget request	NASA:	$4,523,000	Command/service module	$	577,834
including FY 1964 supplemental	Apollo:	2,818,500	Lunar module		242,600
			Guidance & navigation		91,499
			Integration, reliability, & checkout		24,763
Fiscal budget appropriation	NASA:	4,270,695	Spacecraft support		83,663
with FY 1964 supplemental	Apollo:	2,614,619	Saturn I & IB		302,955
			Saturn V		964,924
			Engine development		166,300
			Apollo mission support		170,542
1966					
Original budget request—no	NASA:	$4,575,900	Command/service module	$	615,000
supplemental for prior fiscal year	Apollo:	2,997,385	Lunar module		310,800
			Guidance & navigation		115,000
			Integration, reliability, & checkout		34,400
Fiscal budget appropriation—no	NASA:	4,511,644	Spacecraft support		95,400
supplemental for prior fiscal year	Apollo:	2,967,385	Saturn I/IB		274,985
			Saturn V		1,177,320
			Engine development		134,095
			Apollo mission support		210,385

*Source: *The Apollo Spacecraft: A Chrolonogy*, vols. 1–4, NASA SP-4009 (Washington, 1969–1978).

Fiscal Year				Funding Breakdown (thousands of dollars)	
1967					
Original budget request	NASA:	$4,246,600		Command/service module	$ 560,400
	Apollo:	2,974,200		Lunar module	472,500
				Guidance & navigation	76,654
Fiscal budget appropriation	NASA:	4,175,100		Integration, reliability, & checkout	29,975
	Apollo:	2,916,200		Spacecraft support	110,771
				Saturn IB	236,600
				Saturn V	1,135,600
				Engine development	49,800
				Apollo mission support	243,900
1968					
Original budget request	NASA:	$4,324,500		Command/service module	$ 455,300
including FY 1967 supplemental	Apollo:	2,606,500		Lunar module	399,600
				Guidance & navigation	113,000
				Integration, reliability, & checkout	66,600
Final budget appropriation	NASA:	3,970,000		Spacecraft Support	60,500
including FY 1967 supplemental	Apollo:	2,556,000		Saturn IB	146,600
				Saturn V	998,900
				Engine development	18,700
				Apollo mission support	296,800
1969					
Original budget request	NASA:	$3,677,000		Command/service module	$ 282,821
	Apollo:	2,038,800		Lunar module	326,000
				Guidance & navigation	43,900
				Integration, reliability, & checkout	65,100
Fiscal budget appropriation	NASA:	3,193,559		Spacecraft support	121,800
	Apollo:	2,025,000		Saturn IB	41,347
				Saturn V	534,453
				Manned space flight operations	546,400
1970					
Original budget request	NASA:	$3,168,900		Command/service module	$ 282,821
including FY 1969 supplemental	Apollo:	1,651,100		Lunar module	231,433
				Guidance & navigation	33,866
				Science payloads	60,094
Fiscal budget appropriation	NASA:	3,113,765		Spacecraft support	170,764
with FY 1969 reserve	Apollo:	1,686,145		Saturn V	484,439
				Manned space flight operations	314,693
				Advanced development	11,500
1971					
Original budget request	NASA:	$2,606,100		Flight modules	$ 245,542
	Apollo:	956,500		Science payloads	106,194
				Ground support	46,411
				Saturn V	189,059
Fiscal budget appropriation	NASA:	2,555,000		Manned space flight operations	314,963
	Apollo:	913,669		Advanced development	11,500
1972					
Original budget request	NASA:	$2,517,700		Flight modules	$ 55,033
	Apollo:	612,200		Science payloads	52,100
				Ground support	31,659
				Saturn V	142,458
Fiscal budget appropriation	NASA:	2,507,700		Manned space flight operations	307,450
	Apollo:	601,200		Advanced development	12,500
1973					
Original budget request	NASA:	$2,600,900		Spacecraft	$ 50,400
	Apollo:	128,700		Saturn V	29,309
Fiscal budget appropriation	NASA:	2,509,900			
	Apollo:	76,700			

APPENDIX 3

LUNAR EXPLORATION PLANNING, 1961–1967

Scientific Questions in Lunar Exploration*

Major questions in the exploration of the Moon fall chiefly in three categories of basic problems: 1) structure and processes of the lunar interior, 2) the composition and structure of the surface of the Moon and the processes modifying the surface, and 3) the history or evolutionary sequence of events by which the Moon has arrived at its present configuration. The possibility that ancient rocks and deposits on the Moon's surface may contain a unique record of events related to the formation or accretion of the terrestrial planets gives the scientific exploration of the moon unusual potential significance. There is also the minor possibility of finding prebiotic material. . . .

The major questions are as follows:

Structure and Processes of the Lunar Interior
   (1) Is the internal structure of the Moon radially symmetrical like the Earth, and if so, is it differentiated? Specifically, does it have a core and does it have a crust?
   (2) What is the geometric shape of the Moon? How does the shape depart from fluid equilibrium? Is there a fundamental difference in morphology and history between the sub-Earth and averted faces of the Moon?
   (3) What is the present internal energy regime of the Moon? Specifically, what is the present heat flow at the surface and what are the sources of this heat? Is the Moon seismically active and is there active volcanism? Does the Moon have an internally produced magnetic field?

Composition, Structure, and Processes of the Lunar Surface
   (1) What is the average composition of the rocks at the surface of the Moon and how does the composition vary from place to place? Are volcanic rocks present on the surface of the Moon?
   (2) What are the principal processes responsible for the present relief of the lunar surface?
   (3) What is the present tectonic pattern on the Moon and distribution of tectonic activity?

---

*Listed here are the "Fifteen Questions" that formed the basis for subsequent planning in lunar exploration. Source: National Academy of Sciences–National Research Council Publication 1403, *Space Research: Directions for the Future* (Washington, 1966), pp. 21-22.

(4) What are the dominant processes of erosion, transport, and deposition of material on the lunar surface?

(5) What volatile substances are present on or near the surface of the Moon or in a transitory lunar atmosphere?

(6) Is there evidence of organic or proto-organic materials on or near the lunar surface? Are living organisms present beneath the surface?

History of the Moon

(1) What is the age of the Moon? What is the range of age of the stratigraphic units on the lunar surface and what is the age of the oldest exposed material? Is a primordial surface exposed?

(2) What is the history of dynamical interaction between the Earth and the Moon?

(3) What is the thermal history of the Moon? What has been the distribution of tectonic and possible volcanic activity in time?

(4) What has been the flux of solid objects striking the lunar surface in the past and how has it varied with time?

(5) What has been the flux of cosmic radiation and high-energy solar radiation over the history of the Moon?

(6) What past magnetic fields may be recorded in the rocks at the Moon's surface?

Recommendations of the 1965 Summer Conference on Lunar Exploration and Science, Falmouth, Mass., July 19–31, 1965

The Falmouth conference of 1965 made the first concerted effort to define a systematic 10-year program of lunar exploration, giving primary emphasis to manned exploration. Working groups were established in the disciplines of geodesy/cartography, geology, geophysics, bioscience, geochemistry (mineralogy and petrology), particles and fields, lunar atmosphere measurements, and astronomy. At the conclusion of the conference each disciplinary working group prepared a report, from which a summary was prepared by a coordinating committee. Following are excerpts from the summary, which set forth the major requirements for the program.*

MISSION SUMMARY

INTRODUCTION

In this chapter the major recommendations of the Working Groups are arranged by missions or programs. The suggested priorities, instrument allocations, and mission characteristics for various vehicles are indicated briefly. . . .

Although there was some overlap, most of the recommendations could be divided into these missions: Apollo, Lunar Orbiter, Apollo Extension System–Manned Lunar Orbiter (AES-MLO), Apollo

---

*NASA 1965 Summer Conference on Lunar Exploration and Science, NASA SP-88 (Washington, 1965).

Extension System–Manned Lunar Surface (AES-MLS) and Post–AES. . . . As the scientific instruments and space vehicle characteristics and availability become more clearly defined, the assignment of experiments will become clearer.

OVERALL PROGRAM

The plans and recommendations of the 1965 Lunar Exploration Summer Conference are based on a 10-year program of exploration, beginning with the first manned lunar landing in the Apollo program. The recommendations of this conference are limited to the 10-year period following the first Apollo lunar landings because a decade seems to be the approximate maximum time for which developments can be meaningfully forecast. In addition, the long lead times involved in the development of equipment for use in space flight require that recommendations be made to cover this period of time. In carrying out these recommendations, it will be necessary for the National Aeronautics and Space Administration to conduct its programs in a way that permits a maximum degree of flexibility to meet changing requirements.

The need for flexibility is also important for determining the rate at which missions should be conducted during this ten-year period. It is clearly desirable to schedule flight missions to provide adequate time between missions to react to the findings of one mission by modifying experiments for a later mission. One method of accomplishing this is through the modular construction of individual experiments. It is also desirable to program certain types of experiments so that the time of the operation of the devices used on one mission will overlap with the operating time of devices operated on a subsequent mission. This will provide not only scientific continuity in the experiments, but also simultaneous data from a multiplicity of lunar locations.

In addition, overall program planning considerations dictate the scheduling of missions at relatively close intervals. It is clearly desirable to maintain a certain degree of program momentum, both for psychological reasons and to make certain that the project personnel analyzing the results are provided with definite goals over a relatively long period of time.

On consideration of all of these factors, it was the feeling of the Working Groups that the National Aeronautics and Space Administration should schedule lunar surface missions at a minimum rate of one per year or possibly two through 1974. Lunar orbital missions should be conducted at the rate of one per year. Since many of the lunar surface missions will require two flights each, three to five Apollo/Saturn V vehicles are required annually.

Present indications are that lunar exploration should continue at the same rate in the latter half of the 1970s. However, there seems to be no need now to plan the flight mission assignment schedule for that period of time.

THE EARLY APOLLO MISSIONS

It is assumed that at least the first missions, with durations limited to a day or two and with exploration limited to an area close to the point of landing, will be dominated by operational considerations. Since the highest mission priority is assigned to the safety of the astronauts, the bulk of their time and attention will be devoted to perfecting the procedures of flight. In the relatively short duration of these early missions, the time assigned to scientific lunar exploration as such will be limited.

All experiments should be designed to conserve the astronauts' time, the most valuable scientific commodity on the early missions.

As flight procedures and techniques are perfected, and as improved flight equipment becomes available, plans should be made to gradually increase the duration of stays on the Moon, the distance traveled from the point of landing, and the proportion of the astronauts' time devoted to lunar exploration. A judicious use of "manned" and "unmanned" spacecraft will be required to obtain maximum coverage. Recommendations were made by some groups concerning the specific instrumentation to be carried on each of the first three landings.

Training of the astronauts in sampling techniques and field geology is of the utmost importance to insure the intelligent collection of samples. To assure collection of sterile samples, training is required for the astronauts in the nature of contamination, transfer of contaminants, aseptic transfer procedures, and chemical cleanliness.

Priorities for Experiments

The highest priority activity for the early Apollo landings is to return the greatest number and variety of samples as is feasible. It is desirable that all samples be kept sterile and free of chemical contaminants from such sources as the LEM fuels, the LEM atmosphere or the outgassing or leakage of the astronaut's suit. A variety of easily obtainable samples should be collected, ranging from dust to rock sizes. These samples should be taken as far from the LEM as possible. Both surface and sub-surface samples are required. In the event of a semiaborted or shortened stay on the lunar surface, the astronaut's first scientific duty is the collection of as many samples as possible, without regard to sterility.

The second priority for the Early Apollo is the emplacement of the Lunar Surface Experiment Package (LSEP) by the astronauts. This should be emplaced to attain optimum operating conditions. Next in priority are the lunar geological traverses by the astronauts. If feasible, these should be accurately controlled with automatic procedures and monitoring. The description of topographic and geologic relations along the traverse lines should be supplemented by stereoscopic photographs.

Equipment Requirements

The working groups were asked to consider equipment priority beginning with the most important. Weights assigned are the absolute minimum needed to accomplish the task. In some cases (i.e., sample tools), added weight and complexity would be desirable if weight and space were available. This priority list is as follows:
1. Sample containers (10 lb.) should keep samples sterile and chemically clean. Stainless steel is acceptable. More studies should be completed relative to the use of Teflon in the lunar environment.
2. Sampling tools (10 lb.) should be easily operable, light, and simple; e.g., space hardened rock hammer, rubber mallet, sun compass.
3. Aseptic sample collection tool (10 lb.).
4. Photography (7 lb.): A stereoscopic camera with several filters and polarizing lenses.
5. LSEP experiments: Suggested experiments in order of priority are: a passive seismograph (25 lb.); a magnetometer/particle detector (13 lb.) and a heat flow measurement device (15 lb.); an active seismograph experiment (7–10 lb.); a micrometeoroid detector (15 lb.). The gravimeter should be considered for AES because of its weight and development complexities.

Preliminary Studies

Studies and tests should be started immediately to determine the amounts and effects of the outgassing of the astronauts' suits and the escape of the atmosphere from the LEM. Sterilization of the escaping atmosphere from the LEM should be considered. Analyses of the possible contaminants in the LEM fuel and the effects on sample collection should be undertaken.

Sample Investigations

Upon return of the lunar samples to Earth, they will be prepared at a Lunar Sample Receiving Laboratory (LSRL) for distribution. Here they will be logged in, cataloged, checked for outgassing, measured for low level radiation, and examined for pathogenic agents. Only those tests that must be done immediately will be conducted at the LSRL. The portion of samples to be distributed will be packaged and initial distribution to the selected scientific investigators will be made.

EARLY LUNAR ORBITERS

In the period before the end of the decade two classes of missions are scheduled.

Unmanned Missions

These missions will begin in the period of 1966–1967. The primary function for the approved missions is site selection, and hence, low altitude orbiters are desirable. However, should particles and field experiments be included on later flights, altitudes of 150 to 2000 km are required. Following are some specific recommendations for inclusion on these flights:
1. *Cartography.* Stereophotogrammetric analysis of the photography obtained by the first block of Lunar Orbiters should be carried out to obtain information regarding the character of lunar topography and to gain experience in analyzing lunar photography. It is recommended that later first block Lunar Orbiters, or any second block, be placed in orbits of different inclinations, with priority for a polar orbit.
2. *Particles and Fields.* A three-axis magnetometer and particle package to study day-night changes in particle and field environment should be included. For these experiments the spacecraft should be radioactvely and magnetically clean.

3. *Lunar Atmospheres.* Pressure, flux, and mass measurements for determination of neutral and ionic constituents should be conducted. This study is advantageous for early flights because of the uncontaminated state of the atmosphere.

Manned Missions

Consideration should be given to the inclusion of simple diagnostic experiments to be conducted from the orbiting Command/Service Module (CSM) in conjunction with Apollo experiments on the lunar surface.

AES MANNED LUNAR ORBITER (AES-MLO)

Role of AES-MLO

Since less than 1 percent of the lunar surface will be visited in the near future, a major source of scientific knowledge will come from orbiting spacecraft. Extensive information can be easily obtained for the following reasons:
  1. Variation of orbital inclinations will permit mapping of the entire lunar surface.
  2. Absence of atmosphere allows all regions of the electromagnetic spectrum to be accessible.

Thus, Manned Lunar Orbiters can provide useful additions to the subsurface information obtained from geophysical studies and to the local studies from fixed and traversing surface experiments. Because of the orbiter's nature, however, they cannot provide information with the same precision and detail as the surface instruments. Orbital investigations offer the potential of identifying local areas of the lunar surface with unusual properties that might be of interest for future manned exploration.

It is recommended that one orbiter carrying the remote sensing package be flown before the first AES landings. Polar orbiters are desirable at the earliest time for total coverage of the lunar surface by remote sensors. In the first phase of lunar exploration (complete orbital survey), five or six missions are believed to be appropriate. Subsequently, AES–MLOs will be necessary to support and supplement the AES surface operations, as well as to monitor lunar activity. Launch rates should be approximately one a year.

A systematic program of geologic mapping using orbital data is recommended with the preparation of geologic maps at the following scales:
  1. 1:2,500,000. Synoptic map for general planning and collating of a wide variety of data about the gross features of the Moon.
  2. 1:1,000,000. Complete synoptic geologic mapping.
  3. 1:250,000. Total coverage of the lunar surface. This is a long-range goal.
  4. 1:100,000; 1:25,000. Special purpose, directed toward solution of selected topical problems.

Photography Equipment

For complete photographic coverage of the lunar surface the lunar Orbital Camera System should include the following camera subsystems:
  1. Metric (Mapping) Camera Subsystem. Designed for use in determining the lunar figure and mapping of the lunar surface.
  2. High Resolution Twin Convergent Panoramic Camera Subsystem. Designed for photography of the highest resolution in keeping with coverage of large areas.
  3. Ultrahigh Resolution Camera Subsystem. Designed to produce photography of extremely high resolution and multispectral response of limited areas.
  4. Multiband Synoptic Camera Subsystem. A means for obtaining large areas of multispectral photographic coverage of the reflective properties of the lunar surface in the visible and near-visible portion of the spectrum.

Remote Sensors

Imaging sensors will provide information about surface structure and composition from depths of microns to a few meters. Imaging instruments, including UV [ultraviolet] imagers, IR [infrared] imagers, and high-resolution radars have proven value for surface and near-surface structure and composition studies and for the study of thermal anomalies. Coverage of the entire Moon with these instruments is recommended at an early date.

Nonimaging remote sensors, such as the passive microwave, radar scatterometer, IR, X-ray, gamma-ray, and alpha-particle emission are recommended for inclusion in lunar orbital payloads pending the results of current remote sensor feasibility studies. . . .

Atmospheres, Particles, and Fields Equipment

Atmospheric and ionospheric variability surveys should be conducted. The recommended characteristics of the ion and neutral mass spectrometers are described in the Group Report. Ion traps, solar wind detectors, and pressure gauges can be adapted from the unmanned programs.

Other experiments include the determination of the cosmic-ray albedo, the study of solar and lower-energy galactic cosmic rays (particle telescope), and the search for water using a neutron instrument to detect the hydrogen content of the surface.

Instruments for Subsurface Analysis

There are three orbital instruments that have the potential of obtaining valuable data from depths of kilometers and beyond. The first of these is a magnetometer, which will yield important information on the geology of the lunar crust. Second is a gravity gradiometer yielding details of the variations in the lunar gravitational field. The third is the electromagnetic pulse probe, which has the potential for probing to depth and differentiating the various layers present.

AES MANNED LUNAR SURFACE (AES-MLS)

The AES-MLS is essentially a continuation of the early Apollo missions characterized by longer stay time and larger scientific payloads.

It is suggested that this program can be usefully exploited in five or six missions, extending through 1974. The scientific requirements of this series include stay times up to 14 days and traverses up to 15 km from point of landing.

The longer stay time will probably permit the collection of more material than can be returned to Earth. Hence aids to the selection of samples in the field or prior to return should be provided by analytical equipment that will also measure sample characteristics that may be altered by return or packaging. Sample return is still the most significant achievement in these missions. Local mapping should also have a high priority so that sample location is accurately tied to the local geology.

The capability to off load several LSEP's during each AES mission is also an important aspect. In this way, a small array of stations (LSEP's) could be set into operation, giving important information for revealing the internal properties of the Moon.

The Moon should provide a unique base for astronomers because of its useful environmental characteristics, the most important being the lack of an appreciable atmosphere. However, exhaustive studies of the complete lunar environment are necessary before engineering design can be started.

The primary objective of analytical devices used on the lunar surface should be to extend the power of the observer to differentiate materials that have similar characteristics. The optimum sample return capability would be between 200 and 250 kg (450–600 lb.) per mission. The following basic types of equipment are required for this phase of lunar exploration:

1. Automatic position recording systems. Essential for tracking and recording movements of the astronaut, and the roving vehicle, and knowing the orientation of the camera. The system would automatically telemeter this information back to Earth or to the LEM.

2. Local Scientific Survey Module (LSSM). This surface roving vehicle should have the capability of carrying either one or two suited astronauts and scientific payload of at least 600 lb. An operational range of 8 km radius is a minimum, and 15 km would be more useful. Remote control of the LSSM would also be advantageous both before and after the arrival of the astronauts.

3. Lunar Flying Vehicle (LFV). A LFV would be useful for extending the operational range of the AES and for studying features inaccessible to the LSSM due to topography. It should be able to carry a 300-lb. scientific payload over a distance of 15 km. Continued study should determine how effectively it can be employed in surface operation.

4. Lunar Drills. The development of a 1-inch drill capable of penetrating to a depth of 3 meters in either rubble or solid rock is recommended. It should be operable from a roving vehicle. It is necessary for lunar heat flow studies and for obtaining biological samples.

Because of the liberal weight allowance for equipment delivered to the Moon's surface most Working Groups indicated a wide variety of experiments desired for inclusion in the program. Equipment and experiments include instrumentation for performing gravity surveys, active seismic surveys, magnetic measurements, radioactivity measurements, environmental measurements, and in general instrumentation and supporting equipment for conducting geological-geophysical surveys on the lunar surface.

To obtain maximum output of scientific information from these experiments, astronauts should be given scientific training in specific rather than general areas. The greatest need is for trained geologists; however, specialized training will be required in physics, meteorology, chemistry, and in other fields.

POST-AES

The AES should be followed by a program including long-distance travel, up to 800 km and fixed-site investigation from 2 months to 1 year. These missions should commence about 1975 and proceed at a rate of one per year through 1980. Additional orbital flights also appear desirable during this period so as to conduct simultaneous orbital and surface missions.

A long-range laboratory vehicle for geological and geophysical exploration is required to permit the collection of data to form a broad regional integrated picture of the surface geology and crustal structure. These data will also be essential as a basis for interpretation of the imagery and measurements obtained from the remote sensing orbital vehicle and also to substantiate other investigations.

A series of traverses along the geological belt is suggested, requiring a vehicle with the following characteristics:

1. A minimum range of 800 km.
2. Shelter for a three-person crew.
3. Mission duration capability of up to 2 months.
4. No constraints in returning to starting point.

A Lunar Base is a surface complex that will allow longer stay time, possibly up to a year, than is presently envisioned by the AES concept. Primary needs for a base are visualized to be:

1. The measurement of presently occurring time-varying phenomena; many are geophysical in nature.
2. The study of lunar surface processes.
3. Deep-drilling studies—most important for information on early history, crustal composition, and surface properties of the past. The depth to be reached should probably exceed 300 meters.
4. Detailed study of a critical field area.
5. Construction and manning of large radio and optical telescopes, yet to be defined.

[A 36-page section following this summary in the original report, not reproduced here, is a detailed listing of the requirements formulated by each disciplinary working group.]

Recommendations of the 1967 Summer Study of Lunar Science and Exploration, Santa Cruz, California, July 31–August 13, 1967

The Santa Cruz Summer Study was organized along the same lines as the Falmouth Conference, with five major objectives:

1. To obtain the consensus of the scientific community as to what the future lunar exploration program should be;
2. To prepare detailed science plans for future manned and unmanned lunar missions;
3. To establish the order of priority of experiments to be conducted on all missions;
4. To make recommendations on major hardware items required for the science programs;
5. To make recommendations on the instrument development programs required for the science program and those required to meet Supporting Research and Technology Program needs.

The following is an excerpt from the first section of the conference report,[*] "Summary and Recommendations," pp. 9–29.

---

[*]*1967 Summer Study of Lunar Science and Exploration*, NASA SP-157 (Washington, 1967).

SUMMARY AND RECOMMENDATIONS

The primary purpose of the Santa Cruz Conference was to arrive at a scientific consensus as to what the future lunar manned and unmanned exploration program should be, particularly in the time frame of the Apollo Applications Program (AAP). It was planned that the major results of the conference would include (1) a recommended list of lunar missions with detailed mission plans and priority experiment lists for each mission, (2) a priority list of major hardware items, and (3) recommendations for instrument development and for Supporting Research and Technology Programs.

The details of the findings and recommendations of the working groups are reported in later sections [not included here]. The major recommendations of the conference and the proposed lunar exploration program include (1) systems development, (2) proposed mission sequence, (3) program planning and support, and (4) science mission plans.

SYSTEMS DEVELOPMENT

Lunar Surface Mobility

The most important recommendation of the conference relates to lunar surface mobility. To increase the scientific return from lunar surface missions after the first few Apollo landings, the most important need is for increased operating range on the Moon. On the early Apollo missions it is expected that an astronaut will have an operating radius on foot of approximately 500 meters. It is imperative that this radius be increased to more than 10 km as soon as possible.

To increase surface mobility the following recommendations are made:

1. *It is recommended that a Lunar Flying Unit be developed immediately to be used in AAP and, if possible, on late Apollo flights to increase the astronaut's mobility range.*

   This is the first step toward attaining reasonable mobility. It is expected that the Lunar Flying Unit (LFU) will provide a mobility radius of 5 to 10 kilometers, which is a considerable improvement over the present capability, but not nearly enough. Exploration of lunar surface features such as large craters and their environs will require a range of approximately 25 km or more.

2. *It is recommended that the Saturn V dual-launch capability be developed as soon as possible.*

3. *It is recommended that the dual-mode local scientific survey module (LSSM) be developed on the same schedule as the dual-launch Saturn V.* This wheeled vehicle should be capable of operating in an automatic or manned mode. The primary purpose of the dual-launch system is to carry the recommended LSSM and additional fuel for the Lunar Flying Units. The automated/manned LSSM has a greater capability than the one now planned [i.e., the lunar roving vehicle]. The LSSM used in conjunction with Lunar Flying Units should provide a mobility radius of approximately 25 km.

   The best type of lunar surface mobility system was the subject of considerable debate during the conference. Two different philosophies of exploration on large-scale areas arose.
   1. The Geochemistry Working Group strongly favored the large lunar flying vehicle. A manned vehicle, which provides spot coverage over a wide area, would best afford the opportunity of observation and sample collection.
   2. A substantial, but divided, opinion of the Geology Working Group was that the combination of Small Lunar Flying Units with the LSSM was best for spot coverage and for continuous ground coverage. The Geology Working Group emphasized that the continuous ground observation was essential to solve complex geological problems in areas of limited size. The geologists felt that experience had shown that such studies are critical to solving much larger problems and are necessary to place geochemical and geophysical data in their proper geological context.

At the conclusion of the meeting, a substantial majority of the working groups were in favor of the shorter range, continuous surface traverse using a dual-mode LSSM rather than spot coverage over a large area with a Large Flying Unit. Another reason for the choice of the LSSM is that it is probable that the manned half of a dual launch will carry two Lunar Flying Units. Starting with this, the total mobility system using the LSSM seemed a better choice.

A very important reason for the choice of the LSSM involved the use of an automated mode of operation. Agreement was unanimous on the need for long unmanned traverses on the lunar surface. After the astronauts have returned to the Earth, the LSSM would be sent to a new destination. On its journey, the LSSM would accomplish the following:

1. Stereo TV on the LSSN will permit the LSSM to be controlled from the earth. The LSSM would collect samples along the route. Some of these samples would be aseptically handled and packaged for return to the earth by the next manned lander.
2. The LSSM would conduct a geophysical traverse of a large area using devices such as magnetometers, gravimeters, and radar probes.
3. The LSSM would deploy several small ALSEP-type Remote Geophysical Monitors (RGM) along the traverse. In this way, a network of such units could be built. The RGM would carry instruments such as seismographs, atmospheric mass spectrometers, gravimeters, and magnetometers.

The dual-mode LSSM is more complicated and has greater capability than the vehicle presently planned. The LSSM should have the following characteristics and capabilities:

1. articulation;
2. the ability to pick up rocks;
3. stereo TV with the Apollo bandwidth and a fast shutter;
4. rock analysis (nondestructive) with a storage capacity of 50 pounds;
5. a headlight (night and shadow operation)
6. samples, stowage of approximately 100, maximum weight of 1 kg per sample with some aseptically sealed;
7. the capability to carry and deploy 6 RGMs, each weighing 50 pounds and equipped with instruments such as a gravimeter, a radar probe, and a magnetometer;
8. dead-reckoning navigation with altitude and horizontal ties to known controls;
9. capability for carrying two persons with optional steering modes and capability for carrying 1 or 2 LFUs;
10. relay communications; and
11. the ability to carry a backup portable life support system (PLSS) or an independent life support system.

Block II Surveyor

Other systems working with the automated LSSM are probably needed to develop the geophysical network of Remote Geophysical Monitors on the Moon. This network requires about 10 automated stations distributed over the front face of the Moon with spacings on the order of 1,000 km or more. This system is required to obtain large-scale information about the interior of the Moon.

It is recommended that a Block II Surveyor, or another system, be available in the period from 1970 to 1975, which is capable of deploying experiments such as the following:

1. a passive seismic/tidal gravimeter/tiltmeter (three components);
2. a corner reflector;
3. a gravimeter (geodetic);
4. a mass spectrometer;
5. a total-pressure gauge;
6. a doppler transponder;
7. a facsimile camera;
8. a magnetometer;
9. a plasma probe;
10. low-energy particles;
11. electric field; and
12. a gamma-ray experiment or alpha-counter experiment.

The Block II Surveyor will afford wide geographic coverage for instruments. Special care must be taken in deployment of experiments (magnetometer, gamma-ray, etc.) so that they are deployed in an appropriate environment.

Sample Return Capability

One important, if not the most important, scientific result from the AAP missions will be the return of lunar samples. The amount returned must increase as the capabilities of the vehicles allow.

It is recommended that the total returned payload from the Moon in AAP missions increase to 400 lb. so that a minimum of 250 lb. of lunar samples can be returned.

A consensus of the conference was that a capability to return approximately 50 lb. of refrigerated samples was needed as soon as possible.

Modular ALSEP

It was clearly recognized at the conference that Apollo lunar surface experiments packages or their derivatives such as Emplaced Scientific Stations or Remote Geophysical Monitors would be used on essentially all AAP landing missions. A number of new experiments under development require the ESS or RGM capability, and many of the current experiments should be used on the lunar surface in networks. The capability should be established to include new experiments on an ALSEP prior to a mission.

It is recommended that future ALSEP stations be designed to allow a substantial degree of flexibility to react to new opportunities opened up by new developments or discoveries on the Moon. A modular concept to permit accommodation of new instruments with minimum disturbance of the basic ALSEP system would greatly facilitate such flexibility.

It is expected that the number of candidate experiments for a particular ALSEP mission will exceed the number that can be accommodated on that mission. Flight assignments for the mission should be made as close to the flight time as is practical to reflect the state of knowledge at that time. The experiments would, therefore, be built to meet a standard ALSEP electrical interface and a suitably small choice of mechanical interfaces. This requires that the ALSEP central station have an appropriate number of standard electrical plug-in stations and a central data processor that assigns experiment data rates under the control of a stored program. The processor control program could be stored prior to flight or, preferably, on command from Earth. Remote reprogramming is particularly desirable as it permits the real-time assignment of experiment data rates in response to acquired data.

Telemetry Capability

When several scientific packages are operating on or near the Moon for long periods of time, the capability to handle the increased data return will become a problem. The data return capability is presently a problem in some unmanned programs.

It is strongly recommended that appropriate provision be made to insure continuous telemetry coverage of all scientific packages, both single and simultaneous operations, on and around the Moon. Provision must also be made to recover data continuously from the averted face of the Moon.

Orbital Subsatellites

Subsatellites could be injected from the AAP command and service module (CSM) vehicles into precision orbits to study the lunar environment, including magnetic fields and particle environment and the atmosphere and ionosphere.

It is recommended that a subsatellite system he developed for deploying systems of instruments in close orbit around the Moon.

PROPOSED AAP MISSION SEQUENCE

To understand how the mobility systems fit into the program of flights and why they have been chosen, the proposed program of lunar landing missions should be examined. The program is not in final form and will not be for some time, at least until further mission studies have been made. The program of missions consists of three distinct mission types: (1) manned orbital flights; (2) single-launch Saturn V lunar landing flights; and (3) dual-launched Saturn V lunar landing flights. The proposed program is outlined in Table 1 and Figure 1 [not reproduced here].

Table 1. **Proposed AAP Mission Sequence**

Mission	Mission location	Schedule,[a] (Yr.)	Launch mode
1	Manned orbiter	1st	—
2	Copernicus (central peaks)	1st	Single
3	Davy Rille	2nd	Single
4	Copernicus (walls)	2nd	Single
5	Marius Hills LSSM to the Cobra-Head	3rd	Dual
6	Cobra-Head LSSM to Hadley Rille	3rd or 4th	Dual
7	Manned orbiter	3rd or 4th	—
8	Alphonsus LSSM to Sabine and Ritter	5th	Dual
9	Sabine and Ritter (or end of Alphonsus LSSM mission)	5th	Single
▲	North Pole or South Pole[b]	—	—
■	Tycho[b]	—	—
★	Mare Orientale[b]	—	—
▼	Hadley Rille[b]	—	—

[a]Mission will occur within the program year(s) indicated.
[b]Times, launch modes, and sequence are not established; further study is required.

Manned Orbiter

The first recommended AAP flight is a manned lunar orbital flight with these objectives.
1. AAP landing sites mapping photography (return of film necessary;
2. metric-mapping quality photography with returned film;
3. geochemical remote-sensing using gamma-rays and x-rays on a subsatellite left in lunar orbit or using a directional detector system on the CSM; and
4. flying a family of remote sensors such as passive microwave radiometer, infrared radiometer, radar reflectivity, radio noise survey, magnetometer and plasma probe, multicolor photometry, meteoroid detector, radio radiometer, and fluorescence photometer.

Objective 4 has a somewhat lower priority than the first three.

The only way to obtain adequate data for cartographic purposes is to have the film returned to Earth for analysis. The photography from this flight should provide maps of the Moon in the IR region and (by analyzing the gamma-rays) lunar contour maps of the concentration of potassium and uranium and possibly of other elements. Such data will be valuable in future mission planning.

The first orbiter flight should be followed by a second flight within a time frame so that new remote sensing instruments would be perfected (1) to increase the ability to map the Moon remotely in various electromagnetic frequencies and (2) to obtain greater information on the distribution of the elements.

Single-Launch Mode

The first AAP lunar landers will probably be single launches. Missions 2, 3, and 4 represent the proposed early single-launch landers. Because the crater Copernicus is such a large feature and because information about Copernicus is essential to the understanding of the Moon, two separate missions are proposed: one to the central peaks and one to the crater wall. A proposed science mission plan for the central peaks is presented later in this section.

The durations of this class of missions are flexible. It is desired that the single-launch missions be started as soon as possible; however, a series of useful single launches could be continued for a number of missions. Suggested additional sites for the subsequent single missions are Copernicus H, Gambart, Mösting C, Hyginus Rille, Flamsteed, Dionysius, Hipparchus, the dome near Lunar Orbiter Photographic Site II-P-2, and the Surveyor landing and the Ranger impact sites. The Lunar Flying Unit would be used for mobility on these missions.

Dr. George E. Mueller suggested that the conferees consider what scientific program should be carried out on Apollo flights after the first two or three successful lunar landings. The following ground rules were assumed:

1. Some system constraints will have been removed so that more payload is available on the Apollo CSM.
2. The lunar module (LM) will be able to land at rather rough sites but still in or near the Apollo landing zone of ±10° latitude.
3. No substantial hardware changes will be made in the CSM or the LM.

It was decided that all of the suggested single-launch AAP landing sites would be appropriate for late Apollo, except Copernicus, which requires more mobility. The most appealing sites were Copernicus H, Gambart, Mösting C, and the dome near site II-P-2. The lunar module might land close enough to these sites to allow access to the interesting areas. There was a strong feeling that there should be at least one highland landing site.

Additional mobility should be brought into the program as rapidly as possible. The LFU should be used even if it has only a 1- or 2-km range, which will significantly enhance the scientific return. If the LFU is not ready, the schedule for the Apollo lunar program should be adjusted, after the first few successful landers, to allow the LFU to be brought into these late Apollo missions.

Dual-Launch Mode

The future lunar exploration program must involve mobility systems that require two Saturn V launches. The dual-mode LSSM and a large amount of additional fuel for the Lunar Flying Units will be carried to the lunar surface in an unmanned lander. The manned Saturn V will land nearby later. The present LSSM cannot be carried to the Moon in a manned single-launch system.

A major feature of the dual-launch system is that the unmanned LSSM would make a long traverse (approximately 1,000 km) and arrive at the site of the next manned landing with a collection of surface samples from an extensive region of the Moon. These would be returned to Earth by the next manned lander. The suggested sequence of missions with automated LSSM traverses is shown in Table 1. This technique allows a much larger fraction of the lunar surface to be sampled than could possibly be visited by humans during all of the proposed manned AAP missions (Fig. 1).

PROGRAM PLANNING AND SUPPORT

Fallback Position

If the AAP lunar program level should drop below one dual-launch or two single-launch missions per year, then the Group for Lunar Exploration Planning (GLEP) should meet again to reconsider mission plans. The program at this low level might need significant redirection. For example, in the case of severe budgetary restrictions or other problems that would prevent developing the Saturn V dual-launch capability, an automated system, such as Rover launched by a system less complicated than a Saturn V, might need to be developed.

Solicitation for Experiments

The selection of experiments from the scientific community was a major consideration at the Conference.

To develop a strong science program in AAP, it is strongly recommended that any extension of the Apollo science program (that is, new Apollo hardware, follow-on ALSEP, or AAP), be implemented by open solicitation of experiments from the scientific community. Only in this way can the Manned Space Flight Program build the broad base of scientific support and participation necessary for an active and productive research program.

It is recognized that the time scales involved may pose problems in certain disciplines in implementing the above recommendation. However, there are experiments that could be delivered on a relatively short time scale and thus allow the possibility of a wider NASA scientific program.

Support of Basic Research

The NASA support of basic research programs has been a benefit to the space program and also to universities and other research institutions. This support should not be confined to flight programs but should be continued in all areas of basic science that contribute to the overall NASA objectives of space exploration.

Support of Instrument Development

Many scientific experiments appear very promising for lunar exploration but are not feasible now because (1) detector systems have not been developed to the point of having the desired sensitivity, or (2) theoretical problems have not been fully investigated to insure proper design of the experiment or full interpretation of the results.

Three stages in the development of an experiment and the necessary hardware can be visualized. These are (1) detailed consideration of the importance and feasibility of an experiment, (2) development of necessary scientific tools to implement the experiment, and (3) production of flight hardware.

To carry out the first two stages of producing an experiment listed previously, it is recommended that a strong program in scientific instrument definition and development and a substantial lunar supporting research and technology program be undertaken immediately. The amount of money being invested to produce experiments for AAP flights in the present budget is not compatible with the scope of the program. It is further recommended that adequate time be included in program planning and launch schedules to allow for the scientific development of the appropriate experiments.

Establishment of Project Scientist

It is now apparent that there is a need for more continuous scientific input into the development of scientific flight hardware for lunar missions.

It is strongly recommended that a position of Project Scientist be established within the structure of the Manned Space Flight Program. The responsibilities of the project scientist are to represent the scientific requirements and objectives of the experiment to the project manager and staff and conversely to represent to the Principal Investigator project requirements that may affect the experiment. At least one project scientist should be associated with every MSC project that includes scientific experiments; more than one project scientist may be desirable for a large project in which the number of scientific disciplines is large. Furthermore, it is strongly recommended that project scientists be participating experimenters in the projects for which they are responsible. The position of the project scientist within the organizational structure should be at a level that insures adequate science input into the program.

Astronaut Selection and Training

After basic classroom work and tutorials, the astronaut should be provided the time and opportunity to participate directly in the research and planning activities of the particular missions for which he is selected. This may require one or more thorough refresher tutorial covering the specific topics of prime scientific importance for the mission. In crew selection for any mission, flight-operations ability is, without question, the primary criterion.

It is strongly recommended that ability in field geology be the next most important factor in the selection of the crew members who will actually land on the Moon for the Apollo missions. For some of the complicated scientific missions in the later part of the AAP, the Santa Cruz Conference considers that the knowledge and experience of an astronaut who is also a professional field geologist is essential. In the interest of maintaining career proficiency, astronauts should be provided time to engage in some form of research activity within their professional fields.

SCIENCE MISSION PLANS

As previously stated, the science mission plans in this report are not in final form. Engineering studies must be made and appropriate modifications and plans developed.

It is recommended that an immediate and intensive program of detailed mission analyses be undertaken for all of the prime lunar landing sites and traverses that have been listed by this Conference.

Because of the rapid development of suitable launch capability and the growth of an extensive body of photogeologic maps of the lunar surfce and because of the lead-times required for development of selected systems, the working groups felt that such analyses are urgent. The analyses must be planned on an iterative basis to test the applicability of the recommended plans and of instrument development for the achievement of the general scientific objectives for lunar exploration.

[The "Summary and Recommendations" continues with detailed (though tentative) plans for three AAP missions—to Copernicus, the "Cobra Head" (Aristarchus region), and Alphonsus—including landing sites, mission timelines, and surface activities. The balance of the report consists of the individual reports of eight working groups: Geology, Geophysics, Geochemistry, Bioscience, Geodesy and Cartography, Lunar Atmospheres, Particles and Fields, and Astronomy.]

Lunar Science Objectives:
The Rationale for Apollo Landing Site Selection

Following is a summary of material presented to the Apollo Site Selection Board at its meeting on July 10, 1969, intended to show how the Site Selection Subgroup of the Group for Lunar Exploration Planning arrived at their recommended list of primary and alternate sites (presented at the June meeting of the Board). It relates scientific experiments to the "15 Questions" tabulated above.

Age Dating. Determination of the absolute age of lunar surface materials by radioactive-decay methods was of prime importance. The method dates the time at which a given sample became a closed system for a particular element; for example, the potassium-40–argon-40 method determines the time at which a rock containing radioactive potassium-40 cooled sufficiently to retain the gaseous argon-40 decay product. The age thus determined may reflect a period of volcanism, melting by meteoric impact, or original accretion of lunar material.

Ages of primary interest to lunar scientists were:

(1) the age of the moon's formation or of its oldest crust; this age might make it possible to distinguish among various theories of lunar origin. One key site for locating such material was the Fra Mauro Formation. This widespread blanket of debris is considered to consist of debris from the "Imbrian event," the cataclysmic occurrence that produced Mare Imbrium, the enormous circular mare in the northwestern quadrant of the Earth-

facing side of the moon. The Fra Mauro Formation was given high priority on all lists of potential landing sites.

(2) The time at which the maria became filled with the relatively smooth material, probably once-molten rock, that characterizes them. This date would be relatively easy to establish, since mission planners preferred smooth, level landing sites, most often found in the maria.

(3) The time of significant post-mare events, such as the impacts that created the craters Copernicus and Tycho. Equally interesting was the origin of the sinuous rilles, which appeared to have been formed largely in this later period of lunar history.

Lunar Composition. A primary interest of geochemists was to find "primitive" solar system material to deduce the conditions under which planets and satellites condensed from the solar nebula. Data from analysis of terrestrial and meteoritic samples, along with solar and stellar spectral studies and theoretical physics, comprised the prime source of such information, but both the earth and meteorites have been modified by subsequent heating and weathering, which have obscured the original composition. The moon's small size (possibly resulting in a smaller flow of heat from the interior) and its lack of an atmosphere suggest that it might still have original material on the surface. It was generally accepted that this material could most likely be found in the lunar highlands, which appeared to represent the oldest lunar material.

Second in importance was establishing the bulk composition of the moon. The abundance of the major elements was expected to be important in establishing the moon's origin; if it proved to be totally unlike the earth in composition it could hardly have split off from the earth. The abundance of radioactive elements would indicate how much energy had been available for heat-induced chemical changes in the moon since its formation. To determine the bulk composition of the moon it was necessary to sample as many different geologic units as possible. Particularly important were sites showing evidence of differentiation or the presence on the surface of deep-seated material, such as would be present in and around impact excavations or explosive craters and in blankets of material ejected from craters. Radioactivity could be measured by instruments in lunar orbit, hence the importance of flying experiments in the service module.

Finally, analysis of any present or past lunar atmosphere would give clues to lunar origin and evolution. Gas detectors operating over a long period of time might possibly detect transient events, such as had been reported in the crater Aristarchus. The sinuous rilles (e.g., Rima Prinz) were also likely sites for detecting any emission of gases from the lunar interior.

Major Geomorphic Processes. The study of the processes by which lunar landforms have been created and destroyed was important mainly to second- and third-order questions about the moon, but was essential in holding the first-order questions (above) together. Knowledge of dominant processes would provide the basis for selection of samples and determining their place of origin, as well as providing major clues to past energy expenditure on the moon. Regions of particular interest in this regard were the sinuous rilles and areas of volcanic cratering. Since photography covered a large fraction of the lunar surface, data from a few landing sites would enable geomorphologists to draw conclusions about most of the moon.

Lunar Geophysics. The only source of information on the moon's internal structure was the emplaced ALSEP experiments, which included a seismometer and a heat flow instrument. (Later ALSEPs would include different geophysical instruments.)

Seismic data were expected to yield information on layering in the moon, the rate of release of internal strain, and the number and energy of meteorite impacts. It would be very useful to the seismologists to produce an impact of known energy at a known location, such as by causing a spent S-IVB stage or a lunar module ascent stage to crash on the lunar surface. It was important to have at least four seismic instruments active at one time; they should be about 1,000 km apart with as much angular separation as possible. Thus sites at high latitudes, such as Tycho, were quite important.

Data on heat flow were difficult to interpret but could assist in determining whether the moon had originally been hot or cold.

Lunar gravity and geodesy would determine the extent of the "mascons" (mass concentrations) and whether the moon was in hydrostatic equilibrium. These were well suited to study by orbiting instruments.

The laser retroreflectors were expected to make it possible to determine earth-moon distances within a few centimeters, enabling scientists to measure the librations of the moon with previously unattainable accuracy. Best results would be obtained with widely separated reflectors.

Characteristics of the Recommended Landing Sites. The short list of landing sites for the first 10 lunar exploration missions should include:

(1) two types of mare material, "older" or eastern and "younger" or western;
(2) regional stratigraphic units, such as blanket (ejecta) deposits around mare basins;
(3) various types and sizes of impact craters in the maria and in the highlands;
(4) morphological manifestations of volcanism in the maria and in the highlands; and
(5) areas that may give clues to the nature and extent of processes other than impact and volcanism, which may have acted on the lunar surface.

Following is a description of the 10 prime sites chosen by the Group for Lunar Exploration Planning. The sites were selected in order of their preferred execution on one "G," four "H," and five "J" missions.

(1) Landing Site 2 ("older" or eastern mare). This site is entirely within relatively old mare (Imbrian) material. It includes many large subdued craters 200 to 600 meters in diameter but comparatively few in the size range 50–200 meters, a distribution common to many apparently old surfaces. Determination of the age and nature of this Imbrian mare material was a primary object of landing at this site.

(2) Landing Site 5 ("younger" or western mare) is located within relatively young (Eratosthenian) mare material and displays many craters 50 to 200 meters in diameter and relatively fewer of the larger (200–600 meter) craters. It is surrounded by well developed rays from Kepler, making it likely that it contains material derived from considerable depth. The chief goal of a landing at Site 5 was to determine the age and composition of Eratosthenian mare material.

(3) Fra Mauro Formation. This extensive geologic unit covers large portions of the surface around Mare Imbrium and is thought to be material ejected when Imbrium was formed. Samples from the Fra Mauro Formation would help to understand its nature, composition, and formation and its relation to the "Imbrian event."

(4) Rima Bode II, a single linear rille running close to a fresh, elongate crater and a crater chain, was of interest because both the rille and the crater were possible sources of several dark geologic units most probably of volcanic origin. The site was selected as an example of a region where material of deep-seated origin was expected. An alternative site was Hyginus Rille, similar in characteristics but apparently less fresh. Another site, Littrow, would meet part of the objectives of a mission to Rima Bode II.

(5) Censorinus is a 3.8-km crater located within and near the edge of a highland block south-southeast of Mare Tranqillitatis. It offered the opportunity, early in the exploration program, to sample both highland material and features associated with a fresh impact crater. The proposed site was within the ejecta blanket about 1 km north of the crater rim and allowed investigation of the crater on foot, without mobility aids. If Censorinus presented operational difficulties, Littrow could be considered as an alternative site for this mission.

(6) Copernicus (peak). This bright crater, 95 km in diameter, is the source of visible rays of ejected material extending for several hundred kilometers. Its walls expose a 4-km vertical section of the lunar crust. The floor, some 60 km across, contains multiple peaks with a maximum height of 800 meters. A mission to the central peaks would be mainly a sampling mission with the objective of bringing back material that once lay at considerable depth.

(7) Marius Hills, a group of domes and cones near the center of Oceanus Procellarum west-northwest of the crater Marius, are part of a ridge system stretching some 1,900 km through Oceanus Procellarum. The variety of features in this area and their similarity to terrestrial volcanic structures strongly suggests intensive and prolonged volcanic activity.

(8) Tycho (rim), like Censorinus a fresh impact crater, is in the southern highlands. It is much larger than Censorinus and offers an opportunity to study many features common to large, fresh impact events, including associated volcanism. The proposed landing site was near the *Surveyor VII* spacecraft, offering the option of returning some Surveyor parts. In that area are several generations of flows, a pond or pool, ejected blocks (probably from Tycho), and other ejecta features and structures.

(9) Rima Prinz I, in the Harbinger Mountains northeast of the Marius Hills, is a double sinuous rille—a small meandering rille enclosed within a larger sinuous rille. The origin of the rilles is of great interest because they resemble channels carved by a flowing fluid. A landing near the mouth of Rima Prinz I, selected because of the freshness of its details, would allow examination of the lower part of the eroded valley, sampling the materials and studying the exposed structures. An alternative, Schröter's Valley, displays similar characteristics but appears older than Rima Prinz I.

(10) Descartes. The area of the southern highlands north of the crater Descartes and west of Mare Nectaris is characterized by hilly, groovy, and furrowed deposits reminiscent of terrestrial volcanoes. A mission to a region of intensive and prolonged volcanism within the lunar terrae was considered most important, from both the geological and geochemical viewpoints. An alternative to this site was Abulfeda, just to the southwest.

The Group for Lunar Exploration Planning recommended this list of sites after considering the expected evolution in capability as well as the constraints imposed by operations. Those selected for ''J'' missions, for example, were picked on account of the additional time on the surface that would be available and the increased mobility that would be provided by a powered vehicle. Tycho, accessible only in the early part of the year because of operational limitations, was switched from the third to the second ''J'' mission for that reason, although it was better than the Marius Hills site for exploration on foot.

The Apollo Site Selection Board accepted this list for planning purposes at its meeting on July 10, 1969. As was to be expected, the list underwent considerable revision during the next three years, both as to the choice of sites and the order in which they would be explored, as mission planning became more detailed and operational capability improved.

APPENDIX 4

CHRONOLOGY OF MAJOR EVENTS IN MANNED SPACE FLIGHT AND IN PROJECT APOLLO, 1957–1975

1957

October: The Soviet Union placed the first artificial earth satellite *(Sputnik)* into orbit.

1958

April: The Air Force contracted with the Yerkes Observatory, University of Chicago, to produce a new lunar photographic atlas.

The Air Force published a development plan for its manned space program, which included two exploratory man-in-space projects, a lunar reconnaissance mission and a manned lunar landing and return; the plan envisioned completion of the program in seven years at a cost of $1.5 billion.

June: The Air Force contracted with Rocketdyne to design a single-chamber rocket engine burning kerosene and liquid oxygen and producing 1 to 1.5 million pounds of thrust.

July: President Eisenhower signed the National Aeronautics and Space Act of 1958 (P.L. 85-568) establishing the National Aeronautics and Space Administration.

October: The Special Committee on Space Technology, created in January 1958 and chaired by H. Guyford Stever of MIT, reported its recommendations: development of both clustered- and single-engine boosters of million-pound thrust; vigorous attack on the problems of sustaining man in the space environment; development of lifting reentry vehicles; research on high-energy propellant systems for launch vehicle upper stages; and evaluation of existing boosters and upper stages followed by intensive development of those promising greatest utility.

November: A Space Task Group (STG) was organized at Langley Research Center to implement NASA's first manned satellite project (Mercury). Robert R. Gilruth was named project manager.

1959

January: In a report of the staff of the House Select Committee on Astronautics and Space Exploration entitled "The Next Ten Years in Space, 1959–1969," Wernher von Braun of the Army Ballistic Missile Agency predicted a manned flight around the moon within 8 to 10 years and a manned lunar landing and return a few years later. NASA and industry officials envisioned similar progress.

March: The first F-1 engine was successfully test-fired by Rocketdyne, producing more than one million pounds of thrust.

April: NASA announced the selection of seven pilots for the Mercury program.

NASA created a Research Steering Committee on Manned Space Flight. Over the next several months this committee examined long-term human-in-space problems to recommend future missions and coordination of research programs at the NASA centers. At its May 25–26 meeting the committee recommended the manned lunar landing as a focal point for studies in propulsion, vehicle configuration, structure, and guidance requirements, since a lunar landing would constitute an end objective that did not have to be justified in terms of its contribution to a more useful goal.

November: STG appointed a panel to study preliminary design of a multiperson spacecraft for a circumlunar mission, conduct mission analyses, and plan a test program.

1960

January: NASA presented its ten-year plan to Congress, calling for a program leading to manned circumlunar flight and a permanent earth-orbiting space station to start in 1965–1967 and a manned lunar landing some time beyond 1970. Cost estimates for the plan ran to $1.5 billion annually for five years.

February: NASA approved Project Ranger, a project to send an unmanned, hard-landing spacecraft to the moon to relay television pictures of the lunar surface to earth during the final stages of its flight.

March: The Army Ballistic Missile Agency's Development Operations Division at Redstone Arsenal, Huntsville, Alabama, headed by Wernher von Braun, was transferred to NASA as the George C. Marshall Space Flight Center.

April–May: STG developed guidelines for the advanced manned spacecraft program, including detailed propulsion and spacecraft requirements.

May: A meeting on space rendezvous was held at Langley Research Center to discuss the problems of bringing two spacecraft together in space.

NASA began work on a project (later named *Surveyor*) to send a soft-landing spacecraft to the moon to provide scientific and engineering data on the lunar surface.

July: The House Committee on Science and Astronautics urged NASA to intensify its efforts to send humans to the moon and back "in this decade." In the committee's view, NASA's ten-year plan did not go far enough and the space agency was not pressing forward with enough energy.

July: The name "Apollo" was approved for the advanced manned space flight program.

NASA held its first NASA-Industry Program Plans Conference in Washington to brief industrial management on the overall space program. George

M. Low, chief of NASA's Manned Space Flight program, stated that circum-lunar flight and earth-orbiting missions would be carried out before 1970, leading eventually to a manned lunar landing and a permanent space station in earth orbit.

September: NASA issued a formal request for proposals for six-month feasibility studies for advanced manned spacecraft, to define a system fulfilling STG guidelines, formulate a plan for implementing the program, identify areas requiring long lead-time research and development, and estimate the total cost of the program. In October proposals were received from 14 companies, and in November contracts were awarded to Convair/Astronautics Division of General Dynamics Corp., General Electric Company, and The Martin Company.

November: A program of detailed studies of lunar geology was undertaken by the U.S. Geological Survey, funded by NASA.

STG proposed to organize a number of Technical Liaison Groups to coordinate the activities of NASA centers in research for Apollo.

1961

January: A meeting of the Space Exploration Program Council discussed the manned lunar landing project, with emphasis on three methods of conducting the mission: direct ascent, rendezvous of spacecraft in earth orbit, and rendezvous in lunar orbit. It was decided that all three methods should be explored thoroughly. The Council established a committee headed by George M. Low to define the elements of the project insofar as possible.

February: The Instrumentation Laboratory at the Massachusetts Institute of Technology was selected to conduct a six-month study of a navigation and guidance system for the Apollo spacecraft.

March: The Space Science Board of the National Academy of Sciences recommended that "scientific exploration of the moon and planets should be clearly stated as the ultimate objective of the U.S. space program for the forseeable future."

April: The USSR launched a five-ton spacecraft *(Vostok I)* carrying Major Yuri A. Gagarin on a one-orbit, 108-minute flight.

May: STG proposed a new NASA development center to manage the development of manned spacecraft and projects.

The United States launched its first human into space, Lt. Cmdr. Alan B. Shepard, Jr., who rode a Mercury spacecraft *(Freedom 7)* on a parabolic flight path 116.5 miles high and landed 320 miles down range.

Final reports of the six-month feasibility studies for advanced manned spacecraft were submitted to STG by the three contractors.

President John F. Kennedy addressed Congress on "urgent national needs," which included new long-range goals for the American space program. Kennedy expressed his belief that the nation should adopt the goal, "before this decade is out, of landing a man on the moon and returning him safely to the earth." He requested additional appropriations of $611 million for NASA and DoD for fiscal 1962.

NASA appointed a committee (Lundin committee) to study all possible approaches for accomplishing a manned lunar landing in the period 1967–1970 and to make rough estimates of costs and schedules.

July: Twelve companies were invited to submit proposals for the Apollo spacecraft. A detailed statement of work, based on contractor and NASA design studies, was provided for a three-phase program terminating in a lunar landing.

NASA and DoD created a Large Launch Vehicle Planning Group to study development of large launch vehicles for the national space program.

August: NASA selected the Instrumentation Laboratory of MIT to develop the guidance and navigation system for the Apollo Spacecraft.

September: After a study of several locations around the country, NASA selected a site near Houston, Texas, for its new development center for manned spacecraft. The center would design, develop, and test new manned spacecraft, train astronauts, and operate the control center for manned space missions. In October the Space Task Group, still based at Langley, was formally redesignated as the Manned Spacecraft Center (MSC); personnel would move to Houston starting in 1962.

October: John C. Houbolt and others at Langley Research Center presented to the Large Launch Vehicle Planning Group a study on the use of lunar-orbit rendezvous in a manned lunar landing.

November: After evaluation of proposals from five companies, NASA selected the Space and Information Division of North American Aviation, Inc., Downey, California, to design and build the Apollo spacecraft.

December: MSC announced a new manned program using a two-man version of the Mercury spacecraft, which would test techniques of rendezvous in earth orbit.

1962

February: The first American to orbit the earth, Lt. Col. John H. Glenn, Jr., USMC, completed three orbits in a Mercury spacecraft and returned safely to earth.

February–June: Several groups within NASA were intensively studying the various modes of going to the moon (direct ascent, rendezvous in earth orbit, rendezvous in lunar orbit). The third method required a separate spacecraft to detach itself, land on the moon, and return to lunar orbit to rendezvous with the Apollo spacecraft.

March: At the request of the Office of Manned Space Flight, American Telephone & Telegraph established a group called Bellcomm, Inc., to provide independent analysis of systems and problems in the manned space flight program. For the duration of Apollo, Bellcomm performed many services, including advice on selection of landing sites, for OMSF.

July: NASA Headquarters announced that the lunar-orbit rendezvous mode had been selected for the manned lunar landing project and that requests for proposals would be issued for the second spacecraft (the "lunar excursion module"). MSC invited 11 firms to submit proposals for the lunar excursion module. Nine companies responded; in November NASA selected the Grumman Aircraft Engineering Company to build the module.

August: A summer study conducted by the Space Science Board at the State University of Iowa examined the state of NASA's space research program and made recommendations concerning future efforts. Many scientists expressed objections to Apollo (which was not specifically on their agenda), but the study cautiously endorsed the program's scientific goals.

September: A second group of nine test-pilot astronauts was selected for the manned space flight program.

November: MSC released sketches of the space suit assembly and portable life-support system to be used on the lunar surface.

December: A contract was awarded for construction of a Vertical Assembly Building at NASA's Merritt Island Launch Area, Kennedy Space Center. The $100-million structure would provide space for assembling four Saturn V launch vehicles simultaneously.

1963

February: The President's budget request for fiscal 1964 included $5.712 billion for NASA. $1.207 billion was for Apollo—almost a threefold increase over the previous year.

April: Preliminary plans for Apollo scientific instruments were completed. Emphasis was placed on experiments that promised maximum return for the least weight and complexity and were man-oriented and compatible with weight and volume available in the spacecraft. Experiments would be selected after evaluation of proposals from outside scientists.

May: The Mercury project ended with the 34-hour, 22-orbit flight of astronaut L. Gordon Cooper, Jr., in the spacecraft *Faith 7.*

August: NASA Headquarters approved the Lunar Orbiter project, which would use unmanned spacecraft to take detailed photographs of the lunar surface to be used in selecting landing sites for Apollo.

September: Dr. George E. Mueller became Associate Administrator for Manned Space Flight, replacing D. Brainerd Holmes.

October: Headquarters sent MSC some general guidelines for scientific investigations of the moon. Principal scientific activity was expected to include comprehensive observation of lunar phenomena, collection of geologic samples, and emplacement of monitoring equipment.

October: Fourteen more test pilots were selected as astronauts.

November: A Manned Space Science Division was established in the Office of Space Science and Applications (OSSA), NASA Headquarters, to coordinate the efforts of OSSA and the Office of Manned Space Flight in developing scientific experiments for Apollo.

MSC's Space Environment Division recommended 10 specific areas on the moon for evaluation as landing sites for Apollo. These sites and others would be photographed by Lunar Orbiter, after which some would be selected as targets for Surveyor, a project to land unmanned spacecraft on the moon and study the surface.

President Lyndon B. Johnson announced that NASA's Launch Operations Center at Cape Canaveral (Atlantic Missile Range) would be designated the John F. Kennedy Space Center.

December: An ad hoc group working on Apollo experiments recommended the principal scientific objectives of the program: examination of the surface around the landed spacecraft, geological mapping, investigation of the moon's interior (with instruments), studies of the lunar atmosphere, and radio astronomy from the surface.

1964

March: NASA's Office of Space Science and Applications (OSSA) began organizing groups of scientists to assist in more specific definition of the scientific objectives of Apollo. Outside scientists were called upon to propose experiments in geology, geophysics, geochemistry, biology, and atmospheric science.

April: NASA enlisted the aid of the National Academy of Sciences in preparing a plan to recruit scientists for training as astronauts.

May: The first flight of an unmanned Apollo spacecraft was launched from Kennedy Space Center, demonstrating the compatibility of the spacecraft and the launch vehicle.

July: OSSA announced opportunities for scientists to fly experiments on manned space missions, including the lunar landing missions. The earliest Apollo flights expected to support scientific instruments were the fourth and fifth.

Ranger VII returned the first close-up television pictures of the lunar surface, showing details as small as 1 meter across.

August: MSC proposed to build a special lunar sample receiving laboratory in which lunar samples, protected from contamination, would be received, examined, and issued to qualified outside experimenters. The proposal set off several months of discussion between MSC, Headquarters, and the Space Science Board concerning the requirements for such a laboratory and its best location.

1965

March: First manned flight of a Gemini spacecraft, a three-orbit flight to test spacecraft systems.

May: The Space Science Board recommended that samples and astronauts returning from the moon be quarantined until it could be ascertained that they had brought back no life forms that might contaminate the earth.

The NASA Administrator and the Surgeon General agreed to form an Interagency Committee on Back Contamination to define requirements for biological isolation and testing of material returned from the moon and to advise on the construction and operation of a quarantine facility for samples and astronauts.

June: Six scientists were selected for training as NASA astronauts. Two were qualified pilots; the other four were sent to Air Force flight training school before beginning astronaut training.

July: OMSF established an Apollo Site Selection Board to work with OSSA, MSC, and Bellcomm in choosing the sites where Apollo missions would land on the moon.

The Space Science Board convened a Summer Study at Woods Hole, Mass., to recommend directions for future space research. The agenda included manned exploration of the moon and planets. Conferees drew up a list of 15 questions that should determine the course of lunar research. Following the Woods Hole sessions, another group met at Falmouth, Mass., to formulate specific recommendations for the Apollo and related unmanned projects.

August: Three firms were awarded six-month contracts to design prototypes of an Apollo lunar surface experiments package, which would be left on the moon and would return data by telemetry over a period of time.

September: Meeting with MSC scientists, Public Health Service physicians insisted on rigorous quarantine of astronauts and lunar samples following each lunar mission.

December: Two Gemini spacecraft performed the first space rendezvous, maneuvering to a separation distance of one foot with no difficulty.

1966

February: OSSA selected the experiment complement for the Apollo lunar surface experiments package (ALSEP).

The first Apollo spacecraft, a test version of the command and service module, was launched from Cape Canaveral on a two-stage Saturn I-B rocket.

March: The *Gemini VIII* spacecraft performed a rendezvous with an unmanned target vehicle, then docked with it—the first accomplishment of this critical procedure. The mission was aborted soon afterwards when a small thruster malfunctioned.

NASA Hedquarters selected the Bendix Corporation to build the lunar surface experiments package.

May: *Surveyor I*, the first instrumented spacecraft designed to soft-land on the moon and return scientific data, landed in Oceanus Procellarum.

August: *Lunar Orbiter I*, the first of five photographic satellites to be launched in the following 12 months, returned detailed photographs of nine primary and seven alternate Apollo landing sites.

Contracts were let for the first two phases of construction of the lunar receiving laboratory at the Manned Spacecraft Center.

December: MSC created a Science and Applications Directorate to manage the scientific activities of the center, removing this responsibility from the Engineering and Development Directorate.

1967

January: A Lunar Missions Planning Board was established at MSC.

A flash fire in Apollo command module 012 during preflight simulations at Cape Canaveral killed all three of the astronauts inside. Investigation of the cause of this tragedy by NASA and by Congress revealed serious shortcomings in the design of the spacecraft and management of manufacturing, testing, and manned simulations. Progress in the lunar landing program was drastically slowed; it was later estimated that the fire delayed the first lunar landing by 18 months.

February: MSC announced selection of a scientist, Dr. Wilmot N. Hess, of Goddard Space Flight Center, to head its new Science and Applications Directorate.

March: The Office of Space Science and Applications released the names of 110 principal investigators whose proposals for scientific research on the lunar samples had been accepted.

Eleven scientists were selected for astronaut training, bringing the total number of scientist-astronauts to 15.

May: Prime and backup crews were named for Apollo 7, the first mission to fly after the fire. No launch date was announced, but assignment of crews indicated NASA's confidence that problems uncovered by the fire were on the way to solution.

July: Construction of the lunar receiving laboratory was completed and work was under way to install its specialized scientific equipment.

August: MSC named P. R. Bell, a radiation physicist at the Oak Ridge National Laboratory, to head the lunar receiving laboratory. Bell would report to MSC's Director of Science and Applications.

Wilmot Hess convened a group of NASA and academic scientists at the University of California at Santa Cruz to prepare more detailed plans for lunar exploration based on current expectations for lunar missions. At the end of the conference Hess named a Group for Lunar Exploration Planning to work continuously with MSC in defining the scientific aspects of Apollo missions.

September: A Lunar Sample Preliminary Examination Team and a Lunar Sample Analysis Planning Team, both including outside and NASA scientists, were created to assist the staff of the Lunar Receiving Laboratory in the examination and apportionment of lunar samples.

November: The first test flight of a complete Saturn V was successfully launched from NASA's new facilities at Kennedy Space Center and completed without significant anomalies.

December: OMSF established a Lunar Exploration Office within the Apollo Program Office, merging several program units concerned with lunar exploration. A Systems Development group staffed from OMSF would direct hardware development; a Lunar Science group staffed from OSSA would approve operating plans and scientific objectives, payloads, and principal investigators for specific missions.

During the year: The Interagency Committee on Back Contamination worked out procedures for quarantine and release of lunar astronauts and samples and defined a biological test program to search for extraterrestrial organisms.

NASA and the National Academy of Sciences worked to establish a center for research on lunar and planetary samples adjacent to the Manned Spacecraft Center. The center, to be managed by a consortium of universities, would be the organization through which interested researchers could gain access to the lunar materials for scientific work and would provide office space and other support for visiting scientists.

1968

January: The lunar module was given its first test (unmanned) in an earth-orbiting mission.

August: Plans were set in motion to fly a circumlunar mission on the second manned Apollo flight.

In view of problems in building the instruments and constraints appearing in mission planning, OMSF decided not to fly the lunar surface experiments package on the first lunar landing mission. Instead, a simplified set of instruments (a laser reflector and a passive seismometer) would be developed for the first mission and the more extensive set currently in development would be flown later.

October: Apollo 7, the first manned flight of the Apollo command module, was launched for an 11-day earth-orbital test. All primary objectives of the flight were met.

An operational readiness inspection of the lunar receiving laboratory was conducted and numerous discrepancies were noted. A 10-day simulation of LRL operations similarly uncovered many shortcomings in equipment and procedures.

December: The first flight of a manned mission on a Saturn V was launched on December 21. Apollo 8 flew to the moon, completed 10 orbits, and returned safely to earth on December 27. While in lunar orbit the crew made numerous visual and photographic observations of potential landing sites.

During the year: The Apollo Site Selection Board, working with the Group for Lunar Exploration Planning and Bellcomm, selected five sites as alternatives for the first lunar landing mission. Work continued into 1969 to produce and refine a list from which sites for subsequent exploration missions would be chosen.

1969

March: Apollo 9 checked out manned operation of the lunar module, including rendezvous procedures, in a successful 10-day mission in earth orbit.

May: Apollo 10 carried out all phases of a lunar landing mission except the final descent and landing. The lunar module descended to 50,000 feet (15,000 meters) above the lunar surface, visually verified the approach to the primary landing site for the first landing, and returned to lunar orbit to rendezvous with the command module.

OMSF authorized the Marshall Space Flight Center to proceed with development of a manned lunar roving vehicle capable of carrying two astronauts several kilometers from their landed lunar module. The vehicle would be used on the later Apollo exploration missions.

July: Apollo successfully achieved its primary goal with the landing of the lunar module *Eagle* in the Sea of Tranquility on July 20 and the successful completion of Apollo 11 on July 24. Astronauts Neil Armstrong and Edwin Aldrin spent 2.5 hours on the lunar surface, collected some 50 pounds (23 kg.) of lunar rocks and dust, and emplaced a passive seismometer and a laser retroreflector.

July–August: The Apollo 11 samples were brought to the lunar receiving laboratory, examined, and prepared for issuance to outside scientists. After a three-week stay, the crew was certified free of any biological contamination and released.

August: NASA Headquarters approved a package of experiments for remote sensing of the moon, to be flown in the Apollo service module on missions 12 through 20.

September–November: Lunar samples were released for scientific examination by principal investigators.

October: NASA awarded a contract to the Boeing Company to build the lunar roving vehicle.

November: Apollo 12 performed the first precision landing (within 1 km. of a preselected spot) at a site in Oceanus Procellarum near the spacecraft *Surveyor III.* In two surface excursions (more than 7½ hours spent outside the lunar module) the astronauts emplaced the first complete ALSEP instrument package, collected almost 75 pounds (34 kg.) of samples, and removed several parts from the Surveyor for analysis.

1970

January: The Lunar Science Institute adjacent to the Manned Spacecraft Center was officially dedicated.

Detailed reports on the analysis of samples from Apollo 11 were presented at a Lunar Science Conference in Houston, the first of a series of annual conferences on lunar (and later planetary) science.

Budget restrictions and the need to get on with post-Apollo development forced NASA to cancel Apollo 20 and stretch out the remaining seven missions to six-month invervals.

April: Apollo 13, launched on April 11, was aborted two days later when an oxygen tank containing an undetected defect exploded. Mission Control teams devised emergency procedures to conserve oxygen and electrical power, and the spacecraft and crew were brought back safely to earth on April 17 after looping around the moon. An investigation board concluded that the explosion resulted from a highly unlikely combination of circumstances that were traceable to human oversight.

September: Two more missions, Apollo 15 and 19, were canceled because of budget cuts. The remaining four missions were designated Apollo 14, 15, 16 and 17.

1971

January: Apollo 14 landed at a site of prime scientific interest, the Fra Mauro Formation. During two excursions to the lunar surface the astronauts emplaced a second set of scientific instruments and collected some 92 pounds (40 kg.) of samples, but failed to reach a crater that had been one of their primary objectives. The orbiting CSM carried out considerable photography during the mission, including photography of a landing site proposed for a future mission ("bootstrap" photography).

April: On the recommendation of the Interagency Committee on Back Contamination, NASA discontinued the practice of quarantining returned lunar samples and astronauts. No evidence of viable organisms on the moon had been produced on three lunar landing missions.

July: Apollo 15 carried the first extended lunar module and the first lunar roving vehicle to the moon. The mission landed near Mount Hadley and Hadley Rille and stayed almost 67 hours on the surface—twice as long as any prior mission. The astronauts made three trips from their lunar module, emplaced the third set of experiments (including a seismometer that completed a three-site seismic network on the moon), and drove the "rover" a total distance of 17½ miles (28 km.). The orbiting CSM carried the first scientific instrument module (SIM), which housed sensors that recorded data from the moon's surface. A moon-circling subsatellite was launched to measure particles and fields in the lunar environment. During the trip back to earth the command module pilot retrieved film cassettes from the SIM experiments, the first extravehicular activity conducted during a moon-to-earth voyage.

1972

April: Apollo 16 continued NASA's steady extension of lunar exploration missions, staying 71 hours on the surface, planting the fourth set of instruments, and returning almost 200 pounds (91 kg.) of samples. A second set of SIM instruments was operated, and another subsatellite was launched.

July: A summer study on post-Apollo lunar science outlined priorities for future study of Apollo samples and data. The plan called for two years of organization and preliminary analysis of the data, to be followed by two years of careful examination of those data, after which priority would be given to the key problems that emerged. The study recommended continued support of the curatorial facilities at MSC and collection of data from the lunar surface experiments as long as they produced significant new information.

December: The last lunar exploration mission, Apollo 17, carried the first scientist (geologist Harrison H. Schmitt) to the moon. After landing in the Taurus-Littrow region, the astronauts stayed 75 hours, spent 22 hours outside the lunar module, drove their rover 22 miles (35 km.), and collected nearly 250 pounds (113 km.) of samples.

1973

February: The Manned Spacecraft Center was renamed the Lyndon B. Johnson Space Center.

March: A Lunar Programs Office was established in the Office of Space Sciences, NASA Headquarters, to conduct the Lunar Data Analysis and Synthesis Program. The program would oversee the collection and scientific analysis of data from the lunar surface instruments and the lunar samples.

May: The first post-Apollo manned space flight program began with the launch of *Skylab 1,* a Saturn S-IVB stage converted to a laboratory module capable of supporting three-person crews for long periods in earth orbit. Skylab was the outgrowth of earlier "Apollo Applications" planning intended to use the hardware developed for Apollo to collect scientific data. *Skylab 1* used the last Saturn V rocket ever launched. Crews occupied the laboratory for periods of 28, 59, and 84 days; the last mission ended on February 8, 1974.

August: The Office of Manned Space Flight designated an official to be responsible for the final phasing out of the Apollo project.

1975

July: The Apollo-Soyuz Test Project (ASTP) , the first international manned space mission, was conducted in cooperation with the Soviet Union. An Apollo command and service module fitted with a special adapter docked with a Soviet Soyuz spacecraft to conduct joint operations in earth orbit. After separating from the Soyuz, the Apollo crew carried out a short program of scientific experiments. ASTP marked the last use of the launch vehicles and spacecraft built for the Apollo project.

APPENDIX 5

SUMMARY DATA ON APOLLO MISSIONS

Apollo Manned Missions:
Vehicle, Crew, and Launch Data

Mission No.	Launch vehicle No. [a]	CSM No.	LM No.	Crew [b]	Call signs (CSM, LM)	Date, time [c] of launch	Launch site [d]
Apollo 7	205	101	None	Schirra, Eisele, Cunningham	Apollo 7 (no LM)	11 Oct. 1968 10:02:45 a.m.	ETR LC-34
Apollo 8	503	103	None	Borman, Lovell, Anders	Apollo 8 (no LM)	21 Dec. 1968 7:51:00 a.m.	KSC LC-39A
Apollo 9	504	104	3	McDivitt, Scott, Schweickart	Gumdrop, Spider	3 March 1969 11:00:00 a.m.	KSC LC-39A
Apollo 10	505	106	4	Stafford, Young, Cernan	Charlie Brown, Snoopy	18 May 1969 11:49:00 a.m.	KSC LC-39B
Apollo 11	506	107	5	Armstrong, Collins, Aldrin	Columbia, Eagle	16 July 1969 8:32:00 a.m.	KSC LC-39A
Apollo 12	507	108	6	Conrad, Gordon, Bean	Yankee Clipper, Intrepid	14 Nov. 1969 11:22:00 a.m.	KSC LC-39A
Apollo 13	508	109	7	Lovell, Swigert, Haise	Odyssey, Aquarius	11 Apr. 1970 2:13:00 p.m.	KSC LC-39A
Apollo 14	509	110	8	Shepard, Roosa, Mitchell	Kitty Hawk, Antares	31 Jan. 1971 5:03:02 p.m.	KSC LC-39A
Apollo 15	510	112	10	Scott, Worden, Irwin	Endeavor, Falcon	26 July 1971 8:34:00 a.m.	KSC LC-39A
Apollo 16	511	113	11	Young, Mattingly, Duke	Caspar, Orion	16 Apr. 1972 12:54:00 p.m.	KSC LC-39A
Apollo 17	512	114	12	Cernan, Evans, Schmitt	America, Challenger	7 Dec. 1972 12:33:00 a.m.	KSC LC-39A

[a] 200 number, Saturn IB; 500 number, Saturn V.
[b] Commander, CM pilot, LM pilot.
[c] Eastern Standard Time.
[d] ETR, Eastern Test Range; KSC, Kennedy Space Center; LC, launch complex.

Source for all tables: "Apollo Program Summary Report," JSC-09423, April 1975.

Apollo Manned Missions:
Landing and Recovery Data

Mission	Landing date, time [a]	Mission duration [b]	Landing point, lat., long. [c]	Ocean	Distance from target (naut. mi.) [c]	Recovery ship
Apollo 7	22 Oct. 1968 6:11:48 a.m.	260:09:03	27°38'N, 64° 9'W	Atlantic	1.9	U.S.S. *Essex*
Apollo 8	27 Dec. 1968 10:51:42 a.m.	147:00:42	8° 6'N, 165° 1'W	Pacific	1.4	U.S.S. *Yorktown*
Apollo 9	13 March 1969 12:00:53 p.m.	241:00:54	23°13'N, 67°59'W	Atlantic	2.7	U.S.S. *Guadalcanal*
Apollo 10	26 May 1969 11:52:23 a.m.	192:03:23	15° 4'S, 164°39'W	Pacific	1.3	U.S.S. *Princeton*
Apollo 11	24 July 1969 11:50:35 a.m.	195:18:35	13°18'N, 169° 9'W	Pacific	1.7	U.S.S. *Hornet*
Apollo 12	24 Nov. 1969 3:58:25 p.m.	244:36:25	15°47'S, 165° 9'W	Pacific	2.0	U.S.S. *Hornet*
Apollo 13	17 April 1970 1:07:41 p.m.	142:54:41	21°38'S, 165°22'W	Pacific	1.0	U.S.S. *Iwo Jima*
Apollo 14	9 Feb. 1971 4:05:00 p.m.	216:01:58	27° 1'S, 172°40'W	Pacific	0.6	U.S.S. *New Orleans*
Apollo 15	7 Aug. 1971 3:45:53 p.m.	259:11:53	26° 8'N, 158° 8'W	Pacific	1.0	U.S.S. *Okinawa*
Apollo 16	27 Apr. 1972 2:45:05 p.m.	265:51:05	0°42'S, 156°13'W	Pacific	3.0	U.S.S. *Ticonderoga*
Apollo 17	19 Dec. 1972 2:24:59 p.m	301:51:59	17°53'S, 166° 7'W	Pacific	1.0	U.S.S. *Ticonderoga*

[a] Command module splashdown, Eastern Standard Time.
[b] Hours:minutes:seconds.

[c] Best estimate; may be based on recovery ship position data, CM computer data, or trajectory reconstruction.

Summary of Apollo Lunar Surface Activity

Mission No.	Lunar landing site (lat., long./name)	Date, time [a] of: Lunar landing	Lunar liftoff	Time on lunar surface [b]	Duration of EVAs [c]	Weight of samples collected (kg.)
Apollo 11	0.7°N, 23.4°E Mare Tranquillitatis	20 July 1969 5:17:40 p.m.	21 July 1969 2:54:01 p.m.	21:36:21	2:31:40	21
Apollo 12	3.2°S, 23.4°W Oceanus Procellarum	19 Nov. 1969 1:54:36 a.m.	20 Nov. 1969 9:25:48 a.m.	31:31:12	3:56:03 3:49:15 7:45:18	16.7 17.6 34.3
Apollo 14 [d]	3.6°S, 17.5°W Fra Mauro	5 Feb. 1971 4:18:11 a.m.	6 Feb. 1971 1:48:42 p.m.	33:30:31	4:47:50 4:34:41 9:22:31	20.5 22.3 42.8
Apollo 15	26.1°N, 3.7°E Hadley-Apennine	30 July 1971 5:16:29 p.m.	2 Aug. 1971 1:11:22 p.m.	66:54:53	0:33:07 [e] 6:32:42 7:12:14 4:49:50 19:07:53	---- 14.5 34.9 27.3 76.7
Apollo 16	9.0°S, 15.5°E Descartes	20 Apr. 1972 9:23:35 p.m.	23 Apr. 1972 8:25:48 p.m.	71:02:13	7:11:02 7:23:11 5:40:03 20:14:16	29.9 29.0 35.4 94.3
Apollo 17	20.2°N, 30.8°E Taurus-Littrow	11 Dec. 1972 2:54:57 p.m.	14 Dec. 1972 5:54:37 p.m.	74:59:40	7:11:53 7:36:56 7:15:08 22:03:57	14.3 34.1 62.0 110.4

[a] Eastern Standard Time. Liftoff time calculated from touchdown time plus time on lunar surface.

[b] Touchdown to liftoff, hr:min:sec.

[c] Extravehicular activity in hr:min:sec, LM hatch opening to closing. Multiple EVAs shown separately.

[d] Apollo 13 was aborted following an explosion in the service module 55 hr, 54 min after launch. Intended landing site, on the Fra Mauro formation, was used for Apollo 14.

[e] "Standup" EVA—commander stood up in open upper LM hatch to make visual and photographic observations of the landing area.

Source: JSC-09423, "Apollo Program Summary Report," April 1975.

Apollo Science Experiments

	Experiment	Mission 11	12	13	14	15	16	17
I. Surface Experiments								
S-031	Passive seismic experiment	x	x	(x)	x	x	x	
S-033	Active seismic experiment				x		x	
S-034	Lunar surface magnetometer		x			x	x	
S-035	Solar wind spectrometer		x			x		
S-036	Suprathermal ion detector		x		x	x		
S-037	Heat flow experiment			(x)		x	x	
S-038	Charged particle lunar environment			(x)	x			
S-058	Cold cathode ion gauge		x		x	x		
S-059	Lunar field geology	x	x	(x)	x	x	x	x
S-080	Solar wind composition	x	x	(x)	x	x	x	
S-078	Laser ranging retroreflector	x			x	x		
M-515	Lunar dust detector		x	(x)	x	x		
S-198	Portable magnetometer				x		x	
S-199	Lunar gravity traverse							x
S-200	Soil mechanics				x	x	x	x
S-201	Far-ultraviolet camera/spectroscope				x			
S-202	Lunar ejecta and meteorites							x
S-203	Lunar seismic profiling							x
S-204	Surface electrical properties							x
S-205	Lunar atmospheric composition							x
S-207	Lunar surface gravimeter							x
S-229	Lunar neutron probe							x
II. Lunar Orbital Experiments								
S-158	Multispectral photography		x					
S-176	Command module window meteoroid				x	x	x	x
S-177	Ultraviolet photography, earth and moon					x	x	
S-178	Gegenschein from lunar orbit			(x)	x	x		
S-160	Gamma-ray spectrometer					x	x	
S-161	X-ray fluorescence					x	x	
S-162	Alpha-particle spectrometer					x	x	
S-164	S-band transponder (CSM/LM)		x	(x)	x	x	x	
S-164	S-band transponder (subsatellite)					x	x	
S-165	Mass spectrometer					x	x	
S-169	Far-ultraviolet spectrometer							x
S-170	Bistatic radar				x	x	x	x
S-171	Infrared scanning radiometer							x
S-173	Particle shadows/boundary layer (subsatellite)					x	x	
S-174	Magnetometer (subsatellite)					x	x	
S-209	Lunar sounder							x

(x) not performed (or deployed) on aborted mission.

Source: JSC, "What We've Learned About The Moon," July 1980.

Apollo Lunar Surface Experiments Package Arrays and Status as of April 1975*

Experiment	Apollo 12 Array A	Apollo 13 Array B	Apollo 14 Array C	Apollo 15 Array A-2	Apollo 16 Array D	Apollo 17 Array E
Passive seismic	Short-period Z-axis has displayed reduced sensitivity since deployment.	Not deployed.	Long-period Z-axis inoperative since 3/20/72. Noisy data on long-period Y axis since 4/14/73.	Full operation.	Full operation.	
Active seismic			Mortar not fired. Geophone 3 data noisy since 3/26/71. Geophone 2 data invalid since 1/3/74.		3 of 4 grenades launched. Mortar pitch sensor off scale after 3rd firing on 5/23/72.	
Lunar surface magnetometer	Permanently commanded off 6/14/74.			Permanently commanded off 6/14/74.	Full operation.	
Solar wind spectrometer	Full operation except for intermittent modulation drop in two proton energy levels each lunation since 11/5/71.			Permanently commanded off 6/14/74.		
Suprathermal ion detector	Periodically commanded off to prevent high-voltage arcing at elevated lunar day temperatures since 9/9/72.		Periodically commanded to standby operation to avoid mode changes at elevated lunar day temperatures since 3/29/72.	Periodically commanded to standby operation to avoid mode changes at elevated lunar day temperatures since 9/13/73.		
Heat flow		Not deployed.		Probe 2 not to full depth intended, but experiment provides useful data.	Inoperative since emplacement.	Full operation.
Cold-cathode ion gauge	Inoperative. Failed 14 hours after turn-on 11/20/69.	Not deployed.	Intermittent science data since 3/29/72.	Intermittent science data since 2/22/73.		
Lunar ejecta and meteorites						Thermal control design not optimum for Apollo 17 site. Instrument operated for about 75 percent of lunation.

* The Apollo 14 ALSEP station failed in January 1976. The stations still operating (12, 15, 16, and 17) were turned off September 30, 1977.

Flight Directors for Apollo Manned Missions

Mission	Shift 1	Shift 2	Shift 3	Shift 4
7	Glynn S. Lunney	Eugene F. Kranz	Gerald D. Griffin	—
8	Clifford E. Charlesworth	Lunney	Milton L. Windler	—
9	Kranz	Griffin	M. P. Frank	—
10	Lunney, Griffin	Windler	Frank	—
11	Charlesworth, Griffin	Kranz	Lunney	Windler
12	Griffin	Frank	Charlesworth	Lunney
13	Windler	Griffin	Kranz	Lunney
14	Frank, Lunney	Windler	Griffin	—
15	Griffin	Windler	Lunney, Kranz	—
16	Frank, Philip C. Shaffer	Kranz, Donald R. Puddy	Griffin, Neil B. Hutchinson, Charles R. Lewis	—
17	Griffin	Kranz, Hutchinson	Frank, Lewis	—

Source: memos from Director, Flight Operations, MSC, to multiple addressees, listing personnel assignments for the Mission Control Center for each flight.

APPENDIX 6

PRIME AND BACKUP CREWS, SUPPORT CREWS, AND CAPSULE COMMUNICATORS FOR APOLLO LUNAR MISSIONS*

Apollo 8 Prime Crew

Commander: Frank Borman, Colonel,** U.S. Air Force. Born March 14, 1928, Gary, Indiana. B.S. 1950, U.S. Military Academy; M.S. (aeronautical engineering), 1957, California Institute of Technology. Chosen with the second group of astronauts in 1962; backup command pilot for Gemini IV, command pilot for Gemini VII, the longest manned mission (14 days) up to that time.

Command Module Pilot: James Arthur Lovell, Jr., Captain, U.S. Navy. Born March 25, 1929, Cleveland, Ohio. B.S. 1952, U.S. Naval Academy. Chosen with the second group of astronauts in 1962; backup pilot for Gemini IV, pilot for Gemini VII, backup command pilot for Gemini IX, command pilot for Gemini XII; later served as backup commander for Apollo 11 and commander for Apollo 13.

Lunar Module Pilot: William Anderson Anders, Major, U.S. Air Force. Born October 17, 1933, in Hong Kong. B.S. 1955, U.S. Naval Academy; M.S. (nuclear engineering), 1963, Air Force Institute of Technology. Chosen with the third group of astronauts in 1963; backup pilot for Gemini XI; later served as backup command module pilot for Apollo 11.

Apollo 8 Backup Crew

Commander: Neil A. Armstrong (see prime crew, Apollo 11).
Command Module Pilot: Edwin A. Aldrin (see prime crew, Apollo 11).
Lunar Module Pilot: Fred W. Haise (see prime crew, Apollo 13).

Apollo 10 Prime Crew

Commander: Thomas Patton Stafford, Colonel, U.S. Air Force. Born September 17, 1930, Weatherford, Oklahoma. B.S., 1952, U.S. Naval Academy. Chosen with the second group of astronauts in 1962; backup pilot for Gemini

---

*Source: *Astronauts and Cosmonauts, Biographical and Statistical Data* [Revised May 31, 1978], report prepared for the House Committee on Science and Technology by the Congressional Research Service, Library of Congress, July 1978; Apollo mission reports.
**Military ranks given are those held at the time of the mission.

III, pilot for Gemini VI, backup commander for Apollo 7; later commander, Apollo-Soyuz Test Project.

Command Module Pilot: John Watts Young, Commander, U.S. Navy. Born September 24, 1930, San Francisco, California. B.S. (aeronautical engineering), 1952, Georgia Institute of Technology. Chosen with the second group of astronauts in 1962; pilot of Gemini III, backup pilot for Gemini VI, command pilot for Gemini X; later backup commander for Apollo 13, commander for Apollo 16 (ninth man to walk on the moon), and backup commander for Apollo 17. Later commanded several flights of the Space Shuttle Orbiter.

Lunar Module Pilot: Eugene Andrew Cernan, Commander, U.S. Navy. Born March 14, 1934, Chicago, Illinois. B.S. (electrical engineering), 1956, Purdue University; M.S. (aeronautical engineering), 1961, Purdue and the U.S. Naval Postgraduate School. Chosen with the third group of astronauts in 1963. Pilot for Gemini IX, backup pilot for Gemini XII, backup lunar module pilot for Apollo 7; later backup commander for Apollo 14 and commander of Apollo 17 (eleventh man to walk on the moon).

Apollo 10 Backup Crew

Commander: Leroy Gordon Cooper, Colonel, U.S. Air Force. Born March 6, 1927, Shawnee, Oklahoma. B.S. (aeronautical engineering), 1956, U.S. Air Force Institute of Technology. Chosen with the first group of astronauts in 1959. Backup pilot for Mercury-Atlas 8, pilot for Mercury-Atlas 9 (last flight in the Mercury project), command pilot for Gemini V, backup command pilot for Gemini XII.

Command Module Pilot: Donn Fulton Eisele, Major, U.S. Air Force. Born June 23, 1930, Columbus, Ohio. B.S., 1952, U.S. Naval Academy; M.S. (astronautics), 1960, Air Force Institute of Technology. Chosen with the third group of astronauts in 1963; command module pilot for Apollo 7.

Apollo 11 Prime Crew

Commander: Neil Alden Armstrong (civilian). Born August 5, 1930, Wapakoneta, Ohio. B.S. (aeronautical engineering), 1955, Purdue University. Naval aviator and NASA test pilot, working in the X-15 program at the time of his selection with the second group of astronauts in 1962. Backup command pilot for Gemini V, command pilot for Gemini VIII, backup commander for Apollo 8; first man to walk on the moon.

Command Module Pilot: Michael Collins, Lieutenant Colonel, U.S. Air Force. Born October 31, 1930, Rome, Italy. B.S., 1952, U.S. Military Academy. Chosen with the third group of astronauts in 1963. Served as backup pilot for Gemini VII, pilot for Gemini X; assigned to Apollo 8 but replaced when a bone spur required surgery.

Lunar Module Pilot: Edwin Eugene ("Buzz") Aldrin, Jr., Colonel, U.S. Air Force. Born January 20, 1930, Montclair, N.J. B.S., 1951, U.S. Military Academy; Sc.D. (astronautics), 1963, Massachusetts Institute of Technology.

Chosen with the third group of astronauts in 1963. Backup pilot for Gemini IX, pilot for Gemini XII; second man to walk on the moon.

Apollo 11 Backup Crew

Commander: James A. Lovell (see prime crew, Apollo 8).
Command Module Pilot: William A. Anders (see prime crew, Apollo 8).
Lunar Module Pilot: Fred W. Haise (see prime crew, Apollo 13).

Apollo 12 Prime Crew

Commander: Charles ("Pete") Conrad, Jr., Commander, U.S. Navy. Born June 2, 1930, Philadelphia, Pennsylvania. B.S. (aeronautical engineering), 1953, Princeton University. Chosen with the second group of astronauts in 1962. Served as pilot for Gemini V, backup command pilot of Gemini VIII, command pilot for Gemini XI, backup commander for Apollo 9; third man to walk on the moon. Later served as commander of Skylab 2.
Command Module Pilot: Richard Francis Gordon, Jr., Commander, U.S. Navy. Born October 5, 1929, Seattle, Washington. B.S. (chemistry), 1951, University of Washington. Chosen with the third group of astronauts in 1963. Served as backup pilot for Gemini VIII, pilot for Gemini XI, backup command module pilot for Apollo 9; later served as backup commander for Apollo 15.
Lunar Module Pilot: Alan LaVerne Bean, Commander, U.S. Navy. Born March 15, 1932, Wheeler, Texas. B.S. (aeronautical engineering), 1955, The University of Texas. Chosen with the third group of astronauts in 1963. Served as backup command pilot for Gemini X, backup lunar module pilot for Apollo 9; fourth man to walk on the moon. Later served as commander for Skylab 3 and backup commander for the Apollo-Soyuz Test Project.

Apollo 12 Backup Crew

Commander: David R. Scott (see prime crew, Apollo 15).
Command Module Pilot: Alfred M. Worden (see prime crew, Apollo 15).
Lunar Module Pilot: James B. Irwin (see prime crew, Apollo 15).

Apollo 13 Prime Crew

Commander: James A. Lovell (see Apollo 8 prime crew).
Command Module Pilot: John Leonard Swigert, Jr. (civilian). Born August 30, 1931, Denver, Colorado. B.S. (mechanical engineering), 1953, University of Colorado; M.S. (aerospace science), 1965, Rensselaer Polytechnic Institute, M.S. (business administration), 1967, University of Hartford. Chosen with the fifth group of astronauts in 1966. Originally assigned as backup command module pilot on Apollo 13; took over the prime crew position 72 hours before

launch when Thomas Mattingly (see Apollo 16 prime crew, below) was found to have been exposed to rubella.

Lunar Module Pilot: Fred Wallace Haise (civilian). Born November 14, 1933, Biloxi, Mississippi. B.S. (aeronautical engineering), 1951, University of Oklahoma. Served as an aviator in the U.S. Marine Corps and the U.S. Air Force. Chosen with the fifth group of astronauts in 1966. Served as backup lunar module pilot for Apollo 11 and later as backup commander for Apollo 16.

Apollo 13 Backup Crew

Commander: John W. Young (see Apollo 10 prime crew).
Command Module Pilot: John L. Swigert (see Apollo 13 prime crew).
Lunar Module Pilot: Charles M. Duke (see Apollo 16 prime crew).

Apollo 14 Prime Crew

Commander: Alan Bartlett Shepard, Jr., Captain, U.S. Navy. Born November 18, 1923, East Derry, New Hampshire. B.S., 1944, U.S. Naval Academy. Chosen with the first group of astronauts in 1959, he was the United States' first man in space (Mercury-Redstone 3, *Freedom 7*, suborbital flight March 1961) and backup pilot for Mercury-Atlas 9 *(Faith 7)*. He was grounded because of an inner-ear ailment until May 1969; On returning to flight status he was assigned as commander of Apollo 14 and became the fifth man (the only one of the first group of astronauts) to walk on the moon.

Command Module Pilot: Stuart Allen Roosa, Major, U.S. Air Force. Born August 15, 1933, Durango, Colorado. B.S. (aeronautical engineering), 1960, University of Colorado. Chosen with the fifth group of astronauts in 1966. Apollo 14 was his first assignment; he later served as backup command module pilot for Apollo 16 and Apollo 17.

Lunar Module Pilot: Edgar Dean Mitchell, Commander, U.S. Navy. Born September 17, 1930, Hereford, Texas. B.S. (industrial management), 1952, Carnegie Institute of Technology; B.S. (aeronautical engineering), 1961, U.S. Naval Postgraduate School, Sc.D. (aeronautics and astronautics), 1964, Massachusetts Institute of Technology. Chosen with the fifth group of astronauts in 1966; sixth man to walk on the moon. Later served as backup lunar module pilot for Apollo 16.

Apollo 14 Backup Crew

Commander: Eugene A. Cernan (see prime crew, Apollo 10).
Command Module Pilot: Ronald E. Evans (see prime crew, Apollo 17).
Lunar Module Pilot: Joe Henry Engle, Lieutenant Colonel, U.S. Air Force. Born August 26, 1932, Abilene, Kansas. B.S. (aeronautical engineering), 1955, University of Kansas. Qualified as an astronaut in the NASA-Air Force X-15 project, he was chosen with the fifth group of astronauts in 1966. Apollo

14 was his first and only Apollo assignment; he later participated in the approach and landing tests and the orbital flight tests of the Space Shuttle Orbiter.

Apollo 15 Prime Crew

Commander: David Randolph Scott, Colonel, U.S. Air Force. Born June 6, 1932, in San Antonio, Texas. B.S., 1954, U.S. Military Academy; M.S. (aeronautics and astronautics) and Engineer of Aeronautics and Astronautics, 1963, Massachusetts Institute of Technology. Chosen with the third group of astronauts in 1963. Served as pilot for Gemini VIII, command module pilot for Apollo, 9; seventh man to walk on the moon.

Command Module Pilot: Alfred Merrill Worden, Major, U.S. Air Force. Born February 7, 1932, in Jackson, Michigan. B.S., 1955, U.S. Military Academy; M.S. (astronautical/aeronautical engineering and instrumentation engineering), 1963, University of Michigan. Chosen with the fifth group of astronauts in 1966. Served as backup command module pilot on Apollo 12.

Lunar Module Pilot: James Benson Irwin, Lieutenant Colonel, U.S. Air Force. Born March 17, 1930, Pittsburgh, Pennsylvania. B.S., 1951, U.S. Naval Academy; M.S. (aeronautical and instrumentation engineering), 1957, University of Michigan. Served as backup lunar module pilot on Apollo 12; eighth man to walk on the moon. Resigned from NASA and the Air Force in 1972.

Apollo 15 Backup Crew

Commander: Richard F. Gordon (see prime crew, Apollo 12).

Command Module Pilot: Vance DeVoe Brand (civilian). Born May 9, 1931, Longmont, Colorado. B.S. (business), 1953, B.S. (aeronautical engineering), 1960, University of Colorado; M.S. (business administration), 1964, University of California at Los Angeles. Served as aviator in the U.S. Marine Corps and as flight test engineer and test pilot with the Lockheed Aircraft Corporation. Chosen with the fifth group of astronauts in 1966. Assigned to support crew for Apollo 13; later served as command module pilot for the Apollo-Soyuz Test Project and commanded two flights of the Shuttle Orbiter.

Lunar Module Pilot: Harrison Hagan ("Jack") Schmitt (see prime crew, Apollo 17).

Apollo 16 Prime Crew

Commander: John W. Young (see prime crew, Apollo 10).

Command Module Pilot: Thomas K. Mattingly II, Lieutenant Commander, U.S. Navy. Born March 17, 1936, Chicago, Illinois. B.S. (aeronautical engineering), 1958, Auburn University. Chosen with the fifth group of astronauts in 1966. Served as command module pilot for Apollo 13 until three days before launch, when it was discovered that he had been exposed to rubella (German measles) and had no immunity. Later commanded the fourth orbital test flight of the first Space Shuttle Orbiter, *Columbia*.

Lunar Module Pilot: Charles Moss Duke, Jr., Lieutenant Colonel, U.S. Air Force. Born October 3, 1935, Charlotte, North Carolina. B.S., 1957, U.S. Naval Academy. Chosen with the fifth group of astronauts in 1966. Served as backup lunar module pilot on Apollo 13. Tenth man to walk on the moon.

Apollo 16 Backup Crew

Commander: Fred W. Haise (see prime crew, Apollo 13).
Command Module Pilot: Stuart A. Roosa (see prime crew, Apollo 14).
Lunar Module Pilot: Edgar D. Mitchell (see prime crew, Apollo 14).

Apollo 17 Prime Crew

Commander: Eugene A. Cernan (see prime crew, Apollo 10).

Command Module Pilot: Ronald Ellwin Evans, Commander, U.S. Navy. Born November 10, 1933, St. Francis, Kansas. B.S. (electrical engineering), 1956, University of Kansas; M.S. (aeronautical engineering), 1964, U.S. Naval Postgraduate School. Chosen with the fifth group of astronauts in 1966. Apollo 14 was his first crew assignment; later served as command module pilot for Apollo 17 and backup command module pilot for the Apollo-Soyuz Test Project.

Lunar Module Pilot: Harrison Hagan ("Jack") Schmitt (civilian). Born July 3, 1935, Santa Rita, New Mexico. B.S. (science), 1957, California Institute of Technology; Ph.D. (geology), 1964, Harvard University. At the time of selection as an astronaut Dr. Schmitt was working with the Astrogeology Branch, U.S. Geological Survey, Flagstaff, Arizona, as project chief for lunar field geological methods and was involved in photographic and telescopic mapping of the moon. Chosen with the first group of scientist-astronauts in 1965. Served as backup lunar module pilot on Apollo 15; twelfth and last man, and the only geologist, to walk on the moon.

Apollo 17 Backup Crew

Commander: Fred Haise (see prime crew, Apollo 13)
Command module pilot: Alfred Worden (see prime crew, Apollo 15)
Lunar module pilot: James Irvin (see prime crew, Apollo 15)

Two other missions, Apollo 7 and Apollo 9, were earth-orbiting missions conducted to test the redesigned Apollo (Block II) spacecraft and the lunar module before committing them to a lunar mission.

Apollo 7 Prime Crew

Commander: Walter Marty Schirra, Jr., Captain, U.S. Navy. Born March 12, 1923, Hackensack, New Jersey. B.S., 1945, U.S. Naval Academy. Chosen with the first group of astronauts in 1959; the only member of this group to fly in Mercury, Gemini, and Apollo. Served as backup pilot for Mercury-Atlas 7 *(Aurora 7)* and pilot for Mercury-Atlas 8 *(Sigma 7),* backup command pilot for Gemini III, command pilot of Gemini VI (first mission to conduct a rendezvous).

Command Module Pilot: Donn F. Eisele (see backup crew, Apollo 10).

Lunar Module Pilot (no lunar module assigned to this flight): Ronnie Walter Cunningham (civilian). Born March 16, 1932, Creston, Iowa. B.A., 1960, and M.A., 1961 (physics), University of California at Los Angeles. Chosen with the third group of astronauts in 1963.

Apollo 7 Backup Crew

Commander: Thomas P. Stafford (see prime crew, Apollo 10).
Command Module Pilot: John W. Young (see prime crew, Apollo 10).
Lunar Module Pilot: Eugene A. Cernan (see prime crew, Apollo 10).

Apollo 9 Prime Crew

Commander: James Alton McDivitt, Colonel, U.S. Air Force. Born June 10, 1929, Chicago, Illinois. B.S. (aeronautical engineering), 1959, University of Michigan. Chosen with the second group of astronauts in 1962. Command pilot for Gemini IV. Served as manager of the Apollo Spacecraft Project Office at MSC from 1969 to 1972.

Command Module Pilot: David R. Scott (see prime crew, Apollo 15).

Lunar Module Pilot: Russell Louis Schweickart (civilian). Born October 25, 1935, Neptune, New Jersey. B.S. (aeronautical engineering), 1956; M.S. (aeronautics and astronautics), 1963, Massachusetts Institute of Technology. Chosen with the third group of astronauts in 1963. Later served as backup commander for Skylab 2.

Apollo 9 Backup Crew

Commander, Charles Conrad; Command Module Pilot, Richard F. Gordon; Lunar Module Pilot Alan L. Bean (see prime crew, Apollo 12).

Support crews were assigned for the first time in Apollo, to assume some of the large work load entailed by the greater complexity of the Apollo spacecraft and the lunar missions. Their duties were many and varied: maintaining the flight data file, filling in for prime and backup crew members at design reviews where necessary, and keeping the prime and backup crews informed of changes in procedures (simulations, experiments, etc.). Some support crewmen later served on prime or backup crews but many did not. Following are the members of the support crews for all the Apollo missions.

Apollo 7

Jack L. Swigert (prime crew, Apollo 13)

Ronald E. Evans (prime crew, Apollo 17)

Edward Galen Givens, Jr., Major, U.S. Air Force, born January 5, 1930, Quanah, Texas. B.S., 1952, U.S. Naval Academy. Chosen with the fifth group of astronauts in 1966; killed in an automobile accident June 6, 1967.

Apollo 8

Thomas K. Mattingly II (prime crew, Apollo 16).
Gerald Paul Carr, Lieutenant Colonel, U.S. Marine Corps. Born August 22, 1932, Denver, Colorado. B.S. (mechanical engineering), 1954, University of Southern California; B.S. (aeronautical engineering), 1961, U.S. Naval Postgraduate School; M.S. (aeronautical engineering), 1962, Princeton University. Chosen with the fifth group of astronauts in 1966. No other Apollo crew assignments; later served as commander for Skylab 4.
John Sumter Bull, Lieutenant Commander, U.S. Navy. Born September 25, 1935, Memphis, Tennessee. B.S. (mechanical engineering), 1957, William Marsh Rice University. Chosen with the fifth group of astronauts in 1966. Withdrew from the program in 1968 for medical reasons.

Apollo 9

Edgar D. Mitchell (prime crew, Apollo 14)

Fred W. Raise (prime crew, Apollo 13)

Alfred M. Worden (prime crew, Apollo 15)

Apollo 10

Joe H. Engle (backup crew, Apollo 14)

James B. Irwin (prime crew, Apollo 15)

Charles M. Duke, Jr. (prime crew, Apollo 16)

Apollo 11

Thomas K. Mattingly II (prime crew, Apollo 16)

Ronald E. Evans (prime crew, Apollo 17)

John L. Swigert (prime crew, Apollo 13)

William Reid Pogue, Lieutenant Colonel, U.S. Air Force. Born January 23, 1930, Okemah, Oklahoma. B.S. (education), 1951, Oklahoma Baptist University; M.S. (mathematics), 1960, Oklahoma State University. Chosen with the fifth group of astronauts in 1966. Later served on support crews for Apollo 13 and Apollo 14 and as pilot for Skylab 4.

Apollo 12

Gerald P. Carr (support crew, Apollo 8)

Paul Joseph Weitz, Lieutenant Commander, U.S. Navy. Born July 25, 1932, Erie, Pennsylvania. B.S. (aeronautical engineering), 1954, Pennsylvania State University; M.S. (aeronautical engineering), U.S. Naval Postgraduate School. Chosen with the fifth group of astronauts in 1966. Later served as pilot on Skylab 2.

Edward George Gibson (civilian). Born November 8, 1936, Buffalo, New York. B.S. (engineering), 1959, University of Rochester; M.S. (engineering), 1960, Ph.D. (engineering), 1964, California Institute of Technology. Chosen with the first group of scientist-astronauts in 1965. Later served as scientist-pilot on Skylab 4.

Apollo 13

Vance D. Brand (backup crew, Apollo 15)

Willam R. Pogue (support crew, Apollo 11)

Jack Robert Lousma, Lieutenant Colonel, U.S. Marine Corps. Born February 29, 1936, Grand Rapids, Michigan. B.S. (aeronautical engineering), 1959, University of Michigan; Aeronautical Engineer, 1965, U.S. Naval Postgraduate School. Chosen with the fifth goup of astronauts in 1966. Later served as pilot for Skylab 3 and backup docking module pilot for the Apollo-Soyuz Test Project.

Apollo 14

William R. Pogue (Backup crew, Apollo 11 and Apollo 13)
Bruce McCandless II, Commander, U.S. Navy. Born June 8, 1937, Boston, Massachusetts. B.S., 1958, U.S. Naval Academy; M.S. (electrical engineering),

1965, Stanford University. Chosen with the fifth group of astronauts in 1966. Later served as backup pilot for Skylab 2.

Charles Gordon Fullerton, Lieutenant Colonel, U.S. Air Force. Born October 11, 1936, Rochester, New York. B.S. (mechanical engineering), 1957, M.S. (mechanical engineering), 1958, California Institute of Technology. Transferred to the NASA astronaut corps in 1969 from the Air Force Manned Orbiting Laboratory project (canceled). Later served on the support crew for Apollo 17.

Apollo 15

Karl Gordon Henize (civilian). Born October 17, 1926, Cincinnati, Ohio. B.S. (mathematics), 1947; M.S. (astronomy), 1948, University of Virginia; Ph.D. (astronomy), 1954, University of Michigan. Chosen with the second group of scientist-astronauts in 1967. Principal investigator for ultraviolet astronomy experiments in the Skylab project. Later flew on Space Shuttle missions.

Robert Allan Ridley Parker (civilian). Born December 14, 1936, New York, New York. B.A. (astronomy and physics), 1958, Amherst College; Ph.D. (astronomy), 1962, California Institute of Technology. Chosen with the second group of scientist-astronauts in 1967. Later served as support crewman and mission scientist for Apollo 17 and chief scientist for the Skylab missions.

Joseph Percival Allen IV (civilian). Born June 27, 1937, Crawfordsville, Indiana. B.A. (mathematics and physics), 1959, DePauw University; M.S. (physics), 1961, Ph.D. (nuclear physics), 1965, Yale University. Chosen with the second group of scientist-astronauts in 1967. Later served as mission scientist for Apollo 15; left the astronaut corps in 1975 to serve as NASA's Assistant Administrator for Legislative Affairs; returned in 1978 to participate in the Space Shuttle program.

Apollo 16

Philip Kenyon Chapman (civilian). Born March 5, 1935, Melbourne, Victoria, Australia (naturalized in U.S. May 8, 1967). B.Sc. (physics and mathematics), 1956, Sydney University; M.S. (aeronautics and astronautics) 1964, Ph.D. (instrumentation), 1967, Massachusetts Institute of Technology. Chosen with the second group of scientist-astronauts in 1967.

Anthony Wayne England (civilian). Born May 15, 1942, Indianapolis, Indiana. B.S. and M.S. (geology and physics), 1965, Ph.D. (planetary sciences), 1970, Massachusetts Institute of Technology. Chosen with the second group of scientist-astronauts in 1967. Later mission scientist for Apollo 16.

Henry Warren Hartsfield, Jr., Lieutenant Colonel, U.S. Air Force. Born November 21, 1933, Birmingham, Alabama. B.S. (physics), 1954, Auburn University; M.S. (engineering science), 1971, University of Tennessee. Transferred to the NASA astronaut corps in 1969 from the Air Force Manned Orbiting Laboratory project (canceled). Later flew on two Space Shuttle Orbiter flights.

Apollo 17

Robert A. R. Parker (support crew, Apollo 15).

Charles G. Fullerton (support crew, Apollo 14).

Robert Franklyn Overmyer, Lieutenant Colonel, U.S. Marine Corps. Born July 14, 1936, Lorain, Ohio. B.S. (physics), 1958, Baldwin-Wallace College, M.S. (aeronautics), 1964, U.S. Naval Postgraduate School. Transferred to the NASA astronaut corps in 1969 from the Air Force Manned Orbiting Laboratory project (canceled). Later served as support crewman and capsule communicator in Moscow for the Apollo-Soyuz Test Project and commanded STS Mission 31-B in the Space Shuttle program.

Astronaut Classes Selected Through 1969

First group, selected April 9, 1959:

Selection criteria: under 40 years of age; less than 5 ft., 11 in. tall; hold a bachelor's degree in engineering or equivalent; graduate of test pilot school; qualified jet pilot with at least 1,500 hours of flying time.
Lt. M. Scott Carpenter, USN;
Capt. L. Gordon Cooper, USAF;
Lt. Col. John H. Glenn, Jr., USMC;
Capt. Virgil I. ("Gus") Grissom, USAF (died in AS-204 fire, Jan. 27, 1967);
Lt. Comdr. Walter M. Schirra, Jr., USN;
Lt. Comdr. Alan B. Shepard, Jr., USN; and
Capt. Donald K. ("Deke") Slayton, USAF.

Second group, selected September 17, 1962:

Selection criteria: under 35 years of age; bachelor's degree in a physical or biological science or engineering; experience as a jet test pilot, having graduated from a military test pilot school or attained experimental flight test status in the armed services, the aircraft industry, or NASA.
Neil A. Armstrong;
Maj. Frank Borman, USAF;
Lt. Charles ("Pete") Conrad, Jr., USN;
Lt. Comdr. James A. Lovell, Jr., USN;
Capt. James A. McDivitt, USAF;
Elliott M. See, Jr. (died in T-38 crash, Feb. 28, 1966);
Capt. Thomas P. Stafford, USAF;
Capt. Edward H. White II, USAF (died in AS-204 fire, Jan. 27, 1967; and
Lt. Comdr. John W. Young, USN.

Third group, selected October 18, 1963:

Selection criteria: under 34 years of age; bachelor's degree in science or engineering; experimental flight test status or 1,000 hours of jet flying time.

Maj. Edwin E. ("Buzz") Aldrin, Jr., USAF (Ph.D., astronautics);
Capt. William A. Anders, USAF;
Capt. Charles A. Bassett II, USAF (died in T-38 crash, Feb. 28, 1966);
Lt. Alan L. Bean, USN;
Lt. Eugene A. Cernan, USN;
Lt. Roger B. Chaffee, USN (died in AS-204 fire, Jan. 27, 1967);
Capt. Michael Collins, USAF;
R. Walter Cunningham;
Capt. Donn F. Eisele, USAF;
Capt.Theodore C. Freeman, USAF (died in T-38 crash, Oct. 31, 1964);
Lt. Comdr. Richard F. Gordon, Jr., USN;
Russell L. Schweickart;
Capt. David R. Scott, USAF; and
Capt. Clifton C. Williams, USMC (died in T-38 crash, Oct. 5, 1967).

Fourth group, selected June 28, 1965:
(scientist-astronauts)

Selection criteria: doctor's degree in medicine, engineering, or one of the natural sciences. Applicants were evaluated by a committee of the National Academy of Sciences and those considered qualified were then reviewed by NASA for selection.
Owen K. Garriott, Ph.D. (engineering);
Edward G. Gibson, Ph.D. (engineering);
Duane E. Graveline, M.D. (resigned before beginning training);
Lt. Comdr. Joseph P. Kerwin (M.D.), MC, USN;
F. Curtis Michel, Ph.D. (physics) (resigned 1969); and
Harrison H. ("Jack") Schmitt, Ph.D. (geology).

Fifth group, selected April 4, 1966:

Selection criteria same as those for the third group.
Vance D. Brand;
Lt. John S. Bull, USN (resigned for health reasons, 1968);
Maj. Gerald P. Carr, USMC;
Capt. Charles M. Duke, Jr., USAF;
Capt. Joe H. Engle, USAF;
Maj . Edward G. Givens, Jr., USAF (died in auto crash, June 6, 1967);
Fred W. Haise, Jr.;
Maj. James B. Irwin, USAF;
Don L. Lind., Ph.D. (physics);

Capt. Jack R. Lousma, USMC;
Lt. Thomas Y. Mattingly II, USN;
Lt. Bruce McCandless II, USN;
Comdr. Edgar D. Mitchell, USN (Sc.D., aeronautics and astronautics);
Maj. William R. Pogue, USAF;
Capt. Stuart A. Roosa, USAF;
John L. Swigert, Jr.;
Lt. Comdr. Paul J. Weitz, USN; and
Capt. Alfred M. Worden, USAF.

Sixth group, selected August 4, 1967:
(scientist-astronauts)

Selection criteria same as those for fourth group.
Joseph P. Allen IV, Ph.D. (physics);
Philip K. Chapman, Ph.D. (instrumentation) (resigned, 1972);
Anthony W. England, Ph.D. (awarded 1970, earth and planetary sciences)
    (resigned, 1972);
Karl G. Henize, Ph.D. (astronomy);
Donald L. Holmquest, M.D. (took leave, 1971; resigned 1973);
William B. Lenoir, Ph.D. (engineering);
John A. Llewellyn, Ph.D. (chemistry) (resigned, 1968);
F. Story Musgrave, M.D.;
Brian T. O'Leary, Ph.D. (astronomy) (resigned, 1967);
Robert A. R. Parker, Ph.D. (astronomy); and
William E. Thornton, M.D.

Seventh group, selected August 14, 1969:
(transferred from the canceled Air Force Manned
Orbiting Laboratory [MOL] project)

Maj. Karol J. Bobko, USAF;
Lt. Comdr. Robert L. Crippen, USN;
Maj. C. Gordon Fullerton, USAF;
Maj. Henry W. Hartsfield, Jr., USAF;
Maj. Robert F. Overmyer, USMC;
Maj. Donald H. Peterson, USAF; and
Lt. Comdr. Richard H. Truly, USN.

Capsule Communicators ("CapComs")

All information transmitted to a spacecraft by voice during a mission was passed
up by the capsule communicator or "CapCom." CapComs were picked from the

astronaut corps so that they would be familiar with the spacecraft and the mission and would understand procedures.

Following is a list of CapComs for the Apollo missions.

Apollo 7: Stafford, Evans, Pogue, Swigert, Young, Cernan.

Apollo 8: Collins, Mattingly, Carr, Armstrong, Aldrin, Brand, Haise.

Apollo 9: Roosa, Evans, Worden, Conrad, Gordon, Bean.

Apollo 10: Duke, Engle, Lousma, McCandless.

Apollo 11: Duke, Evans, McCandless, Lovell, Anders, Mattingly, Haise, Lind, Garriott, Schmitt.

*Apollo 12: Carr, Gibson, Weitz, Lind, Scott, Worden, Irwin.

Apollo 13: Kerwin, Brand, Lousma, Young, Mattingly.

Apollo 14: Fullerton, McCandless, Haise, Evans.

Apollo 15: Allen, Fullerton, Henize, Mitchell, Parker, Schmitt, Shepard, Gordon, Brand.

Apollo 16: Peterson, Fullerton, Irwin, Haise, Roosa, Mitchell, Hartsfield, England, Overmyer.

Apollo 17: Fullerton, Overmyer, Parker, Allen, Shepard, Mattingly, Duke, Roosa, Young.

---

*Four civilian non-astronauts served as backup CapComs on this mission: Dickie K. Warren, James O. Rippey, James L. Lewis, and Michael R. Wash.

Accumulated Time in Space for American Astronauts:
Mercury through Apollo

Astronaut	Total hours in space[a]	Missions[b]
James Lovell	715	GT-7, GT-12, A-8, A-13
Eugene Cernan	553.32	GT-9, A-10, A-17
John Young	533.6	GT-3, GT-10, A-10, A-16
Charles Conrad	506.8	GT-5, GT-11, A-12
Frank Borman	477.58	GT-7, A-8
James McDivitt	338.95	GT-4, A-9
Richard Gordon	315.88	GT-11, A-12
David Scott	298.88	GT-8, A-15
Walter Schirra	295.22	MA-8, GT-6A, A-7
Thomas Stafford	290.27	GT-6A, GT-9, A-10
Edwin Aldrin	289.82	GT-12, A-11
Jack Schmitt	288.9	A-17
Ron Evans	288.9	A-17
James Irwin	288.2	A-15
Alfred Worden	288.2	A-15
Michael Collins	266.07	GT-10, A-11
Ken Mattingly	265.9	A-16
Charles Duke	265.9	A-16
Donn Eisele	260.15	A-7
Walter Cunningham	260.15	A-7
Alan Bean	244.6	A-12
Gordon Cooper	225.24	MA-9, GT-5
Alan Shepard	216.29	MR-3, A-14
Edgar Mitchell	216.03	A-14
Stuart Roosa	216.03	A-14
Neil Armstrong	205.98	GT-8, A-11
Fred Haise	142.9	A-13
John Swigert	142.9	A-13
Edward White	97.93	GT-4
Virgil Grissom	5.12	MR-4, GT-3
Scott Carpenter	4.93	MA-7
John Glenn	4.92	MA-6

[a] Liftoff to splashdown.
[b] A, Apollo; GT, Gemini-Titan; MA, Mercury-Atlas; MR, Mercury Redstone (suborbital).

Source: *Astronautics and Aeronautics,* annual summaries of space flights; Apollo mission reports.

APPENDIX 7

CREW TRAINING AND SIMULATIONS*

Even before Project Mercury began, the value of high-fidelity simulation as a training procedure was well established through aircraft flight experience. But whereas aircraft pilots can obtain much of their training during actual flights, the crews for space missions must receive all their training in flight tasks *before* the mission, because a manned space flight is fully committed to its entire mission at liftoff and the crew must be proficient in all anticipated flight operations at the time of launch. Aircraft experience was used in the development of the first space flight simulator, the Mercury Procedures Simulator; subsequent programs drew heavily on experience in actual space flights to design the simulators.

Lack of experience with the environmental factors of space flight gave rise to considerable concern in the early stages of Project Mercury; hence training concentrated on acclimating astronauts to the high acceleration forces of launch and reentry, zero-gravity conditions, heat, noise, and spacecraft tumbling. Once it was established that crews could perform normally under these conditions, subsequent training programs focused more narrowly on the complexities of operating the spacecraft systems.

For Project Apollo (and Gemini before it) a very large effort was devoted to the development of simulators that would duplicate as closely as possible the sights, sounds, and sensations of space flight. It was not possible to produce all these effects simultaneously—in particular, zero gravity or weightlessness could not be sustained for any length of time on earth—hence simulators were broadly divided into two classes: moving-base, in which the crew station could be moved to simulate expected conditions, and fixed-base, in which no motion was imparted to the crew. An outstanding example of a moving-base simulator was the lunar landing training vehicle (see Chapter 9), a free-flying device providing six degrees of freedom and designed to approximate the performance of the lunar module in its final approach to landing on the moon in reduced gravity. Others included the translation and docking simulator, the dynamic crew procedures simulator, and the lunar landing research facility, which allowed training for specific phases of a mission ("part-task trainers"). Most of the astronauts' training time, however, was spent in the fixed-base command module simulator (CMS) and the lunar module simulator (LMS).

Both the CMS and the LMS were built around mock-ups of the respective spacecraft, fully equipped with controls and displays, crew couches, and storage com-

---

*This discussion is based primarily on NASA Technical Note TN D-7112, "Apollo Experience Report—Simulation of Manned Space Flight for Crew Training," by C. H. Woodling, Stanley Faber, John J. Van Bockel, Charles C. Olasky, Wayne K. Williams, John L. C. Mire, and James R. Homer, March 1973.

partments as nearly like the mission-specific spacecraft as possible.* An elaborate optical system projected realistic out-the-window scenes for each stage of the mission, using either films or hand-made mock-ups photographed by high-resolution television cameras. (The preparation of three-dimensional scale models of the lunar surface for use in the simulators was a factor that had to be considered in meeting an assigned launch date.) For solving navigation problems, a star field was projected in the field of vision showing all stars of magnitude 5 or less that appeared in the vicinity of the stars used for determining the spacecraft's position. Switches and controls were designed to have exactly the same "feel" as flight articles and produced responses exactly like those that would be generated during a mission. The CMS system was controlled by four large digital computers and a program of 750,000 words. The analogous LMS required three computers and a 600,000-word program. At the computer control panels, instructors could create every conceivable type of emergency for the astronauts to cope with.

Because two or more crews might be in training at the same time, three command module simulators were provided, one at the Manned Spacecraft Center and two at Kennedy Space Center, where astronauts spent much of their last few weeks of preparation before flight. One lunar module simulator was located at each center. At the height of preparations for Apollo 11 some 175 contractor personnel worked on the development and control of the simulator software; 200 more were assigned to hardware operations and maintenance.

Not only the crews, but also the flight controllers who manned the consoles in the Mission Control Center (MCC) had to become proficient in real-time management of all the systems involved in a flight. Much of their training was conducted independently of the crew, using computer-generated "math models." In the later stages of preparation Mission Control and the spacecraft simulators (either or both) were linked in integrated network simulations requiring both crews and flight controllers to respond to normal and contingency situations. The time devoted to these integrated simulations varied with the complexity of the mission and with experience (see Table 4, following).

Besides their use in crew and flight controller training, the spacecraft simulators were invaluable in working out new operational procedures. During Apollo 13, for example, procedures improvised by flight control teams were checked in real time by astronauts in the simulators before being sent to the crew in the crippled spacecraft.

Zero-gravity training was accomplished in two ways. Short periods (up to about 40 seconds) of partial to null gravity can be achieved in an airplane by flying what was called a "Keplerian trajectory"—a carefully defined parabolic path during part of which centrifugal force offsets gravity. An Air Force KC-135 (the military version of the Boeing 707), structurally reinforced to take the strain of these maneuvers, flew regular missions during which critical operations were evaluated in

---

*In the early days of the project, details of spacecraft design changed so rapidly that it was difficult to keep the simulators up to date. At one point during training for the first manned mission (AS-204), Gus Grissom became so disgusted with the discrepancies between the simulator and the spacecraft that he hung a lemon on the trainer. Courtney G. Brooks, James M. Grimwood, and Loyd S. Swenson, Jr., *Chariots for Apollo: A History of Manned Lunar Spacecraft*, NASA SP-4205 (Washington, 1979), p. 209.

zero gravity. The alternative method used the buoyant effect of water. By attaching weights to various parts of the body of a suited astronaut it was possible to achieve "neutral buoyancy" in a large tank. Many tasks were rehearsed and procedures modified in the neutral-buoyancy facilities at MSC and at Marshall Space Flight Center. The method had the advantage of providing all the time needed; its major disadvantage was that it was less realistic than the aircraft flights on account of the viscosity of the water, which hampered movement. Still, it had its place in zero-gravity training, and astronauts generally found that it gave a conservative estimate of the difficulty of a task. Anything that could be done in the neutral-buoyancy tank could usually be done in space.

Lunar surface simulations were conducted at a site off to one corner of the Manned Spacecraft Center, where a few acres of ground had been pocked with craters and strewn with rocks and gravel to simulate the moon's surface. Here astronauts checked out deployment of the lunar surface experiments and practiced sampling. No attempt was made to approximate the reduced gravity of the moon, although NASA engineers devised a suspension system that offset five-sixths of the astronaut's weight, which was valuable for evaluating techniques of locomotion and manipulation.

Field geology was another important phase of astronaut training. No terrestrial site duplicates the lunar surface, but the rugged conditions expected to be found on the moon have many counterparts on earth. Among the areas visited by astronauts and their instructors were the Grand Canyon in Arizona; volcanic areas in Iceland, Mexico, New Mexico, Alaska, and Hawaii; and the Ries Crater area in West Germany. Each site had specific features applicable to lunar geology, and each provided the opportunity to conduct surface operations analogous to those to be used on the moon.

Besides such "hands-on" rehearsal of mission operations, astronauts sat through many hours of classroom-type work—lengthy briefings on the spacecraft systems, principles of propulsion, guidance and navigation, and orbital mechanics. They also spent many a day in design reviews, crew compartment fit-and-function reviews, and all the other reviews that punctuated the progress of spacecraft from design to delivery and played important roles in the formulation of flight plans.

This discussion does not by any means exhaustively cover the effort that went into simulation and training, but it may indicate the complexity of the simulation program and its importance to the Apollo project. The tables that follow give an indication of the amount of time required to prepare for lunar missions.

Table 1. **Apollo Astronaut Training Summary***

Type of Training	No. of hours	Type of Training	No. of hours
Simulator		**Special Purpose**	
Command module		Lunar science	11,408 [a]
Command module simulator	17,605	Water immersion facility checkout	1,248
CM procedures simulator	1,204	Stowage	993
Simulator briefings	1,195	Extravehicular mobility unit checkout	919
Contractor evaluations	866	Egress	820
Dynamic crew procedures simulator	741	Bench checks	802
Other simulators	156	Walkthroughs	719 [b]
Rendezvous and docking simulator	87	Medical	601
Centrifuge	58	Water immersion facility (zero gravity)	516
MIT hybrid simulator	48	Planetarium	448
Subtotal	21,960	Fire	174
		TOTAL	18,698
Lunar module			
LM simulator	13,317 [c]		
Lunar landing training vehicle	1,130 [d]		
LM procedures simulator	770		
Simulator briefings	533	**Briefings**	
Full mission engineering simulator	179		
Translation & docking simulator	64	Command and service module	4,060
Subtotal	15,993	Guidance and navigation	2,397
		Lunar module	2,130
TOTAL	37,953	Lunar topography	1,458
Procedures		Launch vehicle	656
		Photography	405
Mission techniques	2,730		
Checklist	2,334	TOTAL	11,106
Flight plan	1,987		
Mission rules	1,039		
Design, acceptance	1,011		
Test reviews	814	**Spacecraft tests**	
Team meetings	541		
Training meetings	393		
Rendezvous	288	Command and service module	3,332
Extravehicular contingency transfer	88	Lunar module	1,759
Flight readiness reviews	48	TOTAL	5,091
TOTAL	11,273	PROGRAM TOTAL	84,071

[a] Includes briefings, geology field trips, lunar surface simulations, and lunar roving vehicle trainer operation.
[b] Related to zero-gravity flight operations.
[c] Includes lunar roving vehicle navigation simulator.
[d] Includes lunar landing training vehicle flights (2 hr. per flight), vehicle systems briefings, lunar landing research facility, and lunar landing training vehicle time.

*Tables 1 through 4 adapted from C. H. Woodling et al., ''Apollo Experience Report—Simulation of Manned Space Flight for Crew Training,'' NASA TN D-7112, March 1973.

Table 2. **Apportionment of Training According to Mission Type**

Training Category	Missions before first lunar landing [a]		Early lunar landing missions [b]		Final lunar landing missions [c]	
	Hours	Percent of total	Hours	Percent of total	Hours	Percent of total
Simulators	11,511	36	15,029	56	11,413	45
Special purpose	4,023	13	5,379	20	9,246	36
Procedures	7,924	25	2,084	8	1,265	5
Briefings	5,894	18	3,070	11	2,142	9
Spacecraft tests	2,576	8	1,260	5	1,255	5
Total	31,928	100	26,822	100	25,320	100

[a] Apollo 7, 8, 9, 10.
[b] Apollo 11, 12, 13, 14.
[c] Apollo 15, 16, 17.

APPENDIX 8

THE FLIGHT OF APOLLO 13*

Apollo 13, the third manned lunar landing and exploration mission, had been tentatively scheduled in July 1969 for launch in March 1970, but by the end of the year the launch date had been shifted to April. In August 1969 crew assignments for Apollo 13 were announced: James A. Lovell commanded the prime crew, which included Thomas K. Mattingly II as command module pilot and Fred W. Haise as lunar module pilot. Their backups were John Young, John Swigert, and Charles Duke. The target for the mission was the Fra Mauro Formation, a site of major interest to scientists, specifically a spot just north of the crater Fra Mauro, some 550 kilometers (340 miles) west-southwest of the center of the moon's near side.

On March 24, 1970, during the countdown demonstration test for Apollo 13, KSC test engineers encountered a problem with an oxygen tank in the service module. The spacecraft carried two such tanks, each holding 320 pounds (145 kilograms) of supercritical oxygen. They provided the oxygen for the command module atmosphere and (along with two tanks of hydrogen) three fuel cells, which were the spacecraft's primary source of electrical power. Besides power, the chemical reaction in the cells produced water, which not only supplied the crew's drinking water but was circulated through cooling plates to remove heat from certain critical electronic components. The tanks were designed to operate at pressures of 865 to 935 pounds per square inch (psi) (6,000 to 6,450 kilopascals) and temperatures between –340°F and +80°F (–207°C to +27°C). Inside each spherical tank were a quantity gauge, a thermostatically controlled heating element, and two stirring fans driven by electric motors. The fans were occasionally operated to homogenize the fluid in the tank; it tended to stratify, leading to erroneous quantity readings. All wiring inside the tank was insulated with Teflon, a fluorocarbon plastic that is ordinarily noncombustible. Each tank was fitted with a relief valve designed to open when the pressure rose above 1,000 psi (6,900 kilopascals); the tanks themselves would rupture at pressures above 2,200 psi (15,169 kilopascals) . Both tanks were mounted on a shelf in the service module between the fuel cells and the hydrogen tanks.

The countdown demonstration test called for the tanks to be filled, tested, and then partially emptied by applying pressure to the vent line, thus forcing oxygen out through the fill line. Number one tank behaved normally in this test, but number two released only 8 percent of its contents, not 50 percent as required. Test

---

*This narrative is based primarily on three sources: AIAA Paper No. 70-1260, by Glynn S. Lunney (one of the three flight directors on Apollo 13) , "Discussion of Several Problem Areas During the Apollo 13 Operation," presented at the AIAA 7th Annual Meeting and Technical Display, Houston, Texas, Oct. 19-22, 1970; NASA, "Report of Apollo 13 Review Board," June 15, 1970; and "Apollo 13 Technical Air to Ground Voice Transcription," April 1970. See also the published House and Senate hearings cited in notes 75 and 79, Chapter 11.

engineers decided to proceed with the rest of the test and investigate the problem later. The next day, after KSC engineers had discussed the problem with colleagues at MSC, North American Rockwell (builders of the service module), and Beech Aircraft (manufacturers of the oxygen tanks), they tried emptying the tank again, with no success. Further talks led to the conclusion that the tank probably contained a loose-fitting fill tube, which could allow pressure to escape without emptying the tank.

When normal procedures again failed to empty the tank, engineers decided to use its internal heaters to boil off the contents and applied direct-current power at 65 volts to the heaters. This was successful but slow, requiring eight hours of heating. It was then decided that if the tank could then be filled normally it would not cause a problem in flight. A third test gave the same result as the second, requiring heating to empty the tank.

In view of the difficulty of replacing the oxygen shelf—a job that would take at least 45 hours—and the possibility that other components might be damaged in the process and the launch delayed for a month, NASA and contractor officials decided not to replace the tanks.

The spacecraft was launched on April 11, 1970, and the mission was quite routine for the first two days. At 30 hours and 40 minutes after launch (30:40 ground elapsed time, or g.e.t.), the crew ignited their main engine to put the spacecraft on a hybrid trajectory, a flight path that saved fuel in reaching the desired lunar landing point.* At 46:40** the crew routinely switched on the fans in the oxygen tanks briefly. A few seconds later the quantity indicator for tank number two went off the high end of the scale, where it stayed. The tanks were stirred twice more during the next few hours; and at 55:53, after a master alarm had indicated low pressure in a hydrogen tank, the Mission Control Center (MCC) directed the crew to switch on all tank stirrers and heaters. Shortly thereafter the crew heard a loud "bang" and felt unusual vibrations in the spacecraft. Mission controllers noticed that all telemetry readings from the spacecraft dropped out for 1.8 seconds. In the CM, the caution and warning system alerted the crew to low voltage on d.c. main bus B, one of two power distribution systems in the spacecraft. At this point command module pilot Jack Swigert told Houston, "Hey, we've had a problem here."

Because of the interruption of telemetry that had just occurred, flight controllers in the MCC had difficulty for the next few minutes determining whether they were getting true readings from the spacecraft sensors or whether the sensors had somehow lost power. Before long, however, both MCC and the crew realized that oxygen tank number two had lost all of its contents, oxygen tank number one was slowly losing its contents, and the CM would soon be out of oxygen and without electrical power. Among the first actions taken were shutting down one fuel cell and switching off nonessential systems in the CM to minimize power consumption; shortly after, the second fuel cell was shut down as well. When

---

*The "hybrid" trajectory was designed so that if the main propulsion engine failed, the attitude-control rockets on the spacecraft could change the flight path enough to bring it back to a safe reentry after it rounded the moon.

**This and all subsequent times are in hours and minutes g.e.t. Launch time was 12:13 p.m. Eastern Standard Time on April 11.

the remaining oxygen ran out, the CM would be dead; its only other power source was three reentry batteries providing 120 ampere-hours, and these had to be reserved for the critical reentry period.

An hour and a half after the "bang," MCC notified the crew that "we're starting to think about the lifeboat"—using the lunar module (LM) and its limited supplies to sustain the crew for the rest of the mission. Plans for such a contingency had been studied for several years, although none had anticipated a situation as grave as that of Apollo 13. Many of these studies were retrieved and their results were adapted to the situation as it developed.

Shortly after the accident, mission commander James Lovell reported seeing a swarm of particles surrounding the spacecraft, which meant trouble. Particles could easily be confused with stars, and the sole means of determining the spacecraft's attitude was by locating certain key stars in the onboard sextant. Navigational sightings from the LM were difficult in any case as long as it was attached to the command module, and this would only complicate matters. Flight controllers decided to align the lunar module's guidance system with that in the command module while the CM still had power. That done, the last fuel cell and all systems in the command module were shut down, and the crew moved into the lunar module. Their survival depended on this craft's oxygen and water supplies, guidance system, and descent propulsion engine (DPS). Normally all course corrections were made using the service propulsion system (SPS) on the service module, but flight controllers ruled out using it, partly because it required more electrical power than was available and partly because no one knew whether the service module had been structurally weakened by the explosion. If it had, an SPS burn might be dangerous. The DPS would have to serve in its place.

When word got out that Apollo 13 was in trouble, off-duty flight controllers and spacecraft systems experts began to gather at MSC, to be available if needed. Others stood by at NASA centers and contractor plants around the country, in touch with Houston by telephone. Flight directors Eugene Kranz, Glynn Lunney, and Gerald Griffin soon had a large pool of talent to help them solve problems as they arose, provide information that might not be at their fingertips, and work on solutions to problems they could anticipate farther along in the mission. Astronauts manned the CM and LM training simulators at Houston and at Kennedy Space Center, testing new procedures as they were devised and modifying them as necessary. MSC director Robert R. Gilruth, Dale D. Myers, director of manned space flight, and NASA administrator Thomas O. Paine were all on hand at Mission Control to provide high-level authority for changes.

Soon after the explosion, the assessment of life-support systems determined that although oxygen supplies were adequate, the system for removing carbon dioxide (CO_2) in the lunar module was not. The system used canisters filled with lithium hydroxide to absorb CO_2 as did the system in the command module. Unfortunately the canisters were not interchangeable between the two systems, so the astronauts were faced with plenty of capacity for removing CO_2 but no way of using it. A team in Houston immediately set about improvising a way to use the CM canisters, using materials available in the spacecraft.

Flight controllers, meanwhile, were addressing operational problems. Their first critical decision was to put the crippled spacecraft back on a free-return trajectory, which was accomplished by firing the LM descent engine at 61:30. Mission Control then had some 18 hours to consider the remaining problems; the next was a possible course adjustment to change the spacecraft's landing point on earth. If this was to be done, it was scheduled for "PC + 2"—two hours after pericynthion (closest approach to the moon), after the spacecraft emerged from behind the moon. In the interval, Houston worked out a new flight plan that would minimize the consumption of oxygen, water, and electricity while keeping vital systems operating.

The alternatives for the PC + 2 maneuver were worked out by about 64 hours g.e.t. A major consideration was the total time to splashdown. Left on its free-return course the command module would return at about 155 hours g.e.t. to a landing in the Indian Ocean. Three options would bring it back in the mid-Pacific and could reduce the total mission time to as little as 118 hours. The fifth possibility returned the spacecraft in 133 hours, but to the South Atlantic. For one reason or another, all but one of these choices were discarded. The free-return (no course correction) choice was abandoned, since there was no known reason not to use the LM descent propulsion system. Recovery in either the Atlantic or the Indian Ocean was far from ideal; the main recovery force was deployed in the mid-Pacific and there was not enough time to move it or to make adequate arrangements elsewhere. Two options giving the shortest return time (118 hours) had other drawbacks. Both would require using virtually all of the available propellant, and it was not prudent to assume that no additional course corrections would be required. One of them involved jettisoning the service module, which would expose the CM heat shield to the cold of space for 40 hours and raise questions about its integrity on reentry. After five and a half hours of weighing the choices and their consequences, flight directors met with NASA and contractor officials and presented their findings and recommendations. The decision, made some ten hours before the scheduled engine burn, was to go for mid-Pacific recovery at 143 hours.

During all of these deliberations the atmosphere in the lunar module was gradually accumulating carbon dioxide as the absorbers in the environmental control system became saturated. Members of MSC's Crew Systems Division devised a makeshift air purifier by taping a plastic bag around one end of a CM lithium hydroxide cartridge and attaching a hose from the portable life-support system, allowing air from the cabin to be circulated through it. After verifying that this jury rig would function, they prepared detailed instructions for building it from materials available in the spacecraft and read them up to the crew. For the rest of the mission the improvised system kept the CO_2 content of the atmosphere well below hazardous levels.

The decision to recover in the Pacific fixed the time line for the remainder of the mission and imposed some rigid constraints on preparations for reentry. The final course correction had to be made with the LM engine; command module systems had to be turned on and the guidance system aligned; the service module had to be discarded; and when all preparations had been made, the lunar module would be cut loose. In all these preparations the power available from the CM's reentry batteries was a limiting factor. From the PC + 2 burn until about 35 hours before reentry the sequence of activation of CM systems was worked out, checked in the simulators, and modified. Fifteen hours before beginning reentry the revised sequence of activities was read to the crew, to give them time to review and practice it.

The husbanding of expendable resources, particularly electrical power, paid off on the morning of landing, when it was discovered that power reserves in the LM were adequate to allow use of it in the CM. Some of the early CM activities could then be done at a less hurried pace. The Apollo 13 command module splashed down within a mile of the recovery carrier with about 20 percent of its battery power remaining. Three weary, chilled astronauts came aboard the U.S.S. *Iwo Jima* on April 17 and were flown to Hawaii for an emotional reunion with their families.

Mission Control teams and their hundreds of helpers were no less drained. The usual cigars were lighted up after recovery, but the splashdown parties that evening were subdued: most of those who went quit early and went home to bed. Their efforts were recognized the next day when President Richard M. Nixon, on his way to Hawaii, stopped in Houston to present the Presidential Medal of Freedom, the nation's highest civilian award, to the entire team.

NASA immediately convened an investigation board* to determine the cause of the accident and postponed Apollo 14 until its results were in. Lacking the spacecraft itself—the service module had been jettisoned before reentry, and the crew had been able to take only a few rather poor photographs of it—the board initially had only the data from inflight telemetry to work with. When it became clear that the fault lay in oxygen tank number two, the board carefully reviewed its entire history, from fabrication to launch, as recorded in the detailed documentation that followed every piece of equipment from plant to launch pad. Under the board's direction, MSC and other NASA centers conducted tests under simulated mission conditions to verify its findings. The investigation, which concluded in a few weeks, turned up a highly improbable sequence of human error and oversight that led inexorably to the failure in flight.

Board Chairman Edgar M. Cortright, director of Langley Research Center, explained the board's findings to congressional committees in June. The accident, he reported, was not a random malfunction but resulted from an unusual combination of mistakes as well as "a somewhat deficient and unforgiving design." As the board's report reconstructed the events leading up to the accident, the tank left Beech Aircraft's plant on May 3, 1967, after passing all acceptance tests. It was installed as part of a shelf assembly in service module no. 106 on June 4, 1963, having passed all tests conducted at North American Rockwell during assembly. Design changes in the service module, however, necessitated removing the entire shelf from SM 106 for modification. During removal, which was accomplished by use of a special fixture that fit under the shelf to lift it upward,

---

*Board members were: Edgar M. Cortright, director, Langley Research Center, chairman; Robert F. Allnut, assistant to the administrator, NASA Hqs.; Neil Armstrong, MSC; John F. Clark, director, Goddard Space Flight Center; Brig. Gen. Walter R. Hedrick, Jr. , Hqs. USAF; Vincent L. Johnson, deputy associate administrator for engineering, Office of Space Science and Applications; Milton Klein, manager, AEC-MASA space nuclear propulsion office; and Hans M. Mark, director, Ames Research Center.

workmen overlooked one bolt that held down the back of the shelf, with the result that the removal fixture broke, dropping the shelf two inches. The board concluded that this incident might have jarred loose a poorly fitting fill tube. Subsequent tests did not detect any flaws, and after modification the shelf was shipped to Kennedy Space Center for installation in SM 109, the Apollo 13 spacecraft.

What was not known was that this oxygen tank was fitted with obsolete thermostatic switches protecting its heating elements. Original specifications for the switches called for operation on 28 volts d.c.; in 1965 this was changed to 65 volts d.c. to match the test and checkout equipment at the Cape. Later tanks conformed to the new specifications, but this one, which should have been modified, was not, and the discrepancy was overlooked at all stages thereafter.

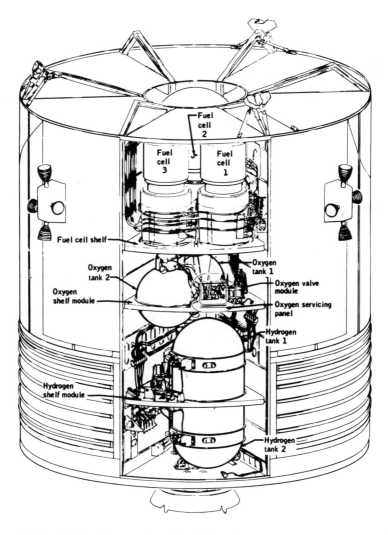

Arrangement of fuel cells and cryogenic systems in bay 4.

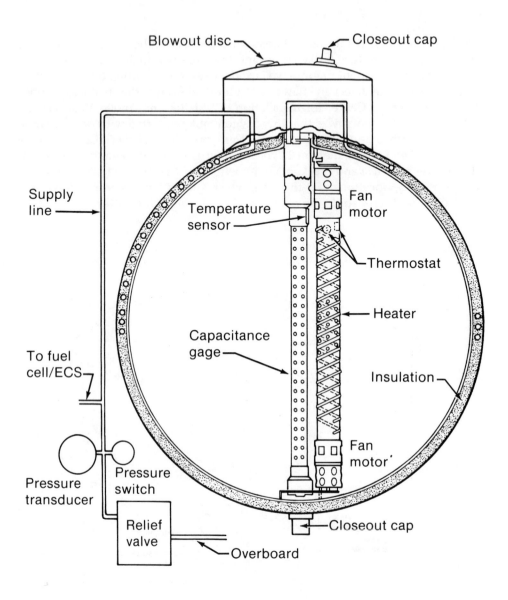

Oxygen tank no. 2 internal components.

In a normal checkout of a normal tank, this would not have mattered, because the switches would not have opened during normal operation. But the improvised procedure used when this tank failed to empty (the result of a loose fitting, as noted above) raised the temperature in the tank above 80°F (27°C), at which point the switches opened. Tests conducted during the investigation showed that the higher current produced by the 65-volt power source caused an arc between the contact points as they separated, welding them together and preventing their opening when the temperature dropped. This went undetected during the detanking procedure at the Cape; it could have been noticed if anyone had monitored the heater current, which would have shown that the heaters were operating when they should not have been. But all attention was on the specific malfunction, and no one was aware that the heaters were on continuously for eight hours on two separate occasions. The result, as tests showed, was that the heater tube reached 1,000 °F (538 °C) in spots, damaging the Teflon insulation on the adjacent fan-motor wiring and exposing bare wire. From that point on, the board concluded, the tank was hazardous when filled with oxygen and electrically powered. Teflon can be ignited at a high enough temperature in the presence of pure oxygen, and the tank contained small amounts of other combustibles as well.

Unfortunately for Apollo 13, the tank functioned normally for the first 56 hours of the mission, when the heaters and the fans were energized during routine operations. At that point an arc from a short circuit probably ignited the Teflon, and the rapid pressure rise that followed either ruptured the tank or damaged the conduit carrying wiring into the tank, expelling high-pressure oxygen. The board could not determine exactly how the tank failed or whether additional combustion occurred outside the tank, but the pressure increase blew off the panel covering that sector of the service module and damaged the directional antenna, causing the interruption of telemetry observed in Houston. It also evidently damaged the oxygen distribution system, or the other oxygen tank, as well, leading to the loss of all oxygen supplies and aborting the mission.

The board pointed out that although the circumstances of the tank failure were highly unusual and that the system had worked flawlessly on six successful missions, Apollo 13 was a failure whose causes had to be eliminated as completely as possible. It recommended that the oxygen tanks be modified to remove all combustible material from contact with oxygen and that all test procedures be thoroughly reviewed for adequacy.

Compared to the AS-204 fire in 1967, Apollo 13 was only a frightening near-miss, and because its cause was localized and comparatively easy to discover, it had fewer adverse effects on the program. Only the skill and dedication of hundreds of members of the often-celebrated "manned space flight team" saved it, however, and the accident served to remind NASA and the public that manned flight in space, no matter how commonplace it seemed to the casual observer, was not yet a routine operation. The same lesson had to be learned once more sixteen years later, when on January 28, 1986, the space shuttle *Challenger* and all seven of its crew were lost a minute after launch. An unforgiving design and the failure of human judgment under pressure combined again to bring a program to a halt while corrective measures were taken.

BIBLIOGRAPHIC ESSAY

The primary source material for the present history was the large collection of documents from the Apollo project maintained, at this writing (1987), in the History Office at the Lyndon B. Johnson Space Center (JSC) in Houston. More than 31,000 documents collected by the JSC historian over a 20-year period occupy nearly 150 linear meters of shelf space in more than 775 document boxes.

Most of these documents came from the "reading files" of JSC engineers and managers. For the most part they are copies of the letters, memoranda, and telexes generated in the management of the Apollo spacecraft project at JSC. Original documents will not be found in this collection; they constitute "record copies," which by law must be retired to Federal Archives and Records Centers—in the case of JSC, that in Ft. Worth, Texas—whence they may be recalled through the records management office at JSC. Thus the History Office collection is an unofficial "historian's source file" rather than an archive. Besides correspondence, these files contain NASA and contractor reports, working papers, minutes of meetings, and other types of documentation. They have been supplemented over the last 20 years with copies of documents obtained from NASA Headquarters and other NASA centers by the compilers of the Apollo chronology and the authors of *Chariots for Apollo: A History of Manned Lunar Spacecraft*. Storage space was always at a premium, and earlier researchers culled the documentation and discarded a considerable amount of material. The basic arrangement of all the Apollo files is chronological, mainly because the documents were in chronological order when they reached the History Office.

By far the largest part of this collection was obtained from the Apollo Spacecraft Program Office at JSC, which directed the design, construction, and testing of the two Apollo spacecraft (command/service module and lunar module). These documents deal largely with the development of the spacecraft themselves, but they contain at least some papers from every organizational element of the project: mission planning and analysis, lunar landing site selection, flight planning, astronaut training, testing programs, etc.

Smaller collections in the historian's source file, separate from the main Apollo chronological file, include:

- 130 boxes of documents pertaining to individual Apollo missions, from the first unmanned test flight of the spacecraft to the last manned lunar mission, Apollo 17. The collection includes premission planning documents, press kits, flight plans, transcripts of onboard and air-to-ground communications, crew debriefings, mission reports, mission science reports, and flight controller logs. The comprehensiveness of these documents varies from mission to mission.
- 30 boxes of documents related to the planning and execution of the scientific experiments conducted during the Apollo missions, in earth and lunar orbit and on the lunar surface.
- 18 boxes of documents on the construction and operation of the Lunar Receiving Laboratory at JSC, from 1964 to 1978, and 1 box relating to the establishment of the Lunar Science Institute.

- 22 boxes of miscellaneous documentation on the command/service module and the lunar module, including systems handbooks, operating manuals, and photographic documentation of individual spacecraft.
- 7 boxes of documents on the lunar landing research vehicle and the lunar landing training vehicle, covering development and operations.
- A small collection of documents on astronaut training, mostly training schedules for specific missions but including outlines of early training programs.
- JSC Apollo Spacecraft Project Office weekly reports, 1962–1966.
- Apollo program quarterly status reports, 1962–1968.
- Apollo working papers and Apollo experience reports. Working papers document the data and assumptions on which mission plans were constructed; experience reports, written after the program was completed, document problems and solutions and lessons learned from the Apollo program.
- Apollo feasibility study proposals (1960) and contractor reports on the feasibility studies, 1960–1961.
- Audio tapes of oral history interviews with more than 300 participants in the Apollo program—NASA and contractor personnel—taken in 1969–1970, plus transcripts of about two-thirds of the tapes.
- A complete collection of Apollo mission reports, plus an Apollo program summary report, containing most of the technical data on each mission.

The most important part of the program not represented in the History Office files is the biomedical. JSC's Life Sciences Directorate has compiled its own historical files, which are presently housed in the Historical Research Center of the Houston Academy of Medicine-Texas Medical Center library, 1133 M. D. Anderson Blvd., Houston, Texas 77030, where they may be consulted by qualified researchers.

The chronological arrangement of the Apollo files makes an index almost indispensable, and part of the historian's task under the contract which supported the writing of this volume was the preparation of an index. Each document in the Apollo files is individually entered in a computer-stored index, except for occasional folders or boxes that were indexed as a unit (e.g., a series of periodic reports). The index entry contains the date of the document, its originator, its title or subject, the name of the person who signed it, and its location (box number) in the file. Using remote terminals in the History Office, the user can rapidly search the entire file in a variety of ways (e.g., by date, originator, subject, or combinations of these categories) and obtain printouts of the retrieved entries. A printout of the complete index, available in the History Office, include tabulations of the number of documents by originator and by subject. This can be useful for quick surveys of the files or for manual searches.

A guide to the Apollo files, available for use in the JSC History Office, includes a more detailed description of the indexing system and an inventory of the Apollo collection. The JSC History Office does not have adequate full-time staff to provide extensive assistance or to conduct document searches, but those who wish to use the files on site can be given some guidance. Arrangements to use the JSC History Office files can be made through the History Office Coordinator, Management Analysis Office, mail code BY, NASA Lyndon B. Johnson Space Center, Houston, Texas 77058.

The value of this documentation is highly variable, as might be expected from the manner in which it was collected. On the whole it represents topics that were important to the Apollo spacecraft program manager at JSC, which encompassed much but by no means all of the Apollo program. More than three-fourths of this material in the Apollo chronological collection originated at JSC or at NASA Headquarters; more than one-third of it is related to four parts of the Apollo project that were the principal concern of JSC: the command and service modules, the lunar module, mission planning and analysis, and test facilities, procedures, and results. In other areas it is sketchy, as in the workings of the Astronaut Office and details of astronaut training.

The chronological arrangement of the files is the major impediment to their use. Correspondence on any given topic is not correlated, except in cases where a secretary collected and copied important background correspondence in a particular case. The index is extremely useful, but it is only a "first cut," limited by the time and funds available for its preparation; cross-referencing, for example, was not attempted. Prior letters, memos, etc., cited in documents can be checked against the index but often are not in the files. Where earlier correspondence is attached, the index entry includes a notation to that effect, but the attached documents are not indexed at the primary entry.

The other major collections (Missions, Lunar Science, and Lunar Receiving Laboratory) were more useful in preparing this history. Documentation of the individual missions is extensive, though variable from mission to mission. It usually includes pre- and postmission documentation, such as flight plans and mission reports; transcripts of inflight communications; press briefings conducted before, during, and after the flights; crew debriefings; and flight directors' logs. Mission files also contain a great deal of material provided to the press by contractors as well as official documentation. The science-related documents seem to have come from the reading files of JSC's director of science and applications, and the science side of the lunar program is well documented, as is the evolution and operation of the lunar receiving laboratory.

Other resources available to researchers at JSC include similar but less voluminous collections for other manned space flight projects (Mercury, Gemini, Skylab, and Apollo-Soyuz). These are housed in the Woodson Research Center (WRC) of the Fondren Library at Rice University in Houston. Under a custodial agreement between Rice and JSC, JSC retains title to the documents while WRC is responsible for their care and preservation and makes them available to qualified researchers. WRC has prepared guides to these collections and is in the process of indexing them in a system compatible with the Apollo index. Arrangements to use these collections can be made through the Woodson Research Center, Fondren Library, Rice University, P.O. Box 1892, Houston, Texas 77001.

The Lunar and Planetary Institute, located adjacent to JSC, has a large technical library devoted to lunar and planetary science. Most of the LPI's holdings are technical in nature, e.g., lunar sample information catalogs, excerpts from transcripts of air-to-ground communications relating to lunar sample collection and documentation. It has a few useful Apollo documents, including collections of minutes of meetings of the Apollo Site Selection Board, 1966–1972, and the Science Working Panel, 1970–72.

Besides these primary documents, the JSC History Office holds a large collection of important secondary sources. This includes a near-complete series of congressional authorization hearings, plus other congressional documents, such as staff reports and a few appropriations hearings, for the period 1958–1972. Another useful source is the weekly publication called *Current News*, a daily compilation of clippings from the national press related to space programs put out by the Office of Public Affairs at NASA Headquarters. For the space flight historian *Current News* serves as a kind of reader's guide to press coverage of major events and an indicator of the state of public opinion concerning the space program. Most of the references to newspaper articles in the notes in this volume were taken from *Current News*, supplemented where possible by consultation of the files of important newspapers available in local libraries.

Oral history interviews with key participants in the Apollo program provided details and insights not often available in the formal documentation. Many of the 300-odd interviews in the JSC History Office files were taken while Apollo was still in progress, many of them before the first lunar landing, and are almost "real-time" discussions of the problems encountered during the program. These interviews were taken by earlier historians with different questions in mind, and consequently they were not as useful for the present work as they might have been. The present author, working more than 10 years after Apollo ended, found fewer of those participants easily accessible within the constraints of his contract but was able to record interviews with a few of them. Their recollections frequently lacked detail because of the passage of time but sometimes yielded information on attitudes and relationships that were useful in reconstructing how and why things happened as they did.

The History Office at Headquarters has a large collection of material on all aspects of the space program. Its holdings are described in *History at NASA*, NASA HHR-50 (Washington, June 1986), available from the History Office, mail code XH, NASA Headquarters, Washington, D.C. 20546. The official files of Headquarters program and project offices have been retired to the Federal Archives and Records Center at Suitland, Maryland. They can be recalled through the History Office in Headquarters, which has an inventory of these holdings. For the scientific aspects of Apollo the papers of Homer E. Newell, associate administrator for space science and applications from 1961 to 1967, were most useful. Newell's own inventory of these documents is more informative than most. Copies of pertinent documents from the Newell files were added to the JSC collection by the present author.

Most of the secondary literature on Project Apollo deals with the accomplishment of the primary objective, the first lunar landing. Considerably less has been published on the second phase of the project, lunar exploration.

The classic study of the origin of Project Apollo is John M. Logsdon's *The Decision to Go to the Moon: Project Apollo and the National Interest* (Cambridge, Mass.: MIT Press, 1970), which deals in detail with the domestic and international political climate in which President John F. Kennedy formulated his challenge to the nation to send people to the moon and back "before this decade is out." An important recent effort to set the entire manned space flight program in a global (and internal American) political context is Walter A. McDougall's . . . *The Heavens and the Earth: A Political History of the Space Age* (New York: Basic Books, Inc., 1985).

The role of Congress, particularly the House of Representatives space committee, is well covered in *Toward the Endless Frontier: History of the Committee on Science Technology, 1959–79,* (Washington: U.S. Government Printing Office, 1980), written by Ken Hechler, a Ph.D. historian and member of the committee for 18 years. Other accounts of the manned space flight program include John Noble Wilford, *We Reach the Moon* (New York: Bantam Books, 1969); Hugo Young, Bryan Silcock, and Peter Dunn, *Journey to Tranquility* (Garden City, N.Y.: Doubleday and Co., 1970); and Richard S. Lewis, *Appointment on the Moon: The Inside Story of America's Space Venture* (New York: Viking Press, 1968). Lewis's *The Voyages of Apollo: The Exploration of the Moon* (New York: Quadrangle Press, 1974) covers some of the same ground as the present volume, discussing each of the Apollo lunar landing missions and its scientific results. Henry S. F. Cooper, space correspondent for *The New Yorker* magazine, has written some good journalistic accounts of the Apollo project in *Apollo on the Moon* (New York: Dial Press, 1969) and *Moon Rocks* (New York: Dial Press, 1970).

Apollo's two accidents, the fatal AS-204 spacecraft fire and the aborted mission of Apollo 13, were the subjects of popular books. The tragedy of Apollo 204 is covered in *Murder on Pad 34*, by Eric Bergaust (New York: G. P. Putnam's Sons, 1968), a sensationalized book that is far from objective. Henry Cooper's *13: The Flight That Failed* (New York: Dial Press, 1973) is an hour-by-hour account of the near-disastrous flight of Apollo 13. Detailed accounts of these grim milestones, including the reports of NASA's investigating boards, are found in published hearings of House committees: U.S. Congress, House Subcommittee on NASA Oversight, *Investigation into Apollo 204 Accident*, 90th Cong., 1st sess. (hereafter 90/1) (Washington: U.S. Government Printing Office, 1967) (3 vols.) , and House Committee on Science and Astronautics, *The Apollo 13 Accident*, 91/2 (Washington: U.S. Government Printing Office, 1970).

NASA's own publications comprise the most complete historical treatments of manned space flight projects now available. A basic chronological reference, based largely on journalistic sources, is the annual series starting with *Astronautics and Aeronautics, 1963: Chronology on Science, Technology, and Policy*, which now covers the years 1963–1977. Chronologies of individual projects include James M. Grimwood, *Project Mercury: A Chronology*, NASA SP-4001 (Washington, 1963); James M. Grimwood and Barton C. Hacker, with Peter Vorzimmer, *Project Gemini Technology and Operations: A Chronology*, NASA SP-4203 (Washington, 1969); and *The Apollo Spacecraft: A Chronology*, NASA SP-4009 (vol. I, by Ivan D. Ertel and Mary Louise Morse, 1969; vol. II, by Mary Louise Morse and Jean Karnahan Bays, 1973; vol. III, by Courtney G. Brooks and Ivan D. Ertel, 1976; vol. IV, by Courtney G. Brooks, Roland W. Newkirk, and Ivan D. Ertel, 1978). These chronologies, based largely on NASA documentation, were compiled during research on project histories.

NASA has published histories of all the projects that contributed to the accomplishment of the lunar landing. Loyd S. Swenson, Jr., James M. Grimwood, and Charles C. Alexander, *This New Ocean: A History of Project Mercury*, NASA SP-4201 (Washington, 1966), gives a full account of the early years of the space age and of NASA along with the development of Project Mercury itself. The second manned space flight project, Gemini, is treated in Barton C. Hacker and James

M. Grimwood, *On the Shoulders of Titans: A History of Project Gemini*, NASA SP-4203 (Washington, 1977), which focuses much more narrowly on the project itself. The Saturn launch vehicles and the launch facilities built at Kennedy Space Center are treated in Roger E. Bilstein, *Stages to Saturn: A Technological History of Apollo/Saturn Launch Vehicles*, NASA SP-4206 (Washington, 1980), and Charles D. Benson and William Barnaby Faherty, *Moonport: A History of Apollo Launch Facilities and Operations*, NASA SP-4204 (Washington, 1978).

Development of the spacecraft for lunar missions is covered in Courtney G. Brooks, James M. Grimwood, and Loyd S. Swenson, Jr., *Chariots for Apollo: A History of Manned Lunar Spacecraft*, NASA SP-4205 (Washington, 1979), which takes the Apollo story through the first lunar landing and foreshadows the subject matter of the present volume in an epilogue. *Chariots* concentrates on spacecraft and mission planning and deals only sketchily with the scientific side of the project.

Besides the official histories, NASA has published *Apollo Expeditions to the Moon*, Edgar M. Cortright, ed., NASA SP-350 (Washington, 1975), a collection of essays written by many of the prominent participants in Apollo. This profusely illustrated volume is a good overview of the program, from conception to completion, for a nontechnical audience.

NASA's space science program is well covered in Homer E. Newell, *Beyond the Atmosphere: Early Years of Space Science*, NASA SP-4211 (Washington, 1980). Several of Newell's chapters are required reading for anyone interested in the complexities of cooperation between government agencies and outside researchers. Newell speaks with authority, having been NASA's director of space science and applications from 1962 to 1967. Although he does not discuss manned space projects in detail, Newell deals with the problems of coordinating Gemini and Apollo with the scientific community outside the space agency and with the highly independent Office of Manned Space Flight and its Manned Spacecraft Center. Some of the same questions are covered in the context of a specific project by R. Cargill Hall in *Lunar Impact: A History of Project Ranger*, NASA SP-4210 (Washington, 1977). Much of the unmanned scientific exploration of the moon was done at the Jet Propulsion Laboratory; Clayton R. Koppes's *JPL and the American Space Program: A History of the Jet Propulsion Laboratory* (New Haven and London: Yale University Press, 1982) is an excellent source on these projects as well as on the management of space science programs. The unmanned Lunar Orbiter project, specifically designed to support Apollo, is the subject of Bruce K. Byers's *Destination Moon: A History of the Lunar Orbiter Program*, NASA TM X-3487 (Washington, 1977).

The Space Science Board of the National Academy of Sciences was the official source of advice and recommendations to NASA from the outside scientific community. Four publications contain the major recommendations concerning the Apollo science program: *A Review of Space Research*, report of the summer study conducted at the State University of Iowa, National Academy of Sciences-National Research Council Publication 1079 (Washington, 1962); *Space Research: Directions for the Future*, report of the summer study at Woods Hole, Mass., NAS-NRC Publication 1403 (Washington, 1966); *NASA 1965 Summer Conference on Lunar Exploration and Science* (Falmouth study), NASA SP-88 (Washington, 1965); and *1967 Summer Study of Lunar Science and Exploration* (Santa Cruz study), NASA SP-157 (Washington, 1967). The general question of NASA's relations with its outside scientific advisory groups has been examined in two monographs: Charles M. Atkins, "NASA and the Space Science Board of the National Academy of Sciences," NASA Historical Note HHN-62, 1966, and Pamela Mack, "NASA and the Scientific Community: NASA-PSAC Interactions in the Early 1960s," (unpublished), 1978. An interesting examination of the relationships between government and the scientific community generally is Daniel S. Greenberg's *The Politics of Pure Science* (New York: New American Library, 1967).

To date, the only NASA insiders to have published accounts of their experiences are several astronauts. Their stories are anecdotal but nonetheless worth reading for insights into the astronauts' side of the manned projects. The best and the most serious of these memoirs is Michael Collins's *Carrying the Fire: An Astronaut's Journeys* (New York: Farrar, Straus and Giroux, 1974). Others include R. Walter Cunningham's *The All-American Boys* (New York: The MacMillan Company, 1977), and Brian T. O'Leary's *The Making of an Ex-Astronaut* (New York: Houghton Mifflin, 1971). O'Leary, an astronomer picked in the second group of scientist-astronauts in 1966, discovered that he had no taste for flying airplanes and resigned from the program a few months after entering it, which somewhat vitiates his claim to be an "ex-astronaut." His book is highly critical of NASA's treatment of scientists in the astronaut program.

Notably lacking in the literature of lunar exploration is any summary of the scientific results of Apollo. This is hardly surprising, for the data from the lunar samples and the emplaced instruments were voluminous and complex, and since only six sites were sampled, extrapolations to the moon as a whole are not likely to be conclusive. The technical literature is staggering in volume. Papers presented at the first eight annual Lunar Science Conferences fill three large volumes each, and more has been published in the scientific journals. The Lunar and Planetary Institute in Houston maintains a computerized bibliography of lunar and planetary science containing some 23,000 entries, of which it is estimated that half deal with lunar science.

It is likely to be years before any consensus emerges among scientists as to the details of the formation and evolution of the moon. Even so, two scientists have undertaken to summarize the gross features of the moon's geological history on the basis of the Apollo data: S. Ross Taylor, who participated in the analysis of the lunar samples, has published *Lunar Science: A Post-Apollo View* (New York: Pergamon Press, Inc., 1975), and Harrison H. Schmitt, the only scientist to get to the moon in Apollo, who presented a summary to an international symposium in 1974, published in *The Soviet-American Conference on Cosmochemistry of the Moon and Planets*, edited by John H. Pomeroy and Norman J. Hubbard, NASA SP-370 (Washington, 1977). Scientific conclusions recorded in the present volume were taken largely from these two sources.

Congressional documents are useful sources for historians of the space program. The Space Act of 1958 required NASA to obtain authorizing legislation for its appropriations, and in annual hearings before House and Senate subcommittees NASA officials summarized the agency's progress during the past year and its plans for the coming year, usually in great detail. Besides these published hearings, reports on specific aspects of the programs were frequently prepared by committee staffs. Especially useful for the lunar science program are the annual authorization hearings before the House subcommittees on manned space flight and space science and applications for fiscal years 1965 through 1969.

From 1958 to 1969 NASA submitted a *Semiannual Report to Congress* summarizing accomplishments of the past six months. A broader view of the nation's space program, encompassing the activities of all government agencies conducting programs in space, is found in the annual *Aeronautics and Space Report of the President*, compiled until 1972 by the National Aeronautics and Space Council and thereafter by NASA.

The author conducted interviews with several participants in the Apollo program and made use of interviews taken by researchers for earlier history projects and filed in the JSC History Office. The major utility of these interviews was to provide details that were sometimes not recorded in documents and to provide explanation of technical points that were not clear. Occasionally they supplied insights from an individual's peculiar point of view.

Persons interviewed by the author:

Alan L. Bean, April 10, 1984, Houston;
Eugene A. Cernan, April 6, 1984, Houston;
Noel W. Hinners, March 16, 1984, Washington;
Joseph P. Kerwin, March 29, 1985, Houston;
Robert O. Piland, October 9, 1984, Houston;
Paul E. Purser, March 10, 1983, Houston;
Harrison H. Schmitt, May 30, 1984, Houston;
John R. Sevier, April 24, 1986, Houston;
Eugene M. Shoemaker, March 17, 1984, Houston;
Donald K. Slayton, October 15, 1984, Houston.

Interviews taken by researchers on earlier projects:

William O. Armstrong, interviewed by Ivan D. Ertel and James M. Grimwood, January 24, 1967;
Alex J. Dessler, interviewed by Loyd S. Swenson, Jr., May 26, 1971 (not transcribed);
John H. Glenn, Jr., interviewed by Robert B. Merrifield, March 15, 1958;
Wilmot N. Hess, interviewed by Robert B. Merrifield, November 7, 1968;
Elbert A. King, Jr., interviewed by Loyd S. Swenson, Jr., May 29, 1971 (not transcribed);
Wendell W. Mendell, interviewed by Loyd S. Swenson, Jr., February 11, 1971;
Russell L. Schweickart, interviewed by Peter Vorzimmer, May 1, 1967;
Donald K. Slayton, interviewed by Robert B. Merrifield, October 17, 1967;
Paul J. Weitz, interviewed by Charles D. Benson, August 19, 1975.

INDEX

ABOUT THE AUTHOR

William David Compton, born in DeLeon, Texas (1927), took B.S. and M.S. degrees at North Texas State Teachers College and a Ph.D. at the University of Texas. He taught chemistry at West Texas State College, Colorado School of Mines, and Prescott College (Arizona). While at Prescott he took a year's leave to study history of technology, earning an M.Sc. from the University of London (Imperial College of Science and Technology) in 1972. In 1974 he was awarded a contract to write NASA's history of the Skylab project in collaboration with Charles D. Benson. *Living and Working in Space: A History of Skylab* was published in 1983 as NASA SP-4208. After three years of working for a Houston engineering consulting firm, Dr. Compton won a second contract from the Johnson Space Center to write the present volume. He also contributed a chapter on NASA and the space sciences to a commemorative volume, *100 Years of Science and Technology in Texas*, published in 1986 by Sigma Xi on the occasion of the society's centenary and Texas' sesquicentennial. In 1983 he wrote the prize essay in the Dr. Robert H. Goddard Historical Essay competition sponsored by the National Space Club of Washington, D.C.

HISTORIES

Anderson, Frank W., Jr., *Orders of Magnitude: A History of NACA and NASA, 1915–1980* (NASA SP-4403, 2d ed., 1981).

Benson, Charles D., and William Barnaby Faherty, *Moonpart: A History of Apollo Launch Facilities and Operations* (NASA SP-4204, 1978).

Bilstein, Roger E., *Stages to Saturn: A Technological History of the Apollo/Saturn Launch Vehicles* (NASA SP-4206, 1980).

Boone, W. Fred, *NASA Office of Delaware Affairs: The First Five Years* (NASA HHR-32, 1970, multilith).

Brooks, Courtney G., James M. Grimwood, and Loyd S. Swenson, Jr., *Chariots for Apollo: A History of Manned Lunar Spacecraft* (NASA SP-4205, 1979).

Byers, Bruce K., *Destination Moon: A History of the Lunar Orbiter Program* (NASA TM X-3487, 1977, multilith).

Compton, W. David, and Charles D. Benson, *Living and Working in Space: A History of Skylab* (NASA SP-4208, 1983).

Corliss, William R., *NASA Sounding Rockets, 1958–1968: A Historical Summary* (NASA SP-4401, 1971).

Ezell, Edward Clinton, and Linda Neuman Ezell, *On Mars: Exploration of the Red Planet, 1958–1978* (NASA SP-4212, 1984).

Ezell, Edward Clinton, and Linda Neuman Ezell, *The Partnership: A History of the Apollo-Soyuz Test Project* (NASA SP-4209, 1978).

Green, Constance McL., and Milton Lomask, *Vanguard: A History* (NASA SP-4202, 1970; also Washington: Smithsonian Institution Press, 1971).

Hacker, Barton C., and James W. Grimwood, *On the Shoulders of Titans: A History of Project Gemini* (NASA SP-4203, 1977).

Hall, R. Cargill, *Lunar Impact: A History of Project Ranger* (NASA SP-4210, 1977).

Hallion, Richard P., *On the Frontier: Flight Research at Dryden, 1946–1981* (NASA SP-4303, 1984).

Hansen, James R., *Engineer in Charge: A History of the Langley Aeronautical Laboratory, 1917–1958* (NASA SP-4305).

Hartman, Edwin P., *Adventures in Research: A History of Ames Research Center, 1940–1965* (NASA SP-4302, 1970).

Levine, Arnold, *Managing NASA in the Apollo Era* (NASA SP-4102, 1982).

Muenger, Elizabeth A., *Searching the Horizon: A History of Ames Research Center, 1940–1976* (NASA SP-4304, 1985).

Newell, Homer E., *Beyond the Atmosphere: Early Years of Space Science* (NASA SP-4211, 1980).

Pitt, John A., *The Human Factor: Biomedicine in the Manned Space Program to 1980* (NASA SP-4213, 1985).

Roland, Alex, *Model Research: The National Advisory Committee for Aeronautics, 1915–1958* (NASA SP-4103, 1985).

Rosenthal, Alfred, *Venture into Space: Early Years of Goddard Space Flight Center* (NASA SP-4301, 1968).

Rosholt, Robert L., *An Administrative History of NASA, 1958–1963* (NASA SP-4101, 1966).